Sound Knowledge

AF759166

Sound Knowledge: The Universal Guide to Audio provides readers with an in-depth understanding of all aspects of sound creation and control, from fundamental studio principles to advanced acoustic techniques.

Beginning with the science of sound and building through microphone techniques, electronics, acoustics and digital audio processing, each chapter provides clear explanations supported by diagrams and real-world applications. Learn professional podcasting setups, understand the intricacies of loudspeaker design, master studio recording techniques and troubleshoot common audio problems. The book covers both analogue and digital domains, exploring everything from vintage recording methods to cutting-edge networked audio systems. Special attention is given to location recording challenges, studio workflows and the art of mixing. With dedicated sections on acoustics, outboard processing and practical problem-solving, this text serves as both a classroom resource and a professional reference. *Sound Knowledge* is designed to sit alongside practical studio time as an aid and guide in sound-based educational courses, yet it also stands alone as a primer for new entrants to prepare to compete for further educational courses or the precious few industry apprenticeships.

Sound Knowledge is the perfect introductory resource for undergraduate audio engineering students, content creators, broadcast professionals and home studio enthusiasts looking to elevate their technical knowledge and production capabilities.

There is a companion website to the book where readers can find further support. Test your knowledge and grow links to other sound professionals at www.SoundKnowledge.org.

Peter R Monnery is BBC (British Broadcasting Corporation) qualified in Technical Operations, as a Producer and an Engineer and has taught audio and broadcasting skills for decades, alongside being a senior player in hands-on audio production and desk operations. Peter has tutored fellow BBC professionals in new disciplines at Broadcasting House, London, including production skills and many facets of operations. Peter has taught in Britain and abroad on leading audio technologies and occasionally lectures at Ravensbourne University in London on specialist subjects.

Sound Knowledge
The Universal Guide to Audio

Peter R Monnery

Routledge
Taylor & Francis Group
LONDON AND NEW YORK

Designed cover image: Cover design created by Peter R Monnery, cover photograph by Peter R Monnery

First published 2027
by Routledge
4 Park Square, Milton Park, Abingdon, Oxon OX14 4RN

and by Routledge
605 Third Avenue, New York, NY 10158

Routledge is an imprint of the Taylor & Francis Group, an informa business

© 2027 Peter R Monnery

The right of Peter R Monnery to be identified as author of this work has been asserted in accordance with sections 77 and 78 of the Copyright, Designs and Patents Act 1988.

All rights reserved. No part of this book may be reprinted or reproduced or utilised in any form or by any electronic, mechanical, or other means, now known or hereafter invented, including photocopying and recording, or in any information storage or retrieval system, without permission in writing from the publishers.

For Product Safety Concerns and Information please contact our EU representative GPSR@taylorandfrancis.com. Taylor & Francis Verlag GmbH, Kaufingerstraße 24, 80331 München, Germany.

Trademark notice: Product or corporate names may be trademarks or registered trademarks, and are used only for identification and explanation without intent to infringe.

British Library Cataloguing-in-Publication Data
A catalogue record for this book is available from the British Library

Library of Congress Cataloging-in-Publication Data
Names: Monnerey, Peter R. author
Title: Sound knowledge: the universal guide to audio /
Peter R. Monnerey.
Description: Abingdon, Oxon; New York, NY: Routledge, 2026. |
Includes bibliographical references and index.
Identifiers: LCCN 2026017773 (print) | LCCN 2026017774 (ebook) |
ISBN 9781041205807 hardback | ISBN 9781041205791 paperback |
ISBN 9781003717065 ebook
Subjects: LCSH: Sound–Recording and reproducing |
Acoustical engineering |
Podcasting | Radio broadcasting | Sound engineers
Classification: LCC TK7881.4 .M636 2026 (print) | LCC TK7881.4 (ebook)
LC record available at https://lccn.loc.gov/2026017773
LC ebook record available at https://lccn.loc.gov/2026017774

ISBN: 9781041205807 (hbk)
ISBN: 9781041205791 (pbk)
ISBN: 9781003717065 (ebk)

DOI: 10.4324/9781003717065

Typeset in Univers
by Deanta Global Publishing Services, Chennai, India

Access the Support Material: Soundknowledge.org

These materials are created and maintained by the author and are not the responsibility of Taylor and Francis.

For Sally, Sarah, Tom, Evie, Archie and Seb who make my World,
And to my close family and friends who help me succeed in it…

Huge thanks for the enduring technical support of Norman Bethray, Andrew Whadcoat and Peter Thomas, resolutely keeping me on track!

Contents

Foreword

I love sound! Nothing else I know can create the simple pure elation of hearing a voice or a tune that reduces me to tears, or dance music that's irresistible, even when my technique is! For the professional there is the thrill of "hands-on" in a live broadcast or concert, where so much can go wrong, and planning so it doesn't! Some may just want an exciting hobby to make a difference away from work. I'm passionate that others join this fantastic world! The text and diagrams in the book are drawn from years of experience lecturing in the workplace and further education. The book is built as a true primer; it does not require any previous experience of sound or audio engineering, and primary school maths is all that is required.

It's understandable in a complex world that it's easier to park students to practise in the studio, rather than take the more difficult and expensive route of formal training in the lecture hall. It's time consuming, and perhaps uncomfortable, to routinely check formal knowledge and to have confidence that the student is progressing as expected. Of course, a combination of practical and theoretical skills is necessary, but at the time of writing I am concerned the pendulum has swung too far for some, and on occasions we are in danger of letting down our future audio professionals with incomplete information, perhaps on a shaky foundation. After a lifetime of loving the job I do, I want others to compete and win at doing the same, to enjoy the confidence and reward that comes from an in-depth knowledge of "Sound" and how to make it work across many disciplines, from simple podcasts to full-on feature film sets.

If you have a creative mind and a determination to succeed, this book is for you. It starts with the essential basics (the foundation of a successful career and so often glossed over!) and grows to a level of understanding in many areas of sound known to few. I make no apology for stripping out the mathematical formulae that engineers love to use as confirmational exclamation marks; sound creation is in the end an art form, and whilst I am the first to say a light touch on physics is essential, it can be couched in friendly, conversational terms that a new reader can easily assimilate. It's like photography; people can have a natural knack to frame a great picture, but without an understanding of aperture, shutter and ISO, the lack of a correct exposure and depth of field will fail the artist when the work is on the exhibition wall…

I had the huge privilege to work in professional paid audio from the age of 15 as a school disco DJ who had to make his own gear. This path led me to leading on stage at corporate entertainment events, including clients like American Express and SmithKline Beecham, as well as supporting major bands of the era. I moved to sound desk operation on BBC Radio programmes at 19, and by age 21 I was left alone to manage complex programmes including a live simulcast between BBC Radio 1 and BBC 1 TV with an audience approaching 20 million. Since then, I have extensively operated on live and recorded network radio programmes, produced music, sport and light entertainment shows on three National networks and an 84-piece Radio Orchestra. I've even been known to work on the repairs of studio equipment as a qualified BBC Engineer. I know that true confidence comes from a holistic understanding of a subject, a personal toolkit that can cure all but the worst of ills in a quick and calm manner. A great sound practitioner should be like a surgeon: skilled, confident, well informed, but, above all, decisive!

This book is not a defining treaty on sound techniques, physics, electronics or acoustics; rather, it is an interlinked careful construction of all to provide a strong foundation to create a confident and competent sound creative. I avidly hope it's an engaging read, fun in places and easy to follow, as it's written for enthusiasts like you to build a rich life doing something you love. I don't claim that every view expressed is of my sole invention; I listen and test information found in daily experience from colleagues, in books and on the internet to find my own path. As a result, I have written about techniques that, tried and tested, have worked for me. Sound is a subject that mixes science and art in equal measure, and I am aware at times there is fashion where people applaud products or techniques which are counterintuitive to a logical understanding of physics. I don't dismiss extravagant claims for certain product strands out of hand, but I do ask that readers use the toolkit provided in this book to make up their own minds rather than following the illogical herd, spending money for nothing (and an empty wallet for free).

I'd like to thank some of our industry's most amazing people, who I admire and still pinch myself I am in their lives! I have been so fortunate through the years to hear the counsel of world-class DJs, producers, musicians, music balancers, audio engineers and fellow operational engineers. Equally, I want to give a huge thank you to the designers of top-flight audio equipment for their encouragement and input to ensure the book is a complete and solid representation of the fundamentals of our industry. That said, no one text can provide every detail culled from every situation, but I believe the knowledge in this book will give you a firm foundation, the confidence and direction for you to take the next steps, wherever your life in sound may take you. Happy reading!

1 Introduction to Sound

Our ambition is to create and control sound, so it's vital to understand what's going on at a fundamental level to give a strong foundation when things get more complex. Understanding these raw basics gives us the toolkit to make great sound in complex situations. So please don't skip this bit as it is the basis of all that comes after it...

MAKING WAVES!

It's amazing how often nature creates predictable outcomes from very different situations. Later in this book you will see electrical calculations used to accurately model room acoustics and sound isolation! For now, let's take an easier comparison. If I throw a rock in a pond, it will make a satisfying *plop*, and although the ripples are beautiful to watch we can't hear them. Yet, in many ways they behave like sound!

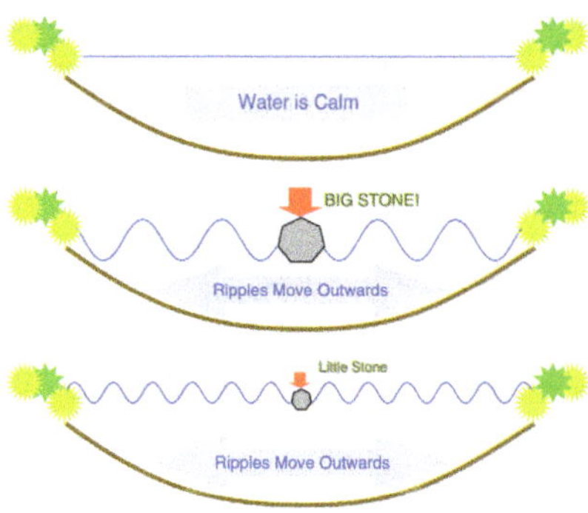

Figure 1.1 A pond making waves with stones

In Figure 1.1 we see rocks make large ripples while pebbles make little ones; yet big or small, the waves will reach the bank at the same time. If I put a cork on the water, it will ride the ripples and stay put, so whilst the waves travel, the water does not! In sound, big movements make large excursions in the air, little ones make small ones; yet equally they travel at the same speed to their destination. A leaf falling through the air may well pass through a sound wave; yet it is undisturbed and simply drops to the ground. Both the ripples and sound obey wave theory rules, yet they are different effects.

Water waves behave like the light waves that make up a rainbow. They are *transverse* waves; their energy is perpendicular (sideways) to the medium they pass through, and whilst our eyes are sensitive to electromagnetic waves, our ears are not. Transverse waves travel through the vacuum of space; sound waves don't; they require a medium. We think of sound movement through air, but it will travel through practically anything, and paradoxically (to me), the thicker the medium gets, the faster the sound moves. Two examples: sound travels almost five times faster in water than in air (great for the whales!), and *18 times* faster in steel! Sound waves can be calculated to give a predictable

DOI: 10.4324/9781003717065-1

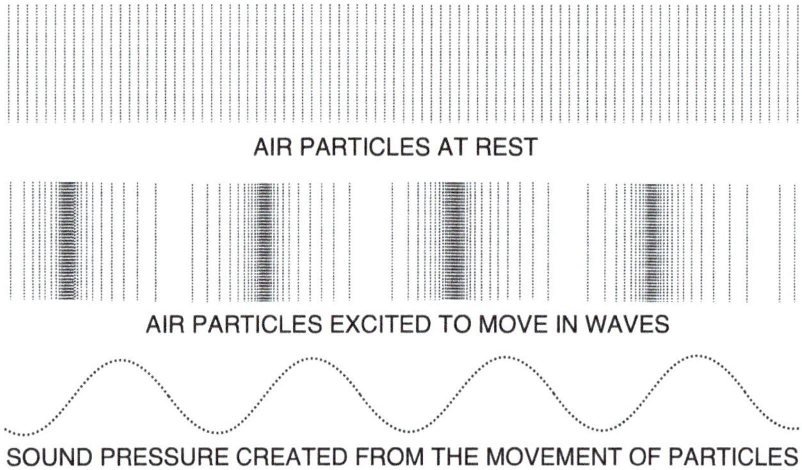

Figure 1.2 Air movement, particles at rest, shown excited and resultant sound pressure

speed through any medium; the more densely packed the particles, the faster the speed.

Sound is not a transverse wave; it is a *longitudinal* wave. It moves *parallel* to the direction of travel. What does that mean? Imagine a rower on the water. Each time there is a pull on the oars the boat shoots forward; each time the oars leave the water, the boat slows. The positive energy of the thrust is mirrored by the negative energy of the rest cycle, yet the boat moves up the river. We don't hear the movement of air particles, we hear the sound pressure that results from their action (Figure 1.2).

In sound waves, the energy of the sound source compresses particles in the medium (normally air), then naturally expands before the next compression. A single cycle of compression and expansion is called a *wavelength* (Figure 1.3). If we count the number of wavelengths that pass in a second, it gives us the *frequency* of the sound. However, it does not give us a clue as to how intense (loud!) it is; that is shown by the force of the compression, which is measured in the *amplitude* of the wave. This is measured between the calm unexcited medium and the maximum deviation by the applied force.

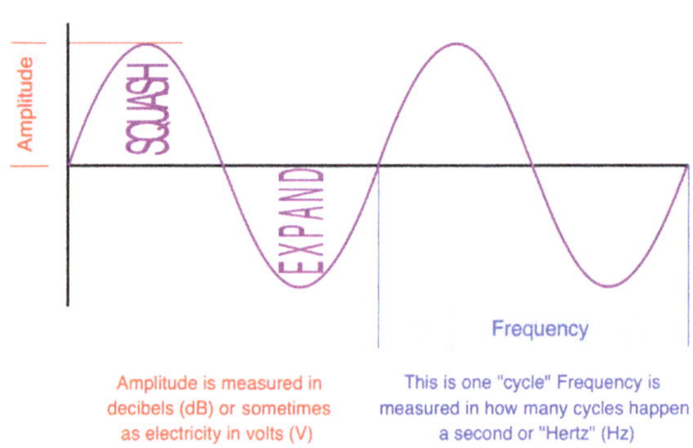

Figure 1.3 A waveform showing wave compression, expansion, frequency and amplitude

Sad as it might be, we can't hear the transverse nature of water ripples, however, our ears are completely brilliant at hearing *longitudinal* waves. When sound is present, the ebb and flow of air particles (*particle velocity*) create changes in *sound pressure* (Figure 1.2) which reach our eardrums and move them in and out in sympathy with the rhythm of the wave. We can capture this movement on other diaphragms such as a microphone and can use instruments to measure the frequency and amplitude of the sound pressure. We see in Figure 1.3 that the frequency is measured in wavelength cycles a second, known commonly as "hertz" (Hz), and the amplitude is measured in 1/10 of a "bel", known as a decibel (dB). When we are young, our ears can hear from very low frequencies of around 20 Hz to very high, 20,000 Hz. Low sound waves below 20 Hz are called subsonic, and high frequencies above 20,000 Hz are called ultrasonic. However most sound we hear sits in the middle range of around 150 Hz to 3,000 Hz.

The speed of sound through the air is much slower than light transmission, but fast enough to be useful to us. The waves carried in the air hit our body, and our ears capture the movement and turn the sound pressure waves into tiny electrical impulses that the brain can decipher. An example of the speed of sound is a thunderstorm (Figure 1.4). If you watch from a distance (a sensible thing to do!) you see the lightning bolt instantly but hear the thunder seconds later. This is the time the sound has taken to reach you at around 343 metres (1,125 ft) every second. If the thunderstorm is a kilometre away, you will see the lightning three seconds before you hear the thunder.

Figure 1.4 A thunderstorm, the lightning is instant, the sound follows

THE SCIENCE OF SOUND

Fortunately, sound obeys the normal accepted rules of physics. As the wavelength increases the frequency decreases. Formulae are intentionally rare in this book, but I think this simple one is worthwhile. As the speed of sound is constant (for a given temperature), what is shown is that as the frequency grows, the wavelength shortens proportionately.

$$\lambda = \frac{c}{f}$$

where:
λ = wavelength in metres
c = speed of sound (343 m/s)
f = frequency in Hz (cycles/second)

My personal guide is to take 1,000 Hz (1 kilohertz; 1 kHz); I know the wavelength is close to 343 mm or around 1/3 of a metre. A halving or doubling of frequency has the opposite effect on the wavelength. Ten times one means 1/10 of the other. Two times frequency is 1/2 the wavelength. Here are some rough examples off the top of my head. At 100 Hz, λ is *about* 3.5 m; 1 kHz λ is therefore approximately 35 mm; 500 Hz λ is 700 mm; 2 kHz λ is 170 mm; and at 50 Hz λ is a massive 7 metres! You get the picture. You might ask why I should care? When we come onto the subject of acoustics, having a few rough figures like this in your head (particularly below 300 Hz) will be very useful!

SOURCES OF SOUND

Many objects produce a natural sound signature; we can recognise birds by their song, a truck as opposed to a car, a flute from a guitar. When we want to create sound, we have choices of what materials we use, however, the sound we hear will come from the medium having a certain length (size) and therefore vibrating (in sound we say resonating) at a particular frequency. Any sound can be characterised by just three parameters: (1) *loudness* (simply amplitude or sound "level"), (2) *pitch* (frequency in physics) and (3) *timbre* (this is known by musicians as overtones or partials, and by engineers as harmonics).

Items that can be made to vibrate at a particular frequency include:

1) Columns of air (organ pipe).
2) Vocal cords (singing).
3) Stretched skins or membranes (drums).
4) Plucked or bowed strings (guitar/violin).
5) Metal or wooden bars (xylophone).
6) Circular plates (cymbals).
7) Electronic oscillator (synthesised music).

This next part on pipe theory is a complex read. *Please don't worry if it's confusing at first*; it will not stop you moving through the book, and trust me, all will come naturally in time! Here we go! If we take the first example from above, a pipe holding a column of air (Figure 1.5), we can form a better perspective of what is going on. An important point, on which we expand later, is that where there is a "fundamental" natural frequency produced by the length of a pipe, there are likely to be harmonic multiples of that base fundamental called "overtones". These add together to provide the "timbre" or characteristic sound of the source.

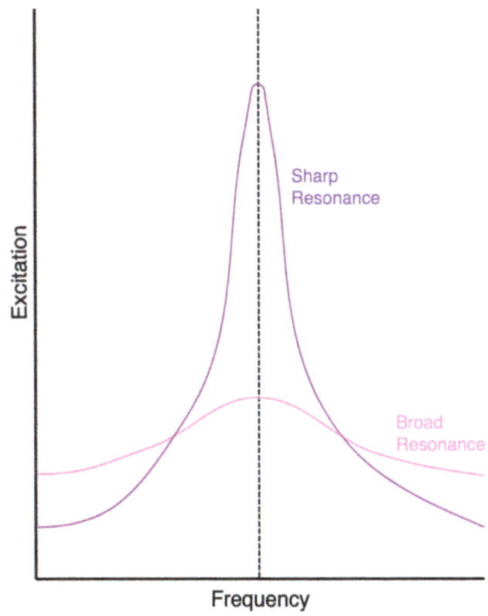

Figure 1.5 Pipe theory showing a resonant frequency peak

Let's close one end of the pipe. The open end is unconstrained and so will resonate at a particular frequency where the sound pressure at its mouth is at its highest. This is called an "antinode". The term for the point of zero pressure is called the "node", and this is what you will find at the closed end wall of the pipe. The longer the length of the pipe, the lower the frequency that will be produced. The distance between a node and the nearest anti-node is 1/4λ, so this must be the "fundamental" length of the pipe when it is first resonating. *The closed-off pipe cannot produce even harmonics*, simply because that would mean a node (zero air pressure) forming at the mouth of the pipe, however, odd harmonics have an anti-node at the mouth of the pipe, and so they are produced. This point of resonance is called a "standing wave".

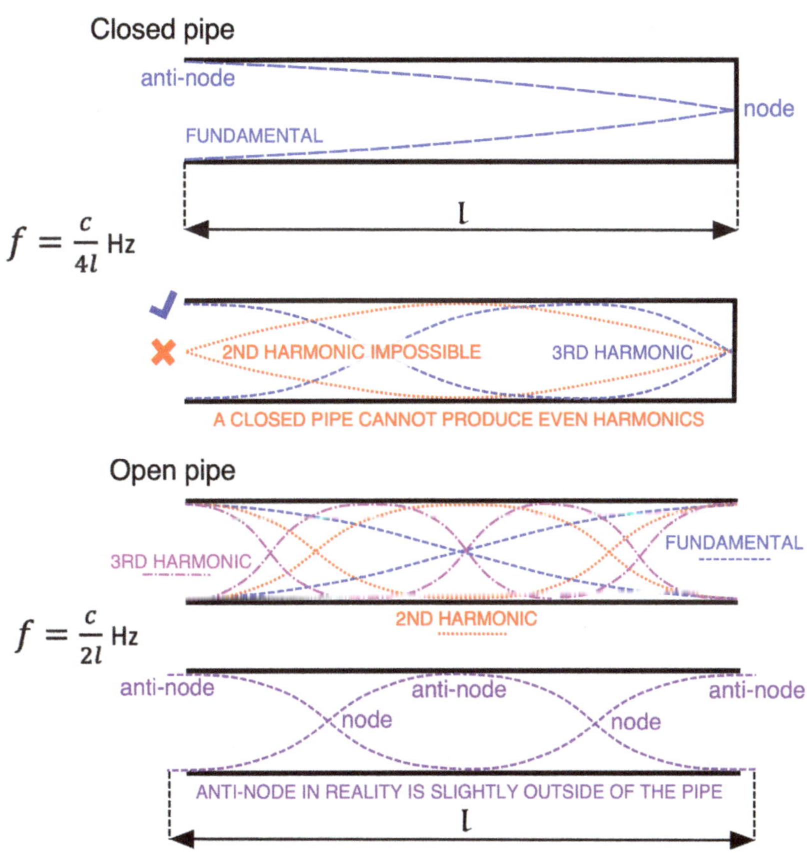

Figure 1.6 Pipe theory showing closed and open pipes and sound behaviour

Pipes with openings at both ends can produce the fundamental and all harmonics of it, whether they are odd or even multiplications of this base frequency. For simplicity we have assumed the antinode is in alignment with the tube end; in reality, it is in the space just outside the opening, extending the effective path length by 0.6R (where R is the radius of a circular pipe). Other shapes have different correction formulae. Figure 1.6 shows closed and open pipe behaviours.

An open pipe will resonate at half the wavelength (1/2λ) when compared to 1/4λ for a closed pipe. This will become relevant when we talk later about ports in loudspeaker cabinets, and absorbers and diffusers in acoustics, and will make more sense practically.

There are scientific proofs for other sources of sound; it's enough for this book to look at one other. Stringed instruments have each end clamped; these are nodes, and the anti-nodes must form in the length. The simplest is one anti-node in the middle of the string, and this is also the fundamental note. In passing, the formula is the same as for an open pipe, however, "*c*" is not the simple velocity of sound. In this case it is the velocity of the wave along the string. This is composed of (1) the length of the string, (2) the tension of the string and (3) the "mass per unit length" (effectively thickness for a given material).

INTRODUCTION TO DECIBELS

We have talked in detail about the length of the wavelength and so its frequency. Time to talk about the amplitude or strength of the sound wave. This is measured in the logarithmic measure of "decibels" (dB). If that sounds mysterious, the dB is our friend as our ears are also logarithmic, so changes in the level of sound (up or down) can be judged by knowing how many decibels have changed. A change of 1 dB is just discernible in comparison of tone or continuous sound (like a violin being bowed). Most people will discern a change of 3 dB on speech, and 6 dB represents a halving or doubling of the sound level. The real-time loss of sound in air, known as the "decay" of a sound wave, follows the inverse square law. For each doubling of distance, the strength of the sound wave is halved as shown in Figure 1.7.

For example, if the loudness of a sound source is 100% at one metre (1 m), it is 50% (or 6 dB down) at 2 m away. At 4 m it is 25% of the original signal strength (or 12 dB down), and so on. Although 6 dB is clearly discernible, it's

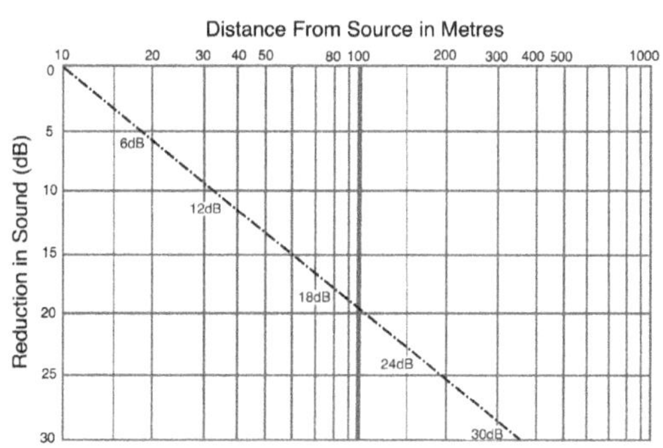

Figure 1.7 Sound loss over distance

not huge. Perhaps two or three presses of the up or down volume button on your TV remote control.

I need to point out the inverse square law applies to sound from one single place, a "point source" such as a solitary loudspeaker, a football supporter or a passing car. But what if there are many supporters or many cars, obviously louder than one, but what else? The sound of many cars travelling on a road (straight or twisty) or a distant crowd describes what is called a "line source", and whilst the proposition of a pure line source is academic, the theory of the distance rule changes. In a pure line source propagation, sound dies away at 3 dB/metre, that is it travels twice as far. In practice multiple source sounds lie between the point and line source calculations. We'll come back to this effect in loudspeakers.

The inverse square law assumes the sound is in free space without any obstacles to modify it. We now understand that sound travels at a finite pace, what happens when a sound wave meets an obstacle? The ground presents a resistance which drags or "grazes" on the sound wave; trees and other "soft" projections more so, just as friction on differing surfaces in physics. The amount of sound attenuation depends on several factors including slope, irregularity in terrain and humidity. Standard formulae exist to find approximate answers and model the outcome, so we can work in various terrains and with different technologies with a high confidence any practical remedial solutions will work as designed on paper.

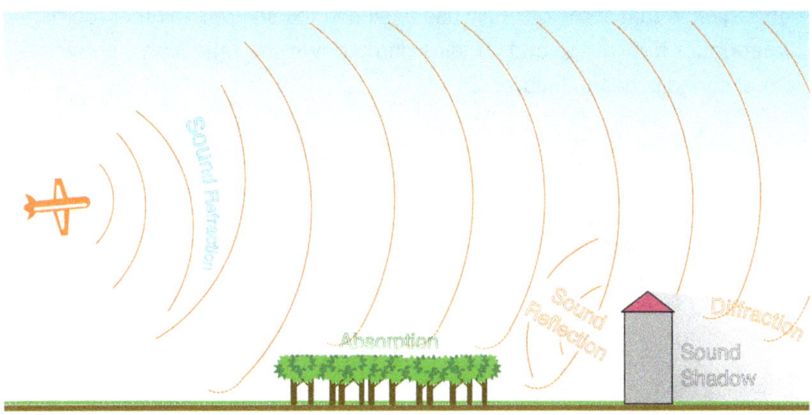

Figure 1.8 Sound grazing over open terrain

I'll now touch on the four ways in which sound can be modified. When sound hits a solid object like a building or a strong fence it reflects off, but more importantly a shadow of high attenuation is formed on the other side. In a perfect world this would be absolute; in reality it is more taking the proverbial wind out of the sails, and as more distance is covered, the sound waves bend around the object (1: diffraction), reducing the effect of the sound losses particularly at low frequencies. If there are several buildings noise can bounce around between them and create a "canyon" effect (2: reflection). The effect of the "sound shadow" is used at the sides of motorways as they pass through urban areas to keep the traffic noise in check. Long tall solid barriers are erected at the sides of the road, and these are often treated with porous absorbing materials on the vehicle side (3: absorption), to keep traffic noise levels lower in the adjacent community. On windy days or damp mornings the sound can bend around the landscape making the barrier less effective. Sound is normally refracted upwards, but from a wind speed of 4 m/s and more, it is refracted back downwards to the ground; it gets noisier, and the sound shadow can be lost. (4: refraction). *So, sound can be altered in its environment by: (1) diffraction, (2) reflection, (3) absorption and (4) refraction.* Surprisingly, sound being "carried along by the wind" is mostly considered negligible in acoustic terms (Figure 1.8).

SOUND IN AN ENCLOSED SPACE

Most of our work with sound takes place in enclosed spaces, rooms we call studios. What happens when a sound wave meets a wall in the room? A sound wave behaves very much like light when it meets a reflective barrier. If the surface is hard like a glass window, the sound glances off the surface at an identical angle to that at which it hit it with little loss in energy. An example in light is where a laser beam hits and reflects off a mirror. Sound reflections can easily be heard by going into a bathroom where the hard tiles reflect your voice, giving a strong echo that some gainfully use to sing in the shower. In other rooms the carpets, soft furnishings and curtains damp down the reflections, and a more natural sound is heard (Figure 1.9).

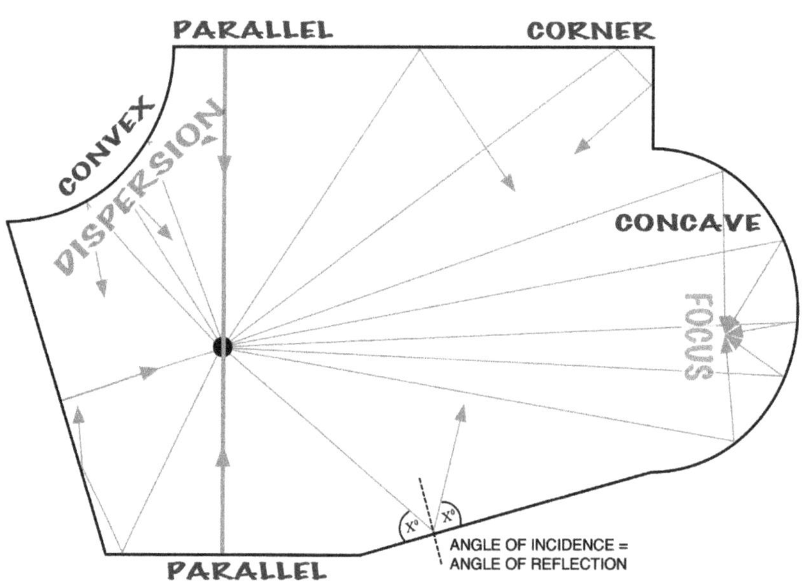

Figure 1.9 Sound waves inside a room

In an empty room sound will bounce off a bare floor, walls and ceiling, and where the surfaces are parallel to each other, this can set up annoying multiple reflections, greatly increasing the sound level at certain frequencies. If you designed a recording studio from scratch, you would avoid any parallel surfaces, corners and any concave surfaces which excite some sounds and suppress others. If we pass light through either convex or concave lenses the light is either focused on a point or scattered outwards. It's the same with sound; in a room's design, dispersion of reflections is useful, and the other trick is to absorb unwanted reflections with a porous finish on sections of the walls, such as wadding, foam or mineral wool. Focus points like window bays or corners are not helpful. We will come back to this subject in much more detail in Chapter 8, "Introduction to Acoustics", but it is worthwhile noting now to get you thinking.

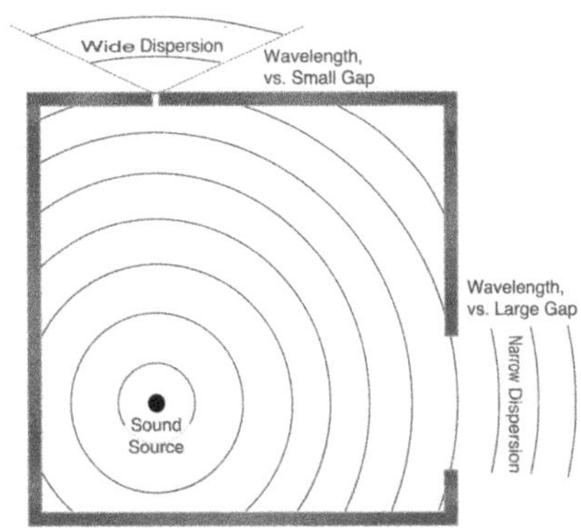

Figure 1.10 Sound waves diffract through large and small room openings

In a final point on sound in spaces, the size of a wavelength and the size of an opening together determine how the sound will distribute through the gap. Given a large opening compared with the wavelength, the sound will "beam" through the gap. A small opening, compared with the size of the wavelength, will lead to a wide dispersion of the sound (Figure 1.10). If you imagine a loudspeaker mounted in a hole in a box, the size of the hole, compared to the wavelength (particularly at the low frequency end), will naturally lead to a wide dispersion of sound in an area.

Earlier, we threw rocks and little pebbles in the pond, to make large ripples and small ones. We noted sound has an equivalent. Big sources of sound make large excursions in the air and produce low frequencies. Small items produce rapid small movements of air and produce high frequencies. Of course, there are many sound sources which sit between the extremes as "middle" frequencies including the human voice. Like the pond, all differing frequencies pass through air at the same speed, so arrive from the source to the ear together. Table 1.1 provides simple examples to get you thinking on the right track.

Table 1.1 An exploration of frequency against common sounds

Sound Source	*Low Frequency*	*Mid Frequency*	*High Frequency*
Percussion (You Hit It)	Timpani	Snare Drum	Triangle
Strings (You Pluck It)	Double Bass	Guitar	Ukulele
Brass (You Blow It)	Tuba	Trombone	Piccolo Trumpet
Engine	Truck	Motorbike	Model Plane
Animal	Bull Elephant	Dog Bark	Mouse

Let's add in a little more detail on some familiar sound sources. In Figure 1.11 the solid line for each instrument gives the range of fundamental frequencies the instrument can play, and the faint extension shows the frequencies the harmonics reach to. I have noted a part of the middle range of frequencies in the bottom scale (pink highlight), which is enough to have clarity in communication. Think old-fashioned phone lines, or ground communication to a plane. It's not high fidelity but people can understand each other without wasting the expensive signal space used by sending a full range of frequencies. We'll look at frequencies in more detail in Figure 1.12.

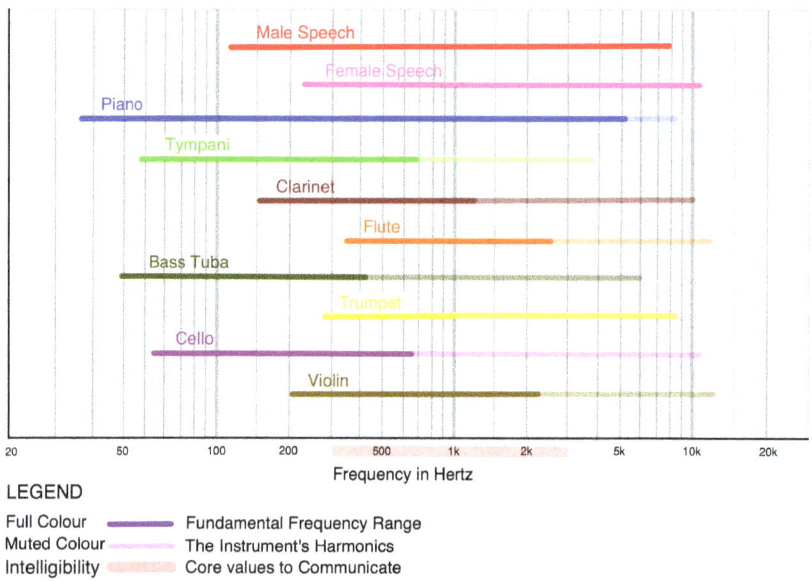

Figure 1.11 Frequency ranges of common sound sources

HARMONICS

So far, we have discussed sound as how loud it is (amplitude), and how many cycles happen in a second (the frequency we hear). A typical audio test tool is a pure single frequency of around 1,000 Hz (1 kHz) tone (also known as a sine wave) to line up equipment to give its best. A tuning fork is a mechanical way of creating a pure sine wave in air. It's unusual in nature to hear just one tone; there is always a primary frequency called the "fundamental", and normally in addition there are multiples of that frequency which add to give a unique sound signature to the source. These additional frequencies are called "harmonics". If we pluck a note on a guitar and blow the same note on a trumpet, they will sound completely different because they have a different set of harmonics.

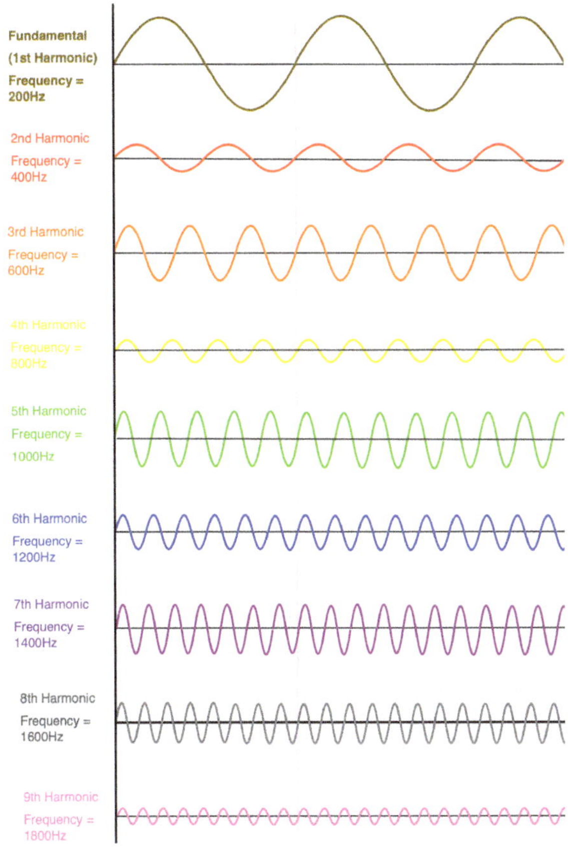

Figure 1.12 The harmonic breakdown of a clarinet

We can describe a note by either naming its frequency or a place in an octave. If we strum an open "A" (2nd thickest) guitar string, we expect the fundamental frequency to be 440 Hz. If you look at the string you can see the fundamental note vibrating, but also see it cycle through other patterns. These are the harmonics at work. The 2nd harmonic is simply double the frequency (880 Hz), and is also an "A", one octave higher in the scale, and so on.

In a final word on harmonics, our ears love "even" harmonics as they add "warmth" to the sound. Conversely "odd" harmonics add an "edge" to the sound. Much-loved vintage valve amps, when turned up till they distort, produce even harmonic distortion. Later "solid state" transistor amplifiers produce odd harmonic distortion. That is the reason stage musicians would rather a badly distorting valve amplifier than a slightly distorting transistor amplifier. Each voice has unique harmonics from the vocal cords, and so two different people singing the same song in the same key create a very different atmosphere. These variations in harmonics make each sound source unique, and there is always something new to be discovered! It is the level of each harmonic, set against the others, that creates the sound. I have said that the harmonics are simple multiples of the fundamental frequency. If the multiples are "even" multiples they will be ×2, ×4, ×6, ×8 and so on. If the harmonics are "odd" they will be ×3, ×5, ×7, ×9, etc. Note the different levels on each harmonic that give shape to the total sound, and when added together, its unique waveform in Figure 1.13.

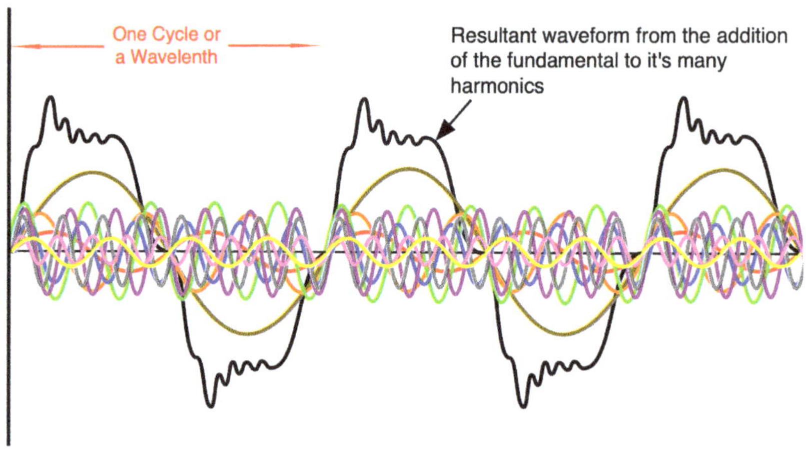

Figure 1.13 Harmonic overlay of a clarinet waveform

THE EAR

So much for sound sources, how do we hear them? If this was a book on painting, we would talk about the perception of colour, light and shade on the eyes. Cordon Bleu cookery? It's all about taste and smell. This book is about sound and there is only one way to access it, through our ears! We still have so much to learn about the ear, and some of what is written here is the result of a lifetime of personal experimentation. I'm not going to please the academics with medical or Latin names, let's just keep it simple. The diagram in Figure 1.14 is correct, but I'm only including those that have a bearing on the needs of a sound practitioner.

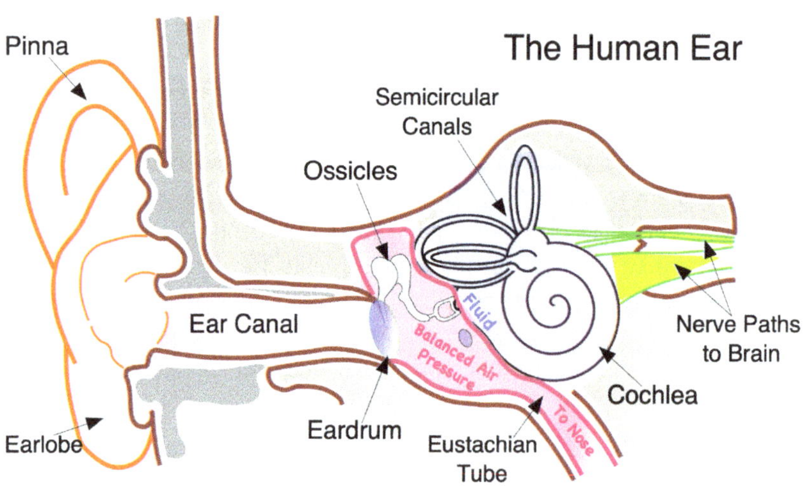

Figure 1.14 The human ear

I'll immediately break my word touching on the semicircular canals as they are neat! Your inner ear doesn't just do hearing; it manages your balance, and just like a smartphone these three loops know which way up you are and make you balance to do apparently simple things like walking (I don't think it's technically so simple!). Sometimes, if you have an inner ear infection like labyrinthitis, these canals can get infected and with gross feelings of seasickness, you must take to your bed for a day or two.

Our ears are very much part of our evolution. Over millennia, our ancestors adapted the jawbone to create an incredibly nuanced hearing. In particular the connection between the eardrum and the cochlea features the smallest bone in our body, the stirrup; and we are still learning how the three bones of the ossicles (hammer, anvil and stirrup) work so well together. But I digress. Sound waves hit our two ears from all directions. A single sound (that is not directly in front or behind us) will reach each eardrum at slightly differing times, and so the wavelength will be in a different part of its phase. This effect is recognised when the two signals are compared in the brain, and the phase difference is turned back into positional information (both direction and distance) with amazing accuracy. You only need to close your eyes and have a friend walk around you, talking all the time, to realise how wonderful this ability is (unless your friend is exceptionally boring!).

They may just be two holes in our head, but we know from our ears with some certainty whether the sound is in front or alternatively above or behind us (Figure 1.15). I believe the shape of the pinna is responsible for our ability to sense this, however, it is not clinically proven. Sound enters each ear canal, and positive pressure pushes the conically shaped eardrum away, flexing a chain of three tiny bones (ossicles) to transfer the energy to the cochlea. The sound wave now contracts in negative pressure, the eardrum is pulled back and again energy is transferred to the cochlea. And that is a single wavelength!

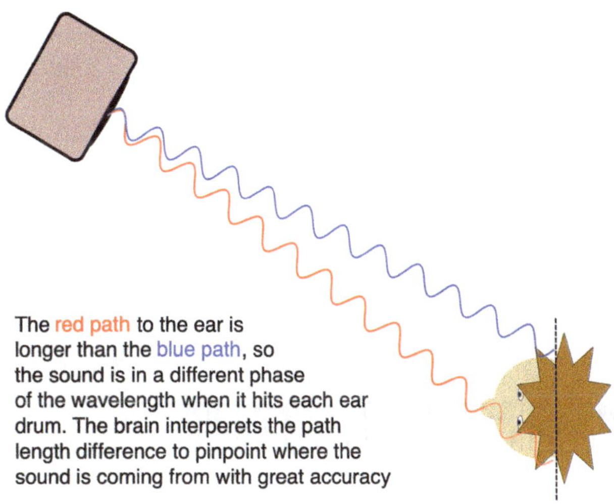

Figure 1.15 Path difference to the ear

The cochlea processes tens of thousands of wave samples a second, passing the frequency content and loudness on to the brain, converted now to faint electrical signals. The brain then compares the signal with the other ear and makes decisions on placement, distance, urgency, emotion and so on, all in a fraction of a second. Now tell me computers are about to take over!

An important point to note is that the inner side of the eardrum is connected to the nose through the eustachian tube to ensure the inner ear is pressure equalised to the outside world (probably the only time you notice this mechanism is when you are in a descending [depressurising] airplane and must swallow to get your hearing back). The conduction from the drum through the three tiny bones of the ossicles, to the cochlea has a mechanical magnification of sensitivity of an amazing 18 times. In short, the ear is a miracle of mechanical biology, capable of tremendous range, yet equally capable of incredible subtlety and nuance, of literally bringing us to tears of joy!

COMPARISON OF WAVELENGTH, FREQUENCY AND OCTAVES

All creatures on earth hear a different range of frequencies. We humans vary, but on average a young person can hear from a very low 20 cycles per second (20 Hz) to a very high 20,000 Hz. That high frequency of 20,000 Hz is normally written as 20 kHz (kilohertz). Below around 20 Hz we feel sound rather than hear it, and our ears simply stop at approximately 20 kHz through evolution. Many other creatures can hear much higher frequencies. Dogs can hear 30 kHz and more which is why the apparently silent "dog whistle" works. Bats create clicks well above our range of hearing and listen to the returning echoes to make sure they don't fly blind. A reminder, frequencies above human hearing are "ultrasonic".

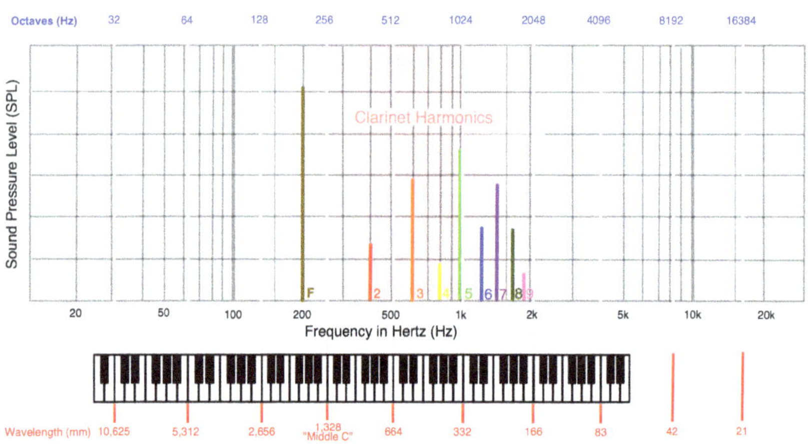

Figure 1.16 A useful comparison of wavelength, frequency and octaves

A range of 20 Hz to 20 kHz is amazing and the ear manages it by hearing frequency logarithmically rather than linearly. Equally the amplitude of the sound range is extraordinary; from the quietest sound to the loudest is a ratio of 1 to 1,000,000,000,000! As said above in the "Introduction to Decibels", the ear hears loudness logarithmically too, so that huge range can be managed in an understandable and practical way. I have to say that there is great controversy around exact note frequencies, forged by competing international interests over hundreds of years. I have stated the "engineers" view of frequency, and whilst A = 440 Hz is the "Pitch Standard" for musicians too, "Middle C" is likely to be tuned to 261.63 Hz rather than the expected 256 Hz. A complex subject in itself!

To recap the ear does not hear loudness or frequency in a normal linear fashion (like we measure distance with a tape or ruler). We need to find ways to express sound which consider the logarithmic nature of the ear.

Where hertz is a linear measure of sound frequency, octaves are the logarithmic version, and where sound pressure (loudness) can be measured linearly (dynes/cm^2) this is hardly ever used. The logarithmic decibel (dB) scale is used to measure both absolute values of loudness and relative changes in sound level. In electrical terms we do use certain voltages as an equipment line-up reference, e.g. BBC line up is 0 dB = 0.775 volts (V) however comparative amplitudes in decibels are normal in equipment too, e.g. raise or lower a signal by 6 dB. We come back to practical use of sound levels and decibels in Chapter 6, "Sound Levels and Connections".

Octaves are a useful default in music notation, as they create a linear notation for notes (like piano keys) but can be confusing in the detailed needs of sound engineering, so we have chosen to stick with frequency (Hz) and instead scrunch the scale up to make the logarithmic plot appear in a linear fashion on the page as shown above in Figure 1.16 and here in Figure 1.17.

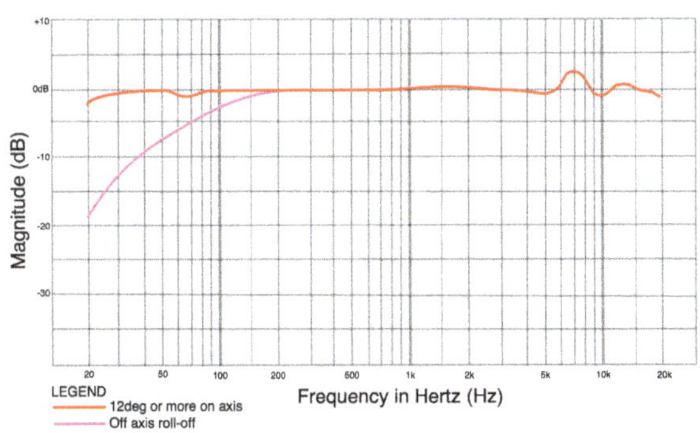

Figure 1.17 A basic frequency response of a microphone

Figure 1.17 shows a simple example of a plot of a microphone against frequency. Here, the "Magnitude" (simply a posh name for amplitude or level) is referenced to the output of the microphone at 1 kHz. From there we can read off positive and negative swings away from the reference. This form of graph is known as a "frequency response" as we can read off the accuracy in decibels at any frequency. For instance, the resonant "lift" at 4 kHz is quite common and will add a bit of sparkle to most voices, but arguably it's not accurate. This often happens in sound, because something sounds great, and perhaps becomes a classic for engineers to use, does not mean it is a high specification. On the road, often quality is sacrificed for robustness, as getting the sound out is preferable to the perfection of silence. Note the off-axis loss of bass when the singer is to the side of the microphone. We will frequently return to the use of a linear measure (frequency in Hz), placed on the graph in a logarithmic axis, the frequency bands compressed to create a scale that mimics how the ear naturally hears.

When we start to talk about different sets of ears (i.e. people in the real world), they will all hear sound levels against frequency a little differently. To take account of this we can add in one further measure, the "phon". It is an identical scale to the decibel; however, we start with a middle test frequency of 1,000 Hz (or 1 kHz) at a level of 60 dB. This is a loudness of 60 phons. If we get a bunch of volunteers to listen to and adjust other frequencies in their headphones to the same level as they hear the 1 kHz sine wave (a pure fundamental tone), we can build a table for an "average" set of ears. Change the level by 20 dB at the 1 kHz centre to create the next "phon" curve and so on. We have Fletcher and Munson (1933) to thank for their work to produce the first graph of "Equal Loudness Contours". These were modified by Robinson and Dadson, and again in 2003 by ISO 226. These tests typically involve young adults using headphones receiving pure tone signals. The "Threshold of Hearing" is commonly judged to be 20 μPa (micro pascals), an imaginary zero sound pressure level or an "SPL of 0 dB", which is interesting to note, but has little use outside of a laboratory. Figure 1.18 is a personal take on the curves; it does not conform to a particular work or study.

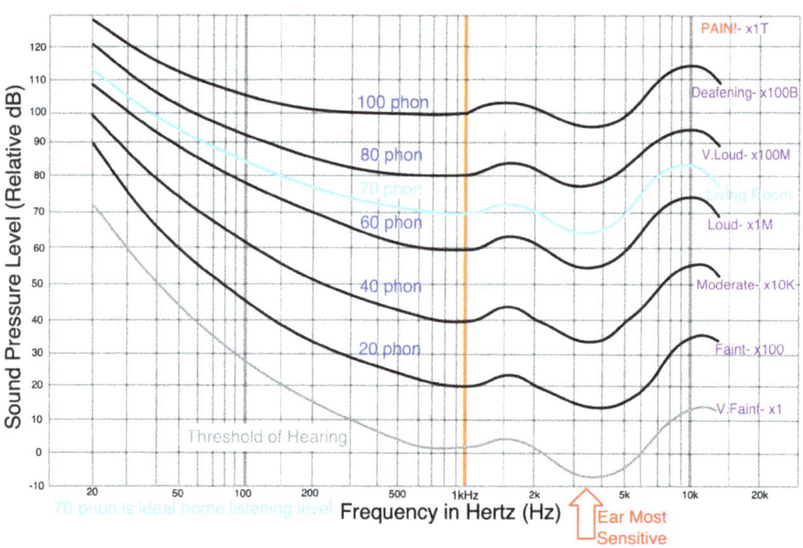

Figure 1.18 A generic Equal Loudness Contours diagram

It's interesting to see the ear is at its most sensitive at around 3.5 kHz at many levels. Over the years people have wondered why, one theory is that it is the frequency of a baby crying or a human scream, and so is an ancient mutual protection for the tribe. We don't perceive these differences; the brain renders all the frequencies "flat". At low volumes we are far more sensitive to the middle frequencies, and our ears flatten out as the volume increases. This effect is capitalised with a "loudness" switch in a mobile phone audio app or a switch on a home high fidelity (HiFi) amplifier which compensates for low listening levels by boosting the bass and top a notch to give the same hearing experience you would enjoy at higher levels. Finally, I have added in a 70 phon curve as that is the level professional bodies estimate most dedicated home listening happens, and so they factor this into sound creation (see digital monitoring levels).

HEARING IMPAIRMENTS

Sadly, as we grow into our mid-20s the pliability of the eardrum starts to reduce, and so the higher frequencies (from 20 kHz) are increasingly lost. At 60+, I am very fortunate to be still able to hear 8 kHz. This natural process is known as presbycusis. But is it the great problem it seems? Look back to the diagram on the range of musical instruments. I can still hear the full range of most instruments; however, I am beginning to lose the subtlety that is in the harmonics of the smaller high-pitched instruments. Like many things as we age, I'm afraid it can only go one way! Other problems can affect our hearing. Listening to LOUD music is OK, listening l-o-n-g is OK but listening l-o-n-g and LOUD will damage your hearing; don't doubt it! There is a loss of sensitivity or deafness often associated with the frequencies of the original sound that damaged your hearing, and you may start to hear phantom sounds which can be temporary or permanent.

Tinnitus is most often a phantom ringing or singing in one ear or both. If you are particularly unlucky the tinnitus can be a deep throbbing hum, or maybe a combination of several pitches at the same time. People have ended their lives for less! A theory is that painter Van Gogh hacked his ear off in an attempt to silence the screaming in his head. I always carry tiny noise-attenuating ear plugs with me (the size of in-ear bud headphones) so I am prepared. With these, the sound you hear is the same quality, just quieter. The same as you don't stare at the sun as you know it will damage your eyes, look after your ears at noisy events, and if you get any temporary tinnitus, it's nature's way of telling you to take better care next time. You have been warned!

Finally on the ear, there is the "Cocktail Party Effect". You may justifiably suppose this is staying in bed till noon the next day, but I'm here to tell you it's the genius of "attentional selectivity". A young brain is amazing at sorting which direction sound is coming from by multiple lightning-fast computations comparing the phases of sound arriving at each ear (and having two good ears is vital), combining them to create a complex sound stage in the brain. Amazingly, a pair of young adult ears placed in a noisy environment can track several voices, lock on and the brain can focus on the sounds that are wanted and reject the sounds that are not. This automatic selection is particularly prone to triggering on hearing your name or, as the textbooks politely put it, "a taboo word".

As we grow older our brains slow, and the surrounding unwanted sound starts to become a problem. This longer latency tarnishes our ability to hear others at the table in a noisy or hard-surfaced environment (busy restaurant), often past the point of being able to pick out the words at all! Further it is the same area of the brain responsible for attention in the first place, so not only is the conversation lost, but also the willingness to participate in the occasion at all! This dual aspect of zoning out is called the "Cocktail Party Problem". It's amazing how many older people can compensate with lip reading, but of course that only works if you can see the lips! Add that into the loss of diction from the high frequencies going south through presbycusis and you'll remember to take gran to a quiet café! We'll come back to the subject of how to get a clear sound to an audience in "Acoustics".

TRANSDUCERS

The ear is a "transducer". It changes air pressure waves to tiny electrical signals which are sent along our nerve pathways to the brain to be deciphered. But what of outside the body? If we want to capture a sound to send it elsewhere for a purpose, we need to turn it into something we can control and manipulate. Luckily, we can use the perfect medium of electricity. In the squash and expansion of air that is sound, we can place a microphone to capture this movement and turn it into a very small voltage. The squash or positive air pressure creates positive volts and the suck of negative air pressure as the air retreats creates negative volts.

This flow of electricity, which moves plus and minus of 0 V is known as "alternating current", an "AC signal", and is (as good as the quality of the microphone will allow) a perfect copy of the waveform produced in the air. Now we have the sound captured as a voltage, we can make it big enough to be useful elsewhere; we can choose just to amplify certain frequencies to modify what we hear, such as bass and treble controls and we can add this signal voltage together with others to make "a mix" of different sound sources, like a sound desk in a band perhaps?

What then? Well, we could amplify it further and put it through another transducer to turn our AC signal back into sound, a mechanism very similar to a microphone but many times larger and requiring a much stronger AC signal. I am sure you have guessed this is a loudspeaker, and it requires a "power amplifier" to make the AC signal strong enough to move a large speaker diaphragm and take our original small sound to a large audience as a public address (PA) system. An alternative would be to use a power amplifier mated with an ultrasonic "carrier" signal to broadcast the signal to many people as a radio channel. Perhaps we want to record the AC signal for use later, either by capturing it in the past on magnetic tape, or now "sampling" the "analogue" voltage many times a second to "store" it as a "digital" facsimile on a computer. All these options will be explored later.

STEREO SOUND

So far, all our theory has been based on a single point source. Life is so much more complex than that; if you stop reading and listen (unless you are in a sound studio yourself!) you will hear so much going on around you: distant shouting from the street, a road drill, the dog next door, a baby crying upstairs and a TV in the next room. Your ears and brain are designed to give a good idea of both direction and distance of all these sources in real time.

This "auditory perspective" was first described by Leopold Stokowski in 1935, and sound engineers quickly realised that the idea of a single point of sound was not only undesirable but also unrealistic. Life is not black and white but endless shades of both subtle and vibrant colour! Early multi-channel sound experiments used headphones rather than loudspeakers, and that should be borne in mind as we still use those early findings today. There are four basic sound configurations: monaural, binaural, monophonic and stereophonic (Figure 1.19).

We don't think twice about whether we listen on speakers or headphones, but the pioneers perceived a great difference when the sound could interact with the acoustics of the room. Delivery at the ear is "aural", and into a space which can modify it is "phonic". A single point sound source is known as "mono", meaning one, and the multipoint source is known as "stereo", meaning many. We have come to expect "stereo" applies to just two speakers, however, that is simply a minimum, and more can be used to create the Stokowski "illusion of the original sound in its environment". However, a step further with

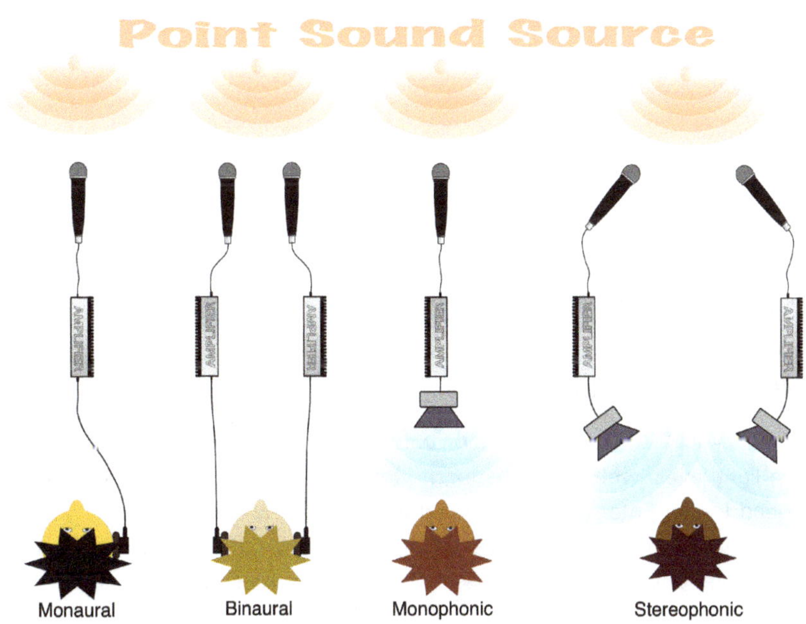

Figure 1.19 Sound reproduction types

multiple speakers, both behind and above, has led to a modern call to say, "but it's more than stereo!" and so the term "multichannel" for cinema-style sound has stuck.

With headphones the sounds on each channel can't interact with the room to recreate Stokowski's ideal, but instruments are placed across a virtual stage in our heads and so at its simplest the illusion is different, but not unenjoyable. Our sound recording systems are routinely "stereophonic" now, two-channel sound is the default for both analogue and digital sources. Competent sound balancers are very aware of creating a sound that meets both the demands of room and headphone listening and modern tools like digital signal processing (DSP) help to improve and rationalise the listening experience for us all.

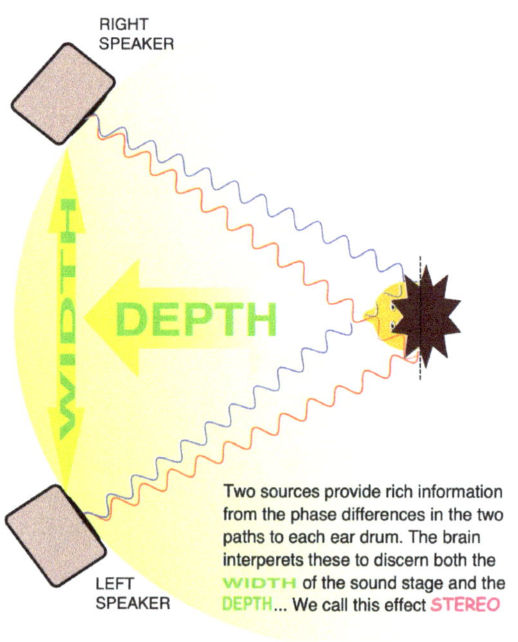

Figure 1.20 Introduction to stereo

The point of stereo is that you don't "hear" the two speakers but are aware of the rich virtual sound stage that is created between them (Figure 1.20). Listening on the most basic of systems should allow an interested listener to pick out the position of each instrument or section of players in that stage, and place which is nearer or further; even higher or lower, from the listening position. If the recording is "live", are you placed in the audience or on the stage? Listen and see! So many techniques, but the great news is there are infinite opportunities for you to invent amazing ways of using the medium of multi-channel sound in new and interesting scenarios.

MEASURING SOUND LEVELS

Just as engineers measure voltage with a voltmeter and photographers measure light levels with a light meter, it would be useful to be able to measure sound levels. Enter the aptly named Sound Level Meter! This tool is mainly used by environmental agencies such as councils and building constructors, to ensure legally stipulated noise levels are not exceeded, however, it does have uses for the sound practitioner. Earlier in this chapter I introduced a graph showing the equal loudness contours of the ear. On this graph we showed the left vertical scale with sound pressure levels (SPL) from the whisper of 30 dB

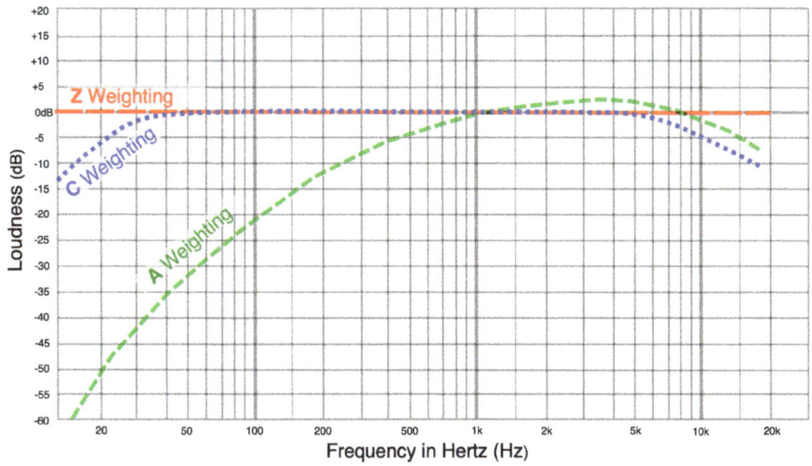

Figure 1.21 Sound Level Meter weighting

to the painful roar of 120 dB; this is set against the horizontal scale showing the human frequency response of 20 Hz to 20 kHz. We noted earlier as sound gets quieter our ability to hear the extremes of the frequency range narrows, and the middle frequencies dominate, and this was shown by curves where decibel levels were corrected by "equal loudness curves" and measured in phons.

The Sound Level Meter takes these changes into account, with "weighting" of the sound levels to match the ear. The most popular is "A Weighting" of the frequency response to give an adjusted view of how annoying background noises might be. The reading is then called A Weighted decibels, dBA or dB(A). This is the default but is often accompanied by a mode for "C Weighting" which is used with loud noise levels, particularly problems such as noise from entertainment venues where the bass thump might be the major nuisance; this is measured as C Weighted decibels, dBC or dB(C). The graph (Figure 1.21) shows why A Weighting would not work for bass problems, and why in any case, the weighting mode during the test must be registered with the reading.

Old Sound Level Meters had a flat frequency setting called Linear or "Unweighted" and this is the most useful for the professional sound engineer wanting to check out their speakers for a flat frequency response in the studio or lounge at home. Many modern meters don't have such a mode, and where they do, it is now called "Z Weighting", although for audio engineers the useful flat response of 10 Hz to 20 kHz is marred by having an allowed variation of measurement of ±1.5 dB in the specification. Clearly "A Weighting" is useless for precision audio work, and if you must use "C Weighting" be aware of the frequency fall off, particularly at high frequencies. One final point, it is unusual in a modern environment to find yourself with zero background noise. A loudness level of around 30 dB can be below your threshold of hearing, so when you see the meter leaping around on the lower decibel levels, it's not faulty; it's just doing its job!

BIBLIOGRAPHY

Architectural and Civil Engineering Department BBC Engineering (1980) *Gude to Acoustic Practice*.

Bronkhorst, Adelbert W. (2000) The Cocktail Party Phenomenon: A Review on Speech Intelligibility in Multiple-Talker Conditions. *Acta Acustica United with Acustica*, 86, 117–128.

Fletcher, H. and Munson, W.A. (1933) Loudness, Its Definition, Measurement and Calculation. *Journal of the Acoustical Society of America*, 5, 82–108.

Mackenzie, G.W. (1964) *Acoustics*. Bungay, Suffolk: Focal Press Limited, Richard Clay and Company Ltd.

Moore, J.E. (1978) *Design for Good Acoustics and Noise Control*. The Macmillan Press Ltd.

NASA.gov (2025) *The Science of Sound*. https://www3.nasa.gov/specials/Quesst/science-of-sound.html

Parkin, P.H., Humphreys, H.R. and Cowell, J.R. (1979) *Acoustics, Noise and Buildings*, 4th edition. London and Boston: Faber and Faber.

Sengpiel, A. and Sengpiel, E. (2025) *Standing Waves*. https://sengpielaudio.com/StandingWaves.htm

2 Professional Podcasting

The information in this chapter is available in more detail in other parts of this book, but such is the passion, variety and sheer quantity of burgeoning podcasters, it makes sense to support this vibrant self-expression with a synopsis of techniques; a shortcut to producing great results quickly and on a tight budget. I want to state, however, that reading the rest of the book will improve your installation, the quality of your work and give you transferable skills in support of growing your hobby into a professional career.

ROOM POSITION

You will find that the chapter and appendices on acoustics provide far greater detail; however, here is a simple set of rules to get you up and running. Choose a room to record on the quiet side of the house, away from traffic. First off, don't be tempted to site your podcasting equipment near a window or wall, particularly near a corner, even though that will seem the practical ideal to fit in with other activities. At these hard "boundaries" of the room the sound waves bounce back, overlapping with other reflections to taint your sound source, often irredeemably colouring the microphone pickup. Some frequencies, particularly at the lower or "bass" end, are lost, or worse, enhanced above normal in "eigentones" to give a boomy sound. A table placed away from the wall is best, and looking into the space is better than staring at a wall. Curtains across a window, particularly thick well-pleated ones, will help to tame the sound in the room.

Pay attention to the fit in the window and door surrounds and take out some time to ensure all the cracks are well sealed with a flexible profile draught excluder to give you much-needed silence. A carpet on the floor, and substantial tapestries to absorb sound on the wall (mounted on a timber batten to make a small air gap of 20 mm to 50 mm) will be amazing, particularly if placed opposite a blank wall. Don't put absorbents opposite each other as it will dampen the room too much. A deep bookcase, filled with a jumble of books on one wall (the long end is best) will help break up discrepancies in the acoustics and produce a superior sound across the frequency range. Soft furnishings are a welcome addition and avoid blocky hard featureless furniture like a sliding wardrobe front or a run of kitchen units. There is a lot more detail in Chapter 8, "Introduction to Acoustics", and for the adventurous, practical help to "Build Your Own" in Appendices 2 and 3 to create simple effective studio absorbers to suit many situations.

DOI: 10.4324/9781003717065-2

WORKSTATION

It's very tempting to place a computer screen behind a microphone so a script can be easily read, and the keyboard accessed for prompt actions. The move to larger screens has not helped here; the flat expanse of the screen reflecting the sound of your voice straight back onto the rear of the microphone and colouring the sound. It's best to place a script on a tablet placed to the side or propped down on the table. The table itself should have a decent cover of thick felt on it, and if you have a script rack for the papers or to prop up a tablet, this also needs to have a felt covering. Avoid placing the microphone on a table stand, as any unconscious knocks to the legs or top will pass up through the stand to the microphone as low frequency thumps.

An Anglepoise microphone stand is so much better as it brings the active parts much closer to the mouth, and even if the stand is anchored to the table, a "shock mount" or "cradle" holding the microphone in place will cut back any structure-borne noise problems. The chair should be comfortable and squeak-free, and one on castors will help you get in the correct position easily.

MICROPHONE CHOICE AND TECHNIQUE

There is an easy choice of "polar" response pattern. Close working "cardioid" is best, and that will likely be a capacitor design (also known as condenser). There is a lot more information in Chapter 4 "Microphones". There are terrific dynamic designs such as the Shure SM7B, but as capacitor microphone prices have crashed, the value for money is clearly on the side of this technology. The next choice will be whether to buy a unit with an analogue or an analogue to digital interface (USB). At the time of writing, it is now possible to have both in a single microphone; the analogue terminates in a solid XLR plug or sometimes a small 3.5 mm jack (for cameras and such like) or a digital interface which terminates in a USB style plug or a proprietary connector for direct digital connection every to a phone or tablet (formally iPhone lightning, now almost all phones USB-C).

Buy a known brand so you can check out the reviews (buy from a reputable dealer as there are a lot of counterfeits of popular microphones); reliability is suspect in the cheaper microphones as they can go noisy or die completely in a year or so. If you are serious about your art, US$100 is an absolute minimum, and the sweet spot of vocal performance versus value is US$250 to US$500. There are fabulous esoteric makes, with passionate advocates, but I'd be tempted for a vocal mic to go to the old school makes where robustness and longevity live alongside reputation and reliable sonic quality. The room will change the sound significantly, so if you can try before you buy, it's well worth doing!

Many microphones have an internal "pop" filter of fine mesh or open foam. It has become common to have an external filter, perhaps for the look of it more than the function! There will be a slight loss of top (crisp) frequencies with each additional filter; however, that must be placed against the disaster that is the "mic blasting" mechanical overload of percussive vocal sounds (mainly Ps and Bs) such as saying "Pop!" or "Blast!" or similar. "Close Miking", bringing your mouth to between 10 mm to 50 mm of the microphone, will enhance the bottom frequencies in an acoustical foible called "bass tip-up", which will suit some as "sexy" and repel others as "boomy". The choice is yours; it's your technique, but clearly close working will make "mic blasting" more likely. As in all this chapter, experimentation and experience will win out to make you, well, *you!*

MONITORING

A rookie mistake is to have the microphone live to the loudspeakers in the room at the time of recording. Even at low volumes the microphone pickup will be coloured by this feedback, and that sound is *never* an improvement. Two choices are left, no listening or using headphones. It's incredibly important to "monitor" your recording as you make it, as a check that all is going OK. This is particularly true if there is more than one person, and much can go wrong if the excitement builds in the conversation! Headphones come in all sizes, prices and technical qualities, however, there are only two acoustic types: those which seal the sound to the ears, and those that don't. Most ear buds seal, over-ear headphones are sold about half and half. "Open backed" allow leakage of sound and "closed back" don't. The latter is *essential* in a live recording environment; the open backed are useful for "airiness" in control areas and living rooms where there are no live microphones, but decent closed back are nearly as good here too, and as a bonus you don't hear the neighbours when listening to your favourite playlist!

So, the choice for podcasting is closed back, and normally the label "Monitor" on the package should suggest the sound is "neutral"; that you will sound the same played on listeners' systems as in your recording room. Sadly, this is often not the case, and whilst these headphones may sound great, they may not be an accurate portrayal of what the microphone recorded. I'd like to tell you reviews would help here, but there is muddled thinking on this subject. My advice is again to buy a branded product, however, know that truly neutral sound is rarely found below US$150 and be prepared to spend more to have confidence in this crucial area. European makes, in general, are still king in telling the truth, but I realise they are not always the most dynamic sounding... But you are not making it just for you to listen to, are you?

MULTIPLE PARTICIPANTS

I've touched on multiple participants above, but let's look a little further. If the podcast is simply informative, a sole presenter, particularly on camera, is a great way to get information across. If the goal is entertainment, the cut and thrust of multiple voices and opinions is absorptive and rewarding. I would recommend from the first that the room design considers whether additional people will take part. It goes without saying that being pushed against a wall is no longer an option, as you will need eye-to-eye contact, probably across a table.

A second microphone should be as the first, cardioid pattern, mounted off the table onto an Anglepoise with a shock mount and a pop filter. However, a simple arrangement of weighted desk stands placed in front of contributors will work *if* the table has 3 mm felt across it and if the mic stands are isolated from the table with (slightly compressed) neoprene foam pads.

Make sure nervous legs don't kick the table, and if you provide headphones to improve the contributor experience, make sure they are on the head and not on the table! After working with many well-known but deaf DJs, I strongly advise individual headphone amps for each person, with individual volume controls. This is doubly so if they bring their own headphones, as different makes have very different sensitivities, and so a given volume setting can produce vastly different sound levels at the ear. If this is not possible and you have to split a headphone output, "in-line headphone attenuators" are available for under US$10 on the internet. I carry one in every headphone case (and a splitter), so I can't be caught out by a deaf guest!

RECORDING AND EDITING

I am assuming that you have grown used to the interface you have with the technical equipment, whether it is to a laptop, or a dedicated device. I can't suggest a single solution, as there are as many products as there are podcasts! I do ask you to practise with your choice and to particularly listen out on the headphones for any distortion that might take place. I advise you to look up the paragraphs on Dynamic Range in Chapter 6, "Sound Levels and Connections" to understand the concept and to keep the model in the back of your mind as you work. Assuming that you are not live, I'm sure you will want to "tidy" your efforts with a spot of editing. Again, this is an area covered elsewhere in this book, but I will say a few precise words. There are several free audio editors for most platforms, and the only difference is likely to be the quality of the "plug-ins" to augment the sound. Even the most basic Digital Audio Workstation (DAW) will allow you to "amplify" your session to maximise the volume and to compress it (make it louder) with a bit of "soft limiting". Other effects are useful, but none more than you do a decent job in the first place!

If you have the luxury of a recording, you will want to streamline it before uploading to make it the best it can be. Straight off the recorder the audio file will have a machine-given file name, like 100723_0125.wav. As soon as you have transferred the file over rename it with a descriptive name. An interview with Alex Jones in January is better named AJones_Jan.wav and better still AJones_Jan_raw.wav. This is to indicate it is the base recording without edits. As soon as I start an edit, I rename the file with "Save As" to AJones_Jan_*edit*

.wav whilst I do the work, and when I have a completed project, I rename to AJones_Jan_*final*.wav so I know that's the one I need to use. The advantage of keeping the raw version is that you can always go back to the source if you lose something in the edit or want to recut the interview for a different purpose; maybe a change of duration or the focus of the interview for, say, an obituary.

My simple rules of editing follow the classic BBC Radio 4 programme "Just a Minute". Avoid (1) hesitation, (2) repetition and (3) deviation. *Keep these in mind and succeed!*

1) Hesitation – people will always let their brain get behind their mouth. For some it is a blank pause, for some the pause is automatically filled with an "uuumm" or an "aaaahhhh". Regardless, it's not a great attention grabber, and is best gone. A simple cut from the beginning of the Um straight hard on to the following word where they have caught up normally works perfectly with their timing, quick and clean. *But* (and it is a big but) *breathing is not a pause*, it is a natural necessity and is expected; otherwise, the brain will be nudging you that something is fishy. Keep it natural; whilst you may remove a few breaths for brevity, check that you have not made your guest appear superhuman!
2) Repetition – this is a lot like the first point... People will "loop" rather than um or ah, filling the hole in their brain freeze, with another go at a point they've just made. Obviously, you don't want or need both, and in my experience the second take is often the more succinct and well put. Before consigning the first take, do give it a listen and ensure you are not throwing any gems in the bin that are worth saving, and be careful of lopping off the joining words. In transferring the phrase, it's easy to leave off a "to" or "and" in the cut, so listen carefully as you excise.
3) Deviation – I hope you have a good reason to press the record button; you are there with a theme, a goal even, to win an ARIA! It's amazing how many people just take whatever they have gleaned, tidy it a bit and shove it out. To be interesting an interview must have a purpose and a flow, and the interviewer has a responsibility to take the guest on that journey. If the season-winning goal scorer breeds budgies with his uncle, I'm sure it has its place, but not today! Keep on track, manage the rhythm of the questions and when your guest goes AWOL, be decisive in bringing them back on track and be ruthless in cutting out the irrelevant answers.

When editing always try to leave and return on a "hard" syllable, even if you must choose to cut in the middle of a word. D, K, O, P, Q and T sounds are good cutting points, F, H, R, S, Y and Z are not. Levels, intonation and energy are your enemies in excessive re-takes, but by careful attention to pauses to maintain a natural flow, most obstacles will be overcome.

ON LOCATION

Figure 2.1 Several portable recorders with pop and wind shields
Photo by the author.

If recording out and about, the microphone within a mobile phone is not yet sufficient on sensitivity and quality. Microphones that plug directly into the lightning (iPhone) or USB-C (Samsung later iPhone and others) can be great, but outside wind can be a killer. Make sure you carry and use a proper furry "wind jammer" (not foamy!). I ensure I buy the right wind jammer each time I buy a new portable recorder (Figure 2.1). As to choice of recorder, my advice is to keep it simple and read the reviews for the current favourite. Some stand-alone recorders will give more choice in recording format and quality at the expense of operational complexity. If it's a simple speech recording, the cheapest stereo recorder from a respected brand will be capable of very similar quality to the more expensive ones. Buying stereo allows you to adjust the levels between you and a guest in the edit.

A few points:

1) Check before and pack any research notes, questions, etc.
2) Research their lifestyle and interests, so you can put the interviewee at ease by finding common points of interest before the record button is pressed.
3) Make sure the recorder has a memory chip in it and that it has enough space on it for the recording.
4) Pack a wind jammer (even if you believe you will be inside).
5) Put in fresh batteries, much cheaper and less embarrassing than setting up a second meeting! (Take a new spare set!) After, throw the partly used ones into a box for remotes, etc.

6) Take a camera (or use your mobile) for a selfie with your guest.
7) Do a test recording and listen back to it to ensure everything works before you meet the interviewee. If you can, listen as you record on headphones (dependent on recording system).
8) Take a recording of the background noise for 30 seconds, so you have authentic atmosphere should you need to patch parts because of other production values.
9) Listen to what's around you, before and during the recording Don't get so absorbed you miss passing motorbikes and distant road workers with jackhammers. Intermittent extraneous noises can make the edit next to impossible.
10) If you use a table microphone mount, check the interviewee doesn't kick the table, putting thumps on your recording. Use a foam pad (neoprene is great) to isolate the mic from the table.
11) Don't stop the recorder for pauses; you risk forgetting to start it again to continue, and each time there will be a new file name to patch together. If you must break the recording, be sure to pick up properly so you don't leave any gaps in the story It's easy to leave out crucial linking words in second takes.

Chapter 13, "Planning for a Location Recording", gives detailed help for location work if that is core to your podcast activities.

A CHOICE OF KIT

A minimum location kit is shown in Figure 2.2. Inside buildings, I have the choice of handheld (careful not to move your fingers) or table mount on a tripod. I have shown a 150 mm × 150 mm neoprene pad under the tripod to stop structural noise. The recorder is shown vertical, as it would be if your guest was seated the other side of the table. This gives good eye contact. Clearly (and I have seen it done!), the microphones point up towards faces, not 90 degrees out to the sides! If you are side by side (better to handhold in this situation), you would angle the recorder towards the pair of you, and if you are recording a room, place the recorder on a shelf, using the pad and point the recorder at the action (vaguely horizontal I would suspect). If you are close working nearer to the microphones (less than 200 mm [8 inches]), then use the foamy pop filter to stop mic blasting. You might as well use it anyway, just in case, and it does look more professional!

Outside you should use the wind jammer as even if there is no wind, you can guarantee a breeze will build while you are recording. I've shown new spare batteries in the photo as a reminder to always take them, and I normally have a battery checker in the glovebox whenever I go on location. The kit shown in the photo will fit neatly in a child's pencil case, and I often use one for this purpose.

Figure 2.2 An example of a kit for a location recording Photo by the author.

VIDEO BLOGGERS – MICROPHONES IN SHOT

There are two current schools of thought in video recording of podcasts. The first is to make a point of placing the microphone close in shot in front of the blogger as a valued prop; this is loved by technologists and is almost bulletproof in creating a good sound. That may not suit your style and gets awkward with more than one contributor. If there is a "round table" setting table stands are great as they leave the face unobstructed yet are close to get a good sound. Be sure to put each mic stand on an "isolation pad" of neoprene foam (or similar) to stop table leg kickers destroying your masterpiece.

There are many great lapel-mounted wireless microphone kits on the market, and these either have a receiver that connects directly to the camera or a dedicated sound recorder. Don't place the mic too high on the clothes so it is under the shadow of the chin; otherwise, you will lose the higher frequencies. Place it on the outermost clothing so nothing can brush over the face of the microphone when moving, as it causes unwanted rustling on the recording. If the lapel mic is too bright, simply mount it upside down. If indoors remove the obligatory wind jammer so the microphone is far less obtrusive. Again, make sure any battery-powered units are fully charged before the day begins.

Noisy venues, with interviewing, are best managed by a single cardioid, directed closely at the person who is talking at the time; but this is an extreme, and it is better to keep the microphone between you and stationary so nothing is missed, adjusting the microphone distance between you only when one person is clearly louder than the other.

A chapter which includes going out and about would be remiss without mentioning safety. Ensure any cables on the floor are routed so they don't become a trip hazard, use mains extensions sparingly, and ensure they are in good condition. Take breaks to get fresh air, keep any drinks well away from your equipment and if walking whilst interviewing watch your step and don't get so lost in the interview you don't notice annoying background sounds on your route. It is common to find a waterproof mobile phone, but much rarer to find a waterproof camera, microphone or laptop, so please look after your precious investments!

A reminder that these topics are expanded in other chapters of the book; I hope you take up the offer and enjoy!

BIBLIOGRAPHY

Alton, F. (1984) *Acoustic Techniques for Home & Studio Everest*, 2nd edition, 4th Printing. United States of America: Tab Books Inc.

Architectural and Civil Engineering Department BBC Engineering (1980) *Guide to Acoustic Practice*. BBC Engineering Training Department.

Mackenzie, G.W. (1964) *Acoustics*. Bungay, Suffolk: Focal Press Limited, Richard Clay and Company LTD.

Walker, R. (1990) *The Design and Application of Modular, Acoustic Diffusing Elements*. BBC Research Department, Engineering Division.

3 Introduction to Electronics

Why is electronics in a book about sound? As soon as we want to make sound bigger or take it outside the room, we must change it into an electrical signal so it can travel on wires. This AC voltage (Chapter 1, "Introduction to Sound") lives in an environment where it needs to be managed so it doesn't get corrupted by unwanted signals and is not damaged by reckless changes, so it comes out the other end in a pristine state. The knowledge gained here will help you solve problems in many sound situations and allow you to produce recordings with confidence. *I promise I'll keep the maths simple (and that's all that's needed!).*

Electricity universally exists in a ring of wires, a circuit, an "electrical circuit". If we break the ring the circuit stops working. When the circuit is complete, electricity flows around it, using some sort of mechanical or chemical pump. In a torch the pump is chemical, namely a battery or batteries joined together, and practically we place in a switch so when the light is not needed, we can break the circuit and save the energy of the batteries.

DIRECT AND ALTERNATING CURRENT

Figure 3.1 shows a real torch. It uses two 1.5 volt batteries which added together make 3 volts (V) to power the 3 V bulb. A sprung strip on the left joins one battery to the other (note they need to point in the same direction so plus goes to minus) and a plastic peg pushes two brass sprung contacts together to complete the circuit (see red ring) and make the bulb light. The circuit diagram of the torch in each state is shown in Figure 3.1.

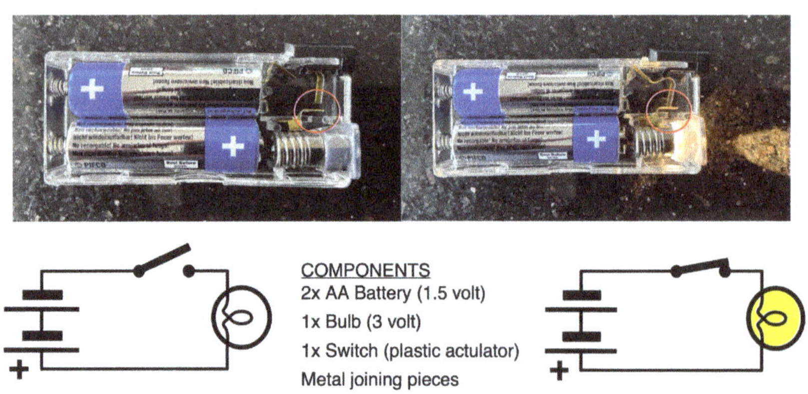

Figure 3.1 Torch, unlit and lit, and the circuit diagram equivalents Photos by the author.

DOI: 10.4324/9781003717065-3

The electricity from the battery flows in one direction, (traditionally) from the positive or "**+**" to the negative "**–**". This is called "direct current" or "DC". If the light is in our home and it's turned on by a wall switch, *it is still a circuit*, just like the torch. However, the mechanical generator for the power that comes to our home reverses its direction of current many times a second, simply because it's easier to make and transport electricity in this way.

This is called "alternating current" or "AC". You will be pleased to know there are just two types, AC and DC. The DC needed for our circuits comes from one of two sources: chemical reaction (a battery) or AC converted to DC by using a device called a rectifier. But how do we make the AC? This is a simple physical process you probably learnt in school.

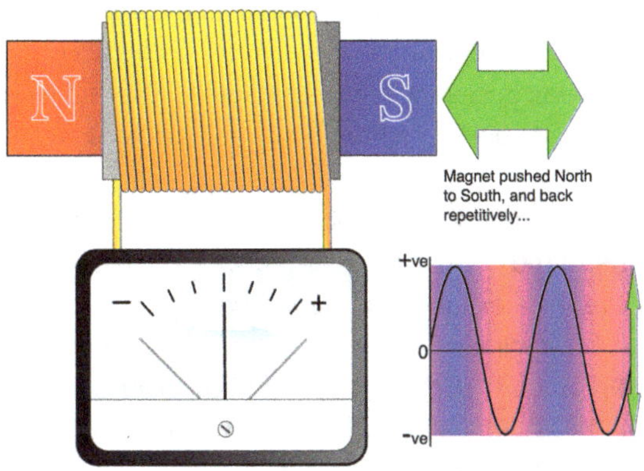

Figure 3.2 Simple AC generator with meter display

Figure 3.2 shows a simple bar magnet placed into a coil of wire. It is then physically moved in and out of the coil. In Figure 3.2, moving the magnet from north to south creates positive volts; moving it back again creates negative volts. If the magnet stands still, there is no output. If we could create a smooth cyclic motion back and forth, a sine wave would be produced. In a power company generator, the magnet spins on a shaft, presenting north then south as it rotates over adjacent coils; the principle is the same, just a different application. In our simple generator, I'm sure you can see we could equally hold the magnet still and move the coil up and down it, from north to south and back… We can choose whether we have a moving-magnet or a moving-coil system.

As discussed before, when sound is converted to electricity it is "AC" and follows all the same physics. A microphone is simply a very small, very weak AC generator. Take a magnet, and place over it a coil. Attach the coil to a thin diaphragm which moves in sympathy with the incoming sound pressure. We have made a "moving-coil microphone".

PRACTICAL ELECTRONIC CIRCUITS

Let's get into some practical circuits! We'll start with the simplest of circuits, the one battery torch shown in Figure 3.3. When the switch is closed the bulb receives the battery volts across its terminals. The bulb draws the amount of electricity it needs to light; this is the current flow.

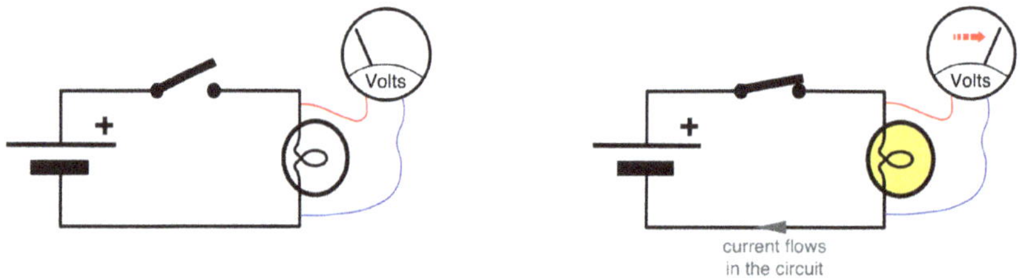

Figure 3.3 Simple bulb circuits, one on, one off

What is weak and what is strong? I guess it's all relative. A super soaker can reach you the other side of the garden with a little water. A full bucket poured over your head is a lot of water, but it can't travel far. Work is being done in both cases. You are wet, and the water has travelled to reach you. The super soaker is an example of a high pressure, low volume delivery. The bucket is a low pressure, high volume delivery. In electricity the pressure or "potential difference" is measured in "volts" (symbol **V**), and the flow (amount) of "current" (symbol **I**) is measured in "amps".

The combination of volts and amps does an amount of work which is called "power" (symbol **P**) and is measured in "watts". In the above examples a similar number of watts could have been expended by very different combinations of values of volts and amps, but their combination creates the power necessary to do the job (**V** × **I** = **P**). There is another basic concept to introduce. In the torch, the bulb used the battery power to change electricity to light; the bulb drew power from the circuit. This loss of electricity is called "resistance". Resistance in "ohms" (symbol **R** or **Ω**) is measured in reverse which makes the maths easy. The *smaller* the value of resistance, the *bigger* the load on the power supply (zero ohms [0 Ω] = a short circuit). The current **I** *times* the resistance **R** equals the number of volts **V**: **I** × **R** = **V** and this rule is called "Ohm's Law". The two formulae are shown in Figure 3.4.

If you know any two, you can work out the third value. It sounds complex; but hang on, in time it's as natural as putting on your socks!

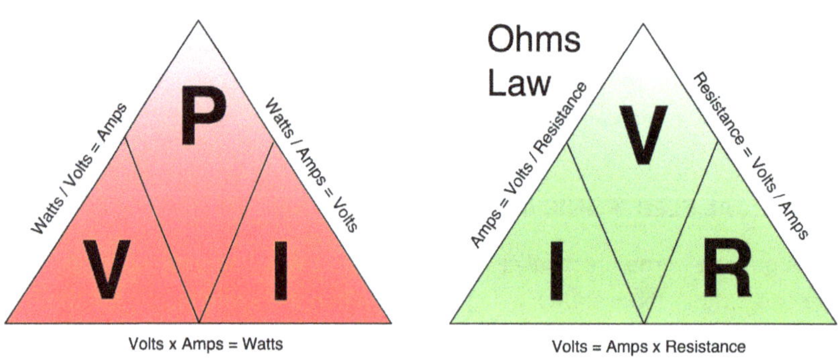

Figure 3.4 These formulae allow us to find a missing value if we have the other two

SERIES AND PARALLEL

Our torch with its old-fashioned bulb is about the simplest circuit you can get. Let's take it to the next level... We want a brighter torch! Let's add another identical bulb. We could (1) strap the two wires across the existing bulb, or (2) break into the circuit and add in another bulb like a daisy chain (as we do for batteries) and then reconnect. In either case we need double the power, yet there is a fundamental and important difference in the setup shown in Figure 3.5.

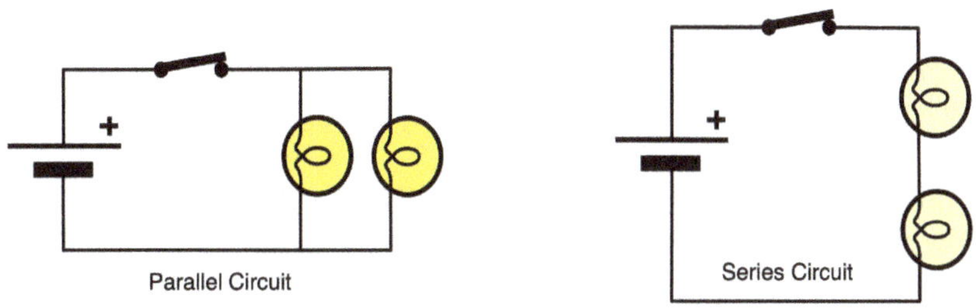

Figure 3.5 Simple series and parallel circuits

In Figure 3.6 we see the first one has the *same* voltage across both bulbs (parallel connection) but the *current* now flows through two routes, doubling the current drawn from the battery. In the second the voltage is now split across each bulb and so each one "sees" half. The current flow (as the resistance of each is identical) is halved and is the same flow through each bulb. This is called a "series" connection. To make the torch work properly we must double the available power. In the series circuit we need to double the voltage (another battery!), or the bulbs will be dim; in the parallel circuit we need a battery capable of supplying double the current or it will be flat in half the time.

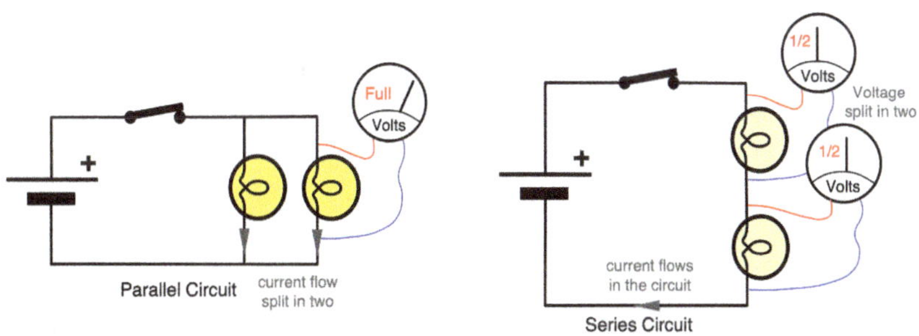

Figure 3.6 Simple series and parallel circuits showing voltage and current

ELECTRICAL SYMBOLS

A bit by the back door we have introduced the idea of electrical symbols used in "circuit diagrams" to represent different components. I'm sure you are now familiar with the battery, the bulb and the switch. Let's introduce a new one, the pure load of the "resistor".

Figure 3.7 Resistor design standards

Different international standards are used in defining this simple component, the UK Institution of Engineering and Technology allows both "the box" and the "squiggle" (Figure 3.7). If you look online at international designers of circuits each gets used about the same amount, but I prefer the "squiggle" symbol when sketching out a new design on scrap paper, and I think it is less likely to be misinterpreted to be anything else. Other components also have variants of their symbol but in my experience, they normally look so like each other that you will be able to make a confident guess. Note, in Figure 3.7 of two identical resistors in series, I have put the resistor's value of 100k (also written as 100 kilo-ohms or 100,000 Ω) in "the box". Again, this convention is popular, but it's also OK to place it outside the perimeter like we have with the "squiggle". I have used my favourite convention in this book, as the content is intentionally international, but please don't be put off by "the box" resistor design when you see it.

POTENTIAL DIVIDERS

In a series circuit of the torch, we see that the voltage has been divided between the two bulbs. If the bulbs are substituted with resistors, if like the bulb, they are the same value, it will split the voltage in half. However, if they are different values, we can calculate the voltage split between the two simply by comparing their values. This is called a "potential divider". Resistors can be almost a short circuit of just a few ohms to almost a completely open circuit with values in the mega-ohms; however, useful values are mostly in the middle of the range of hundreds or low thousands of ohms (kilo-ohm) to many tens of thousands of ohms. Layouts can seem confusing, and one resistor in series with another might not be clear at first look. Have a look at Figure 3.8.

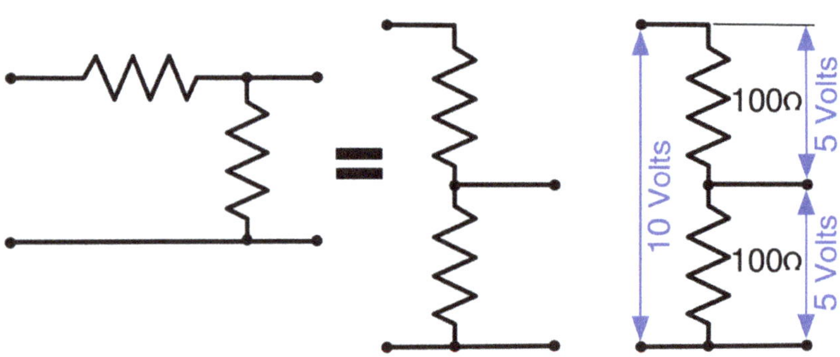

Figure 3.8 Introducing the potential divider

Figure 3.8 shows how the eye can be fooled, and as long as the source resistance of the battery is low and the load resistance applied across the output of the divider is high (ten times greater or more) the accuracy of resistor ratios splitting the voltage can be very high, to points of a per cent. Resistor values are shown by three (or four) coloured bands on the body which roughly follow the colours of the rainbow. Black and brown for "0" and "1" respectively are followed by red (2), orange (3), yellow (4), green (5), blue (6), violet (7), topped up with grey (8) and white (9), and an extra band indicates a guaranteed accuracy to within ± so many per cent.

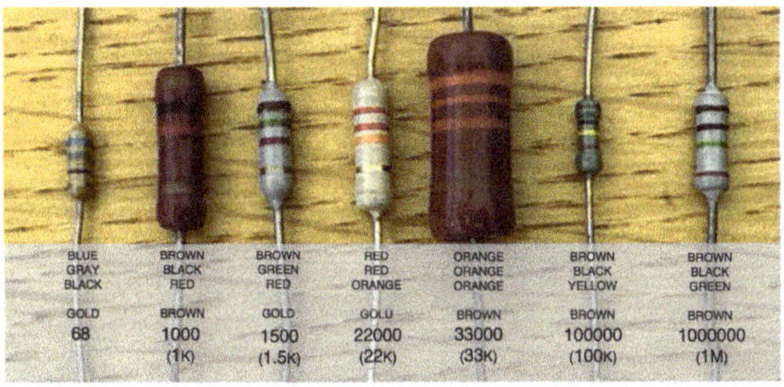

Figure 3.9 Resistors
Photo by the author

A silver band for within 10% of the stated value, gold for 5%, red for 2% and a brown ring for 1%. In Figure 3.9 I show a few from the junk box. What accuracy does each one have? Remember, it's not the absolute values that matter in a potential divider, but the ratio between them. If the load resistance is much higher than the resistor it bridges the accuracy will be fine! In Figure 3.10 there are a few more potential divider examples including an introduction to the "potentiometer" on the right, which varies the voltage split across a single resistor by a sliding or rotating contact. This could be, say, for a volume control or fader.

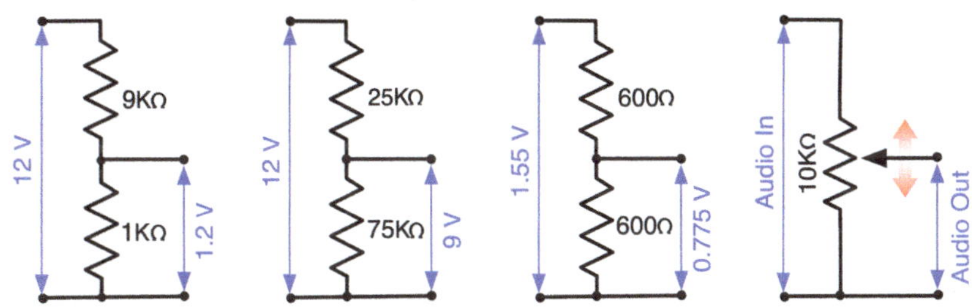

Figure 3.10 Four potential dividers show how voltage is divided by resistor values

Much of the time the internal resistance of the generator is insignificant. The car battery does not notice you have put the interior light on; it is unchanged... Let's take that same car on a cold autumn morning. The internal resistance of the car battery has risen from almost nothing as a natural consequence of the drop in temperature. Further our battery is old and has damage which has increased the resistance in each of its cells. This resistance may still be comparatively low, yet the engine starter (a big fat motor) is also low resistance (therefore high load!). If the internal resistance of the battery is, say, the same as the starter, the voltage is split like a potential divider, and so the 12 volt battery can only deliver 6 V to the starter as the other 6 V is lost in the internal resistance of the battery – sadly not enough volts to start the car and do the school run!

So sometimes the generator's internal resistance does become important, and in my experience no more than in sound engineering. Where the generator resistance becomes significant, we can show it as a resistor in series with the "load" circuit resistance (Figure 3.11), and the simple rules of series circuits apply.

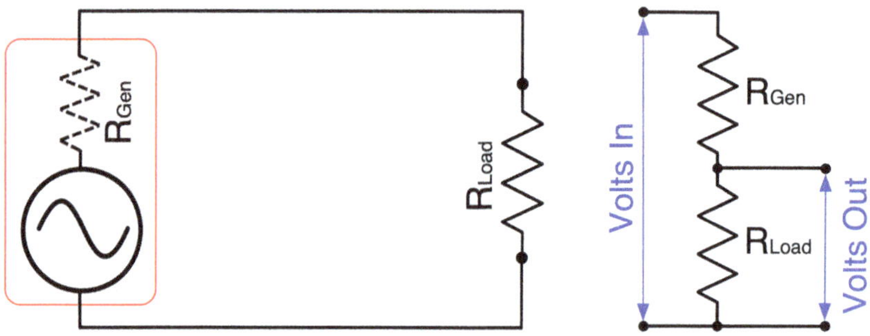

Figure 3.11 Generator internal resistance is in series with load

IMPEDANCE

If life was simple, anything that impeded the flow of electricity in a circuit would be pure "resistance". There are components we can use that change their resistance with what life is doing around them. We have specialist resistors which change value with the amount of light falling on them, which are useful for making night lights; those which change resistance with temperature, which we use as home thermostats and those particularly useful to us in sound, which change their resistance with the frequency of the audio signal passing through them. This can be annoying as we might not want the frequencies taken away by a bad circuit, but it can also be useful when we want to change or restrict what is happening in a circuit to make it do as we want!

COILS AND PLATES

A popular microphone, called a "dynamic" microphone, uses a fine coil of wire (attached to a diaphragm which responds to the movement of air) moving across a magnet to produce a tiny alternating current voltage. The blanket term for any coil of wire is an "inductor". This inductor has a residual resistance in the many turns of wire, but it also has another parameter to it. The AC resistance of the coil changes with frequency, so we have a unique term to explain the behaviour of the coil in an AC circuit. Instead of resistance we use the term "inductance".

Another sort of microphone uses the properties of spacing two plates of metal opposite each other with a thin gap between them. This is a type of capacitor, but there are many others for different tasks. The gap mostly has a form of insulator between the metal surfaces so we can roll up the plates (like a Swiss roll, jam separating the sponge layers) to save space. The "capacitor" or (old term) "condenser" microphone has one flexible plate and a close parallel fixed one with an air gap in between the two. Again, a capacitor changes its resistance with frequency; this time we call this unique capacitor created resistance, "reactance". We know resistors are measured in ohms (**Ω**). Component capacitors (**C**) are measured in farads (**F**) and in circuit inductors (**L**) are measured in Henrys (**H**).

Each of these two new units is very large so, like we saw, "bels" (B) of sound being impractical and so measured as *deci*bels (dB), you will find we work with fractions of Henrys and farads: milli (m) 0.001, micro (μ) 0.000001, nano (n) 0.000000001 and even pico (p) 0.000000000001! Table 3.1 helps to understand the values you are most likely to encounter.

Table 3.1 The most encountered component multipliers

Multiple	*Descriptor*	*Symbol*	*Use ([**bold**] indicates major uses)*
1,000,000	Mega-	M	*Resistor*
1,000	Kilo-	k	**Resistor**
1	–	–	**Resistor**
1/1,000	Milli-	m	**Inductor**, *Resistor, Capacitor*
1/1,000,000	Micro-	μ	**Capacitor**, **Inductor**
1/1,000,000,000	Nano-	n	**Capacitor**, *Inductor*
1/1,000,000,000,000	Pico-	p	*Capacitor*

The combined effect on an *AC* circuit of resistance, inductance and reactance is known by one simple overarching term: *impedance* (**Z**). All our AC audio circuits are bound to have a complex relationship because of the range of frequencies we need to manage and the physical structure of the transducers to turn sound to electricity and back again, so the source generation and load losses are *always referred to as the impedance* (***Z***) of the circuit.

We have talked about AC circuits, as that is what audio signals are; however, capacitors and inductors have uses in DC circuits too, where their attributes are used to achieve different ends. For instance, a capacitor in a powered DC circuit builds a store of electricity (a charge) like the inflation of a balloon on the end of a pipe. When power is removed, the balloon metaphorically deflates, this energy returning power to the circuit. Capacitors are used in all power supplies and in virtually all electronic timing circuits, from police car lights to kitchen egg timers.

Inductors in a DC circuit are like a flywheel; they take energy from the circuit, metaphorically spinning up a wheel, which, when the current flow stops, keeps spinning for a while and the energy (current) flow momentarily continues. This impressive DC feature is not made much use of in sound; however, in AC circuits, inductors make great microphones, loudspeakers, transformers and are highly useful in loudspeaker crossovers (see Chapter 5, "Loudspeakers").

The circuit symbols for resistors, capacitors and inductors are shown in Figure 3.12.

Figure 3.12 Resistor, capacitor and inductor symbols

AUDIO FILTERS

We can crack on and immediately make a simple circuit given what we have learnt. We know a capacitor changes its resistance with frequency; if we place it in a potential divider with a normal resistor, the output voltage of the network will change as the frequency changes. The "AC source" could be any sound captured as a voltage, and on the other side, we have some sort of resistive "load" the signal passes to. The graph in Figure 3.13 is a simple plot of level in decibels against frequency (low left and high right as usual).

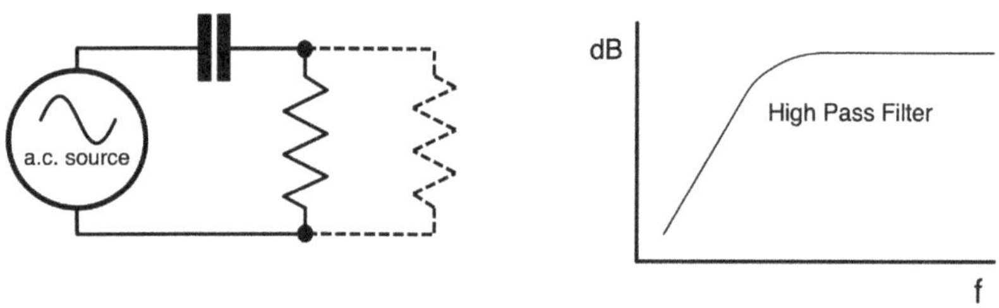

Figure 3.13 High pass filter (HPF) with capacitor in the input

The circuit in Figure 3.13 is used at the input of many audio amplifiers. From the graph you can see we lose the bottom frequencies, but the good news is that in DC terms the circuit does not let the damaging power supply volts through from the preceding circuit to harm the circuit beyond. The capacitor acts as a "DC block" which protects the circuit from external faults. So, what of the low frequencies we have sacrificed? The larger the value of the capacitor, the lower the frequency that can pass through, so a large capacitor will limit the loss of bottom frequencies to an acceptable level.

Note the term "high pass" to denote high frequencies pass unaffected, but the low frequencies are attenuated. In the last century this sort of circuit was used with a record turntable to lose the noisy low frequency rumbling of the bearings in the platter, so it was called a "Rumble Filter".

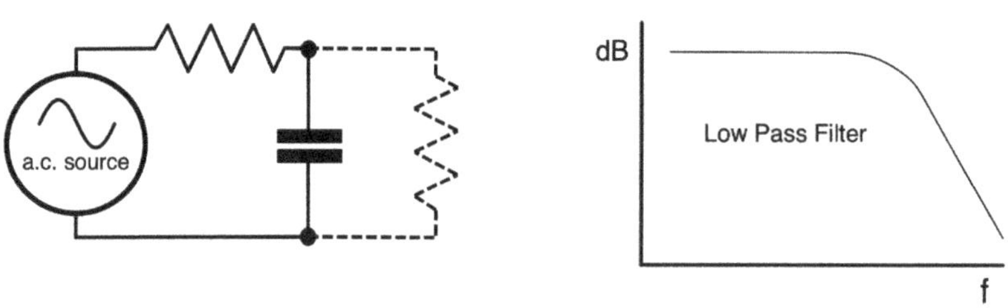

Figure 3.14 Low pass filter with capacitor across the load

A simple reversal of the resistor and capacitor creates a circuit that will pass the low frequencies (including DC current), yet it will roll off the high frequencies (Figure 3.14). A low pass filter. How could this be useful? Two thoughts come to mind. In the last century old vinyl records became worn through many plays and would be "scratchy". Many believed rolling off the high frequencies made the sound better, a "scratch filter". Another thought is that frequencies don't stop just because they are too high for us to hear them. Frequencies continue to rise through to ultrasonic waves and then to radio waves. It makes sense to roll off the high frequencies to include audio at 20 kHz but not much higher; otherwise, the radio waves can get in the wiring and interfere with our perfect audio. A current large problem in older sound desks is the intrusion of a nearby mobile phone "polling" with the nearest mast, a sort of "I'm still here!" – a ga-dic, ga-dic, ga-dic, ga-dic sound. Without taking steps to shield this signal from our circuit, the audio gets ruined. It's also worth mentioning that these two circuit configurations are essential in the formation of "equalisation" (EQ) the Bass and Treble controls found on home and professional sound equipment alike.

Coils can be used in a similar way to capacitors, but just reversed with the resistor to achieve the same result. Where the capacitor is in series, the inductor would be in parallel. Where the capacitor is in parallel, the inductor would be in series. Capacitors are used more often simply because they are cheaper than inductors to make…

Let's take a special use of inductors, the transformer (Figure 3.15). If we take two coils of wire and place an iron-based coupler (like a ferrite rod) through both centres, an AC current will pass from one coil to the other. We assume 100% efficiency and state "Power in" = "Power out". If I put in 1 mW, I expect to get 1 mW out. If the two coils of wire are the same, volts in = volts out, and current in = current out. So, what have we achieved by installing a transformer?

The transformer has completely blocked the transit of DC so like the simple series capacitor above, a transformer is a useful "DC block" to isolate the power supplies between connecting circuits. In addition, the tie to the common earth path is broken, output voltages "float", the output being the difference or "differential" voltage so any nasties in the source current are cancelled out by "balanced line working". However, it has one more very useful feature. If we change the number of the coil turns on one side, so they are now unequal, the *power transfer remains the same*; however the voltage and impedance seen at the terminals changes with the ratio of the windings.

As an example, let's take a transformer where the number of turns of wire on the coils is ten times bigger on one side than it is on the other. If we put 1 V AC in, we will get 10 V out (but note the current rating will drop by approximately the same ratio), and the impedance is a *square* of the number of coils. So, if the impedance on the low side is 100 Ω it will be 10 kΩ on the high side. This can be very useful in providing "free" noiseless gain in a microphone circuit. We will come back to audio transformers in Chapter 4, "Microphones", using them to modify microphone inputs and in balanced line working.

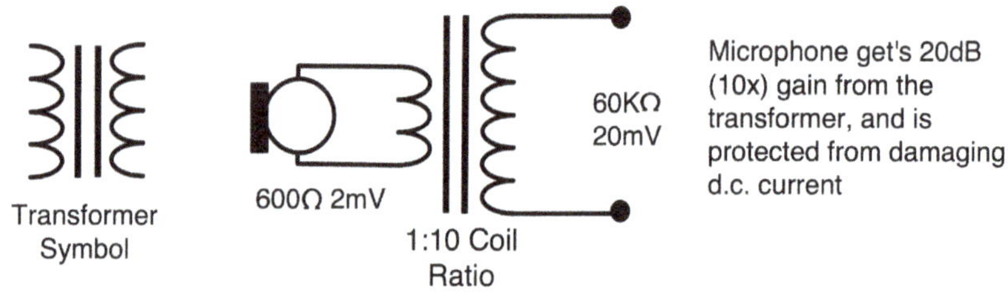

Figure 3.15 Microphone with transformer

SOURCE AND LOAD IMPEDANCE

There are two approaches to stringing audio equipment together, the old-fashioned way is to match the source impedance to the load impedance, which passes the maximum amount of *power*, but loads down our signal to half the voltage (a 6 dB loss). This was the original British "General Post Office" way, using a source and load impedance of 600 Ω (Figure 3.16).

Technically 1 mW of power into a load of 600 Ω creates an odd voltage level of 0.775 V, yet this remains a standard UK line-up level historically known as 0 dBm (0 dB into 1 milliwatt) or Peak Programme Meter (PPM) 4. Note, there is more practical detail on levels and decibels in Chapter 6, "Sound Levels and Connections".

The BBC adopted this system early in the last century as it used the telephony systems available in the day; however, it does create a problem. The "series" circuit shows how this worked. Half the volts are lost in the source. The problem is if you wish to string another 600 Ω "load" across to send the audio signal to two places (a parallel connection) the audio is further loaded down by another 3.5 dB, so the expected "line up level" is lost...

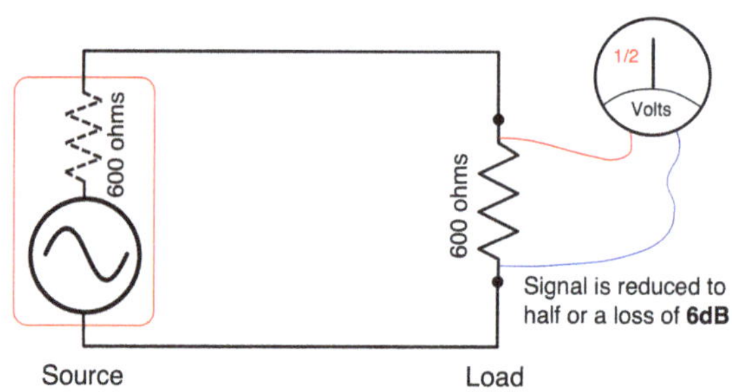

Figure 3.16 600 Ω source and 600 Ω load

Another more practical approach is to make the source impedance much lower than the load impedance. Back to the car battery not noticing if the internal light goes on: if the source impedance is *ten times* or more lower than the load impedance, the source will not be affected by the load (Figure 3.17). This approach has been used in home HiFi equipment for more than half a century, and increasingly in the sound industry, impedance matching is ignored in professional equipment, using the term dBu in levels (u = unspecified) rather than the impedance conscious dBm. *Note, impedance matching remains key with balanced microphone inputs and sending audio desk outputs over analogue lines in a studio centre.*

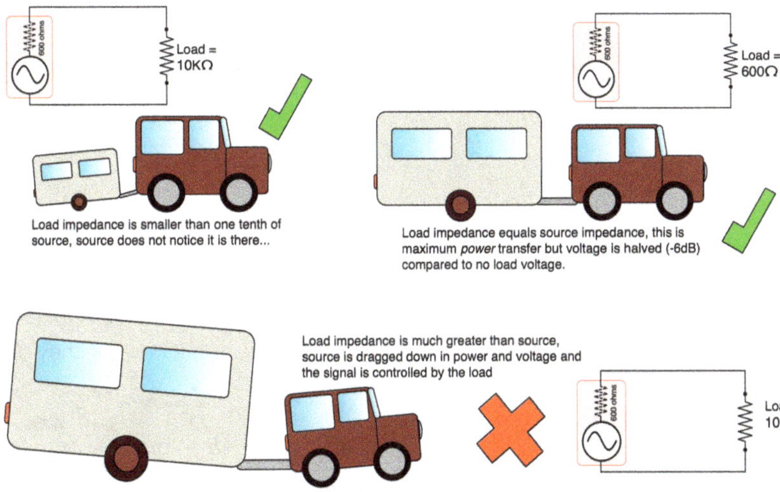

Figure 3.17 Car and caravan impedance loading

In desk line inputs, or where external equipment is plugged into a sound desk, expect that a low impedance source feeds one or more high impedance destinations. Sometimes a combination is used in that a 600 Ω mixer output is bridged by a load of 10 kΩ or more, so a nod is given to both systems. Regardless, you need to keep an eye on it and learn to understand what the metering and your ears are telling you. A final point, discussed in detail in Chapter 6, "Sound Levels and Connections", is that there are two standards of levels: the lower domestic (consumer) level of around 300 mV and the professional level of around 1–2 V. Thought must be given to a mix and match between these two systems.

So why is any of this important? The two absolutes of any audio system are the point where the signal is so high it distorts or so low the background system noise becomes invasive. Having a known reference and using it wisely allows us to position our signal in a sweet spot between the two: a decent level without fear of clipping the signal!

SERIES AND PARALLEL IN COMPONENTS

Earlier, we tried making a brighter torch with an extra bulb in either series or parallel. What happens if we place resistors, capacitors and inductors in simple networks?

Addition

Starting with the simpler addition, the values of several series resistors or inductors, or two *parallel* capacitors, can be simply added together to gain a total (Figure 3.18).

$$R = R + R + R$$

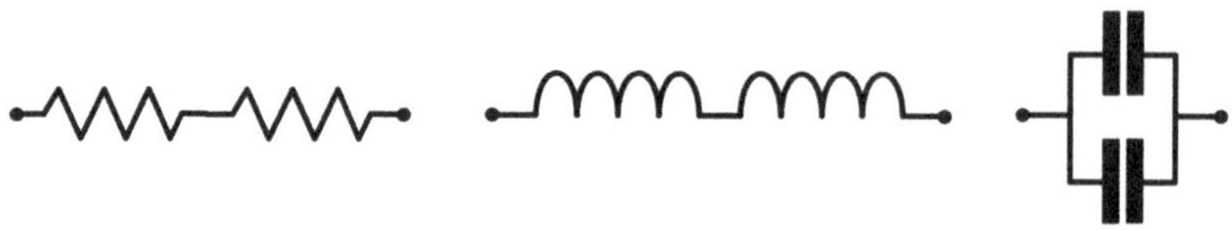

Figure 3.18 Addition of components

Reduction

Placing resistors or inductors in parallel, *or again the reverse* – capacitors in *series* – creates a reduction of value (Figure 3.19). If the two components are the same, the new value is simply half what it was, but what if they are different?

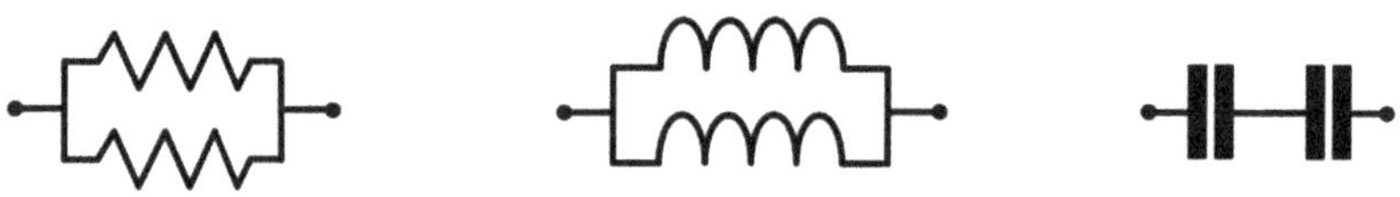

Figure 3.19 Reduction of components

This equation is given for interest, and as a route to further study; however, it's not one you need to recall. **R** can be substituted by **L** for inductor, **C** for capacitor or equally **Z** for impedance. Additional components can be added; simply invert and add as the others.

$$\frac{1}{R}=\frac{1}{R}+\frac{1}{R}+\frac{1}{R}$$

Resonance

In the previous paragraphs we have learnt that either capacitors or inductors can be placed with a resistor to create a filter to remove either the low frequencies or the high ones. We have also learnt about series and parallel circuits.

What about using capacitors and inductors in the same circuit? Each store energy; what if they could pass that storage back and forth between them like a pendulum in an old clock? I'm sure you have guessed that's exactly what happens, and like the weight and length of a pendulum set up a frequency of movement that is exact enough to accurately time a clock mechanism, capacitor and inductor pairs create a resonance at a particular frequency, greatly amplifying or attenuating the signal around that value (Figure 3.20).

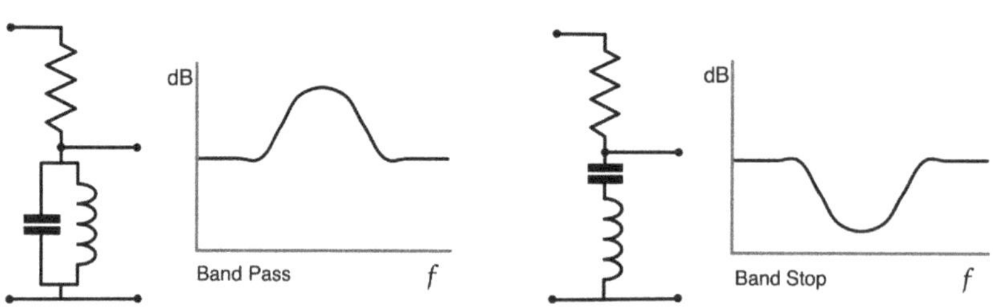

Figure 3.20 Resonance in RLC circuits

If "L" and "C" are in parallel, frequencies of signals are amplified (Band Pass Filter), and if they are placed in series (Band Stop Filter) the result is a loss of audio signal around the core frequency. In theory the peak (or trough) is sharp and tight, however, the inherent resistance of the components dampens the sharpness. We can choose (and normally do) to place a resistor in series with the LC network to widen and flatten the resonant effect, to make it useful in a chosen application. The sharpness (and therefore width of frequencies affected) is known as the "**Q**"; the higher the Q, the sharper the peak (or trough).

Note a band pass filter can be pushed further so that it turns amplification into oscillation, and that is how we can create sine wave test tones to line up our audio equipment. We can easily take a few more components to create other useful audio test tones including square wave, triangle wave and sawtooth wave as shown in Figure 3.21 (see Chapter 11, "Sound Synthesis").

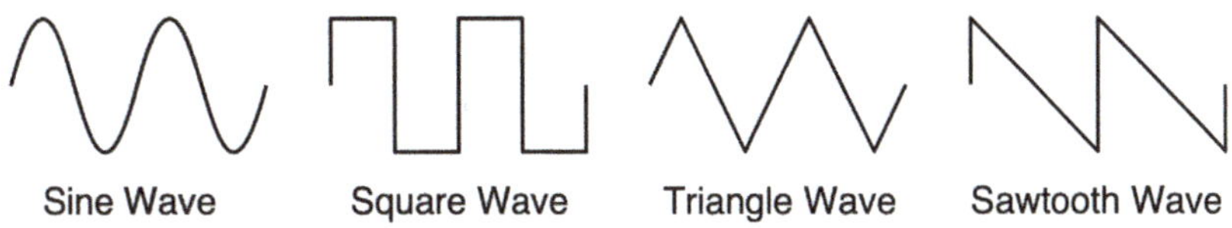

Figure 3.21 Four popular AC waveforms

Two uses of resonant circuits:

1) On a mixer, we want to boost the mid frequencies to enhance the sound of a voice, and so we use a *band pass* configuration at a low Q. This is called a "Mid" or "Presence" control.
2) We have an annoying buzz on a recording. It turns out interference from the power supply has been induced in a bad audio lead. We can use a tuned *band stop* filter at a high Q to remove this troublesome frequency while leaving the rest of the signal intact. There are numerous combinations of RLC circuit to create other effects, but these are beyond the need or scope of this book.

A PRACTICAL USE OF A CAPACITOR

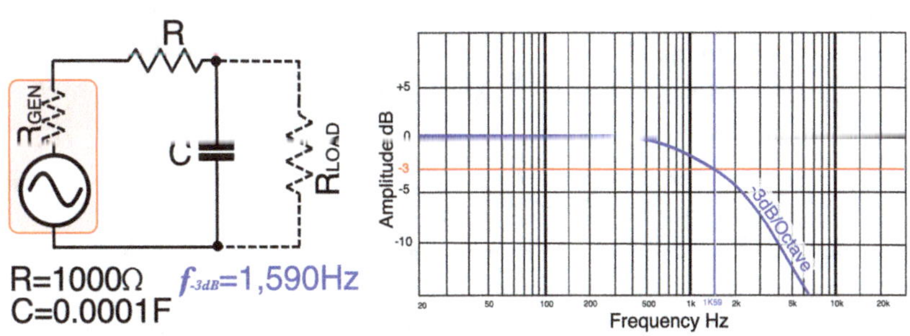

Figure 3.22 Practical low pass filter

The circuit in Figure 3.22 is a practical use of a resistor and a capacitor. R & C have been used to make a sound filter that passes through all the low frequencies, until at a calculated point of around 1.6 kHz the high frequencies slope away cutting them off at a rate of 6 dB/octave (a halving of voltage for every doubling of frequency). This is known as a low-pass filter, as only the high frequencies are attenuated. Simply reversing the resistor and capacitor's positions makes it a high-pass filter; again, with a slope that attenuates the low frequencies from 1.6 kHz at 6 dB/octave.

As we learnt above, the circuit assumes the source impedance is low and the load resistance is high, so they won't skew the result of the equation. Please note this RC circuit is given for interest only, and as a route to further personal study. At the calculated value of our chosen frequency (**f**), the sound level (amplitude) has already dropped by 3 dB, and is known as the "–3 dB breakpoint". The components we have examined in this chapter are called "passive" as they don't need a power supply to work. However, they do have losses, so if we need to make the AC signal bigger (louder) again to make practical use of it, we must "amplify" it by a calculated number of decibels. This addition is called "makeup gain".

SEMICONDUCTORS

There are "active" components called semiconductors, and a key one is the three-legged silicon transistor. A small AC signal voltage is applied to the base and mirrors through to change the resistance between the collector and the emitter. This is used in a potential divider with a resistor in the circuit and so the much larger output voltage of the power supply swings in sympathy with the small input signal creating amplification. A DC power supply taken from a battery or from the house supply is used to create the AC "gain" of the circuit (the opposite of loss). An amplifier can provide voltage gain or current gain, or as in a "Power Amplifier" for a loudspeaker, a combination of the two.

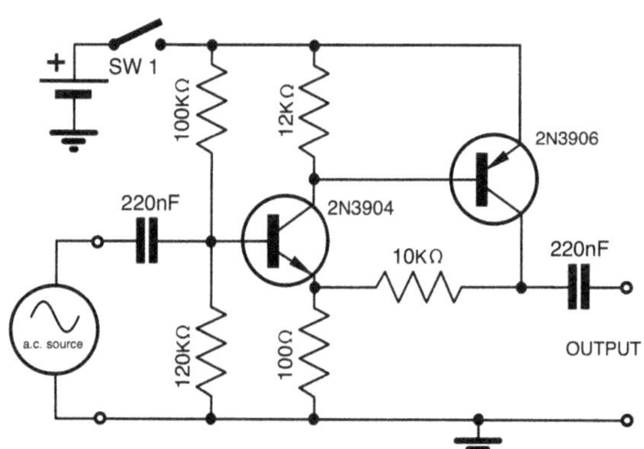

Figure 3.23 Practical two transistor amplifier

For interest only, Figure 3.23 shows an amplifier circuit diagram featuring two transistors. The positive of the battery goes through the power switch to the circuit, the negative of the battery goes to "ground" which is the common rail or "0 V" (common to DC *and* AC sources). Capacitors at each end block the DC from entering or leaving the circuit and resistors set up the transistors to work properly. Spot the potential divider! A circuit where you can see and touch each component is called a "discrete" circuit, however, many modern circuits are complex, and the components we have talked about are made very small and placed on a silicon chip called an "integrated circuit" (IC). These miniature building blocks are highly flexible and can make complex diverse solutions in tiny packages.

THE INTEGRATED CIRCUIT

Whilst a knowledge of ICs is not directly relevant to a career in sound, I have been persuaded that a light touch on the subject would be useful. An "IC" is simply a normal discrete circuit made much smaller. Apart from the obvious benefit of taking up less room, ICs or "chips" (microchips) show a great cost and power saving over the discrete equivalent. There are two main types of chips: those which process the computation of numbers (data) are called Digital ICs, and those which manage continuously varying signals like audio are Analogue ICs. There are crossover devices which manage both, an obvious one being a chip to turn analogue sound into digital code (analogue to digital converter; ADC) and to reverse the process to create sound again, the digital to analogue converter or DAC.

Early Digital ICs would contain a few dozen transistors, each one operating as a DC switch, creating a larger function. Over time, the complexity grew so these circuits might process the data, save the data (memory) or a combination of both. The quest over the years has been to make each transistor smaller, so more can be packed on the tiny silicon wafer; and to work faster, so the time to process is shorter. At the time of writing, artificial intelligence (AI) is in the ascendancy, and this requires serious parallel processing power. A chip developed by Graphcore in Bristol (UK) is said to contain 60 billion transistors and be capable of processing 1,500 functions concurrently! In the early days of digital chips, simple tasks such as logic gates would be performed by dedicated ICs. Computer processing power is now such that complex logical tasks can be written into software and performed as a simulation of the old chips, so a single computer workstation can perform many tasks simply by choosing the appropriate software. A modern computer can record multiple channels of high-quality audio by digitising it in real time and carry out audio signal manipulation formerly done using large expensive analogue boxes strapped into the studio audio mixer. This is now so advanced we can make useful changes to a sound file that would have been impossible before computers. These mathematical sound modifications are called digital signal processing or simply DSP.

SIMPLE OPERATIONAL AMPLIFIERS

Analogue ICs have taken a different path to digital ones. It's about quality rather than quantity. Parameters such as signal distortion hardly matter in the digital world, and power supplies are mostly adequate at just 5 V. Audio signals need a large voltage swing to avoid clipping off the peaks, and good accuracy and quality in the components to keep any additional noise

Figure 3.24 An 8 pin IC

generated as low as possible. The IC that fits the bill is an operational amplifier or "op-amp" is shown in Figure 3.24 and in diagram form in Figure 3.25. These can cost cents, with maybe four circuits in one IC or many dollars, each using laser trimming at the time of construction to ensure the component precision that brings great quality. The op-amp can do many analogue tasks including comparing levels, inverting signals, amplifying voltage, creating a current source, precision rectification and so on. It does not care if the input is DC or AC. The op-amp has two inputs, one inverting the phase and one non-inverting (where the output phase matches the input). The output can drive low impedances, much lower with certain devices; for example, 600 Ω is not a stretch for the NE5534. I've shown a typical 8 pin out op-amp; the amplifier itself is designed to deliver as much gain as possible; the better ones can do so over a wider bandwidth.

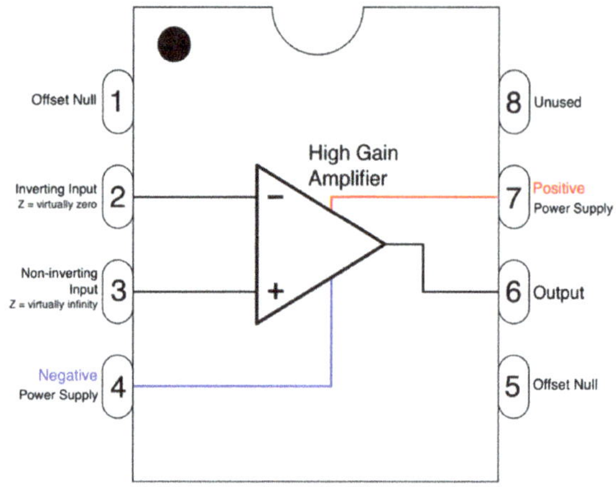

Figure 3.25 An 8 leg IC pin layout

At its simplest, there are two configurations of op-amp audio amplifier: one which uses the inverting input (pin 2) and the other, the non-inverting (pin 3). The decision of which one to use in each situation is the designer's choice. The op-amp will happily amplify DC, so like the transistor amplifier above, we put DC blocking capacitors at either side to make it an AC only amplifier. It needs powering (pins 7 and 4), to allow a decent voltage range; +15 V and –15 V are common, giving a "swing" between the two of 30 V. This means the common rail for supply and the signal (input and output) is the joining of the two at 0 V. Note that this supply rail connection is shown by the "ground" or earth symbol. We have said the amplifier has huge gain; this must be brought down to a useful predetermined level, and we do this by choosing the right value of resistors in the surrounding circuit to adjust how much of the output signal is fed back to the input. The more the feedback, the less the gain; however, usefully, the lower the signal distortion. The two configurations are shown in Figure 3.26.

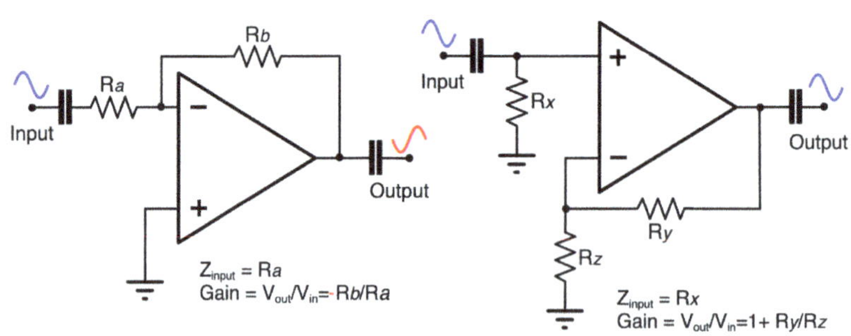

Figure 3.26 Two simple op-amp designs

In the left "inverting" amplifier if R*a* = 10 kΩ and R*b* = 100 kΩ (a ratio of ten) the gain is ×10 or 20 dB. That simple! As the "**–**" input is virtually at earth potential (a short circuit), R*a* is also the input impedance of the circuit. Note the output is out of phase with the input, it is *inverted*.

In the right "non-inverting" amplifier the gain equation is slightly more complex. If R*y* = 18 kΩ and R*z* = 2 kΩ, the gain is (18/2) +1 or ×10. Again, a 20 dB amplifier. The "**+**" input is very high impedance, so we use R*x* as a parallel resistor to set the impedance to the value we want.

In closing, ICs are used extensively in modern sound desks for cheapness and space saving, and with good design, the specifications can be excellent. In particular, the two op-amp inputs can be configured together (usually with other op-amps) to create a "differential" input, which mimics a balancing transformer at a fraction of the price (see Chapter 4, "Microphones"). However, they also bring their own problems, and it remains amazing to me the audio quality that can be gained from a simple well-designed discrete circuit, be it valve or transistor!

AC SIGNALS AND VOLTAGE RAILS

An amplifier uses a DC power supply to multiply an AC source, which we have explained will swing positive and negative, with respect to a voltage point set by the potential divider at the input (and other small factors such as voltage drop across components); the limit of gain being when either a positive voltage swing or a negative one approaches the supply rail where it can go no further and so is restrained at that point, distorting the waveform.

The voltage swing is described as Vpp (voltage peak to peak) or we might choose to look at just one half of the signal, which is Vp (peak voltage). However, when we want to say how high a signal is in level, we use neither of these values. Instead, we measure the AC signal at a value where the voltage *does the same work* as the equivalent DC voltage. If this seems strange, it comes from a time when early electrical light bulbs and heaters needed to be able to equally work off an AC or DC supply depending on geography.

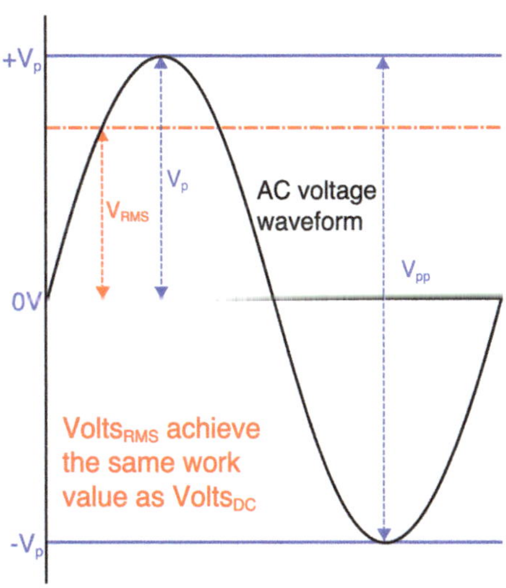

Figure 3.27 AC voltage waveforms

A mains equivalent of around 230 VAC is an RMS value (root mean square), the peak to peak measurement is over 600 volts! It is the same for audio meters; the default for measuring an audio level is not the *peak to peak* or *peak* voltage; it is the much lower V_{RMS} as shown in Figure 3.27.

This is as far as we need to go to be a successful sound practitioner. I sincerely hope you have been able to stay with this explanation, and believe me, we will come back to these principles at various points in the book, so your efforts have been well worthwhile! Electronics is a fun hobby, so if I've made you curious, you will find huge support to go further on the internet.

BIBLIOGRAPHY

Carter, B. and Mancini, R. (2018) *Op Amps for Everyone*, 5th edition. Newnes.

Hartley Jones, M. (1988) *A Practical Introduction to Electronic Circuits*, 2nd edition. Cambridge University Press.

Horowitz, P. and Hill, W. (1989) *The Art of Electronics*, 2nd edition. Cambridge: The Press Syndicate of the University of Cambridge.

Jolly, W.P. (1978) *Electronics*. Teach Yourself Books/Hodder and Stoughton.

4 Microphones

Many find the subject of microphones the most contentious in the field of sound. Explanation can be tricky. Too simple and some professionals may scoff for glazing over points, too complex or impractical and it does no good as people don't follow and learn, so please have patience with me. I can only ask that you have read Chapter 3, "Introduction to Electronics", before attempting this one as it will smooth your path in the coming paragraphs.

I hope it makes sense that there must be a tensioned diaphragm to move in sympathy with the sound pressure that impinges on it, like it does on your own ear drum. Then there must be a mechanism to turn the diaphragm movement into electricity; this is called a transducer. The important point to grasp from the first is that you must separate the *transducer* (the internal guts that turns sound to electricity) from the *directionality* (in which directions the microphone can "hear"). So here we go.

TECHNOLOGY AND DIRECTIVITY

In the beginning, over 100 years ago, there were many types of technology employed to turn sound into electricity. Some were cheap but good enough (like carbon-based ones in telephones) and others were high end and prohibitively expensive. Early broadcasters (even though the initial medium of radio on "AM" was very poor quality) were engineering biased and so wanted the best! Two quality early designs won through: "dynamic" where sound landing on a diaphragm moves a coil of wire in a magnetic field and so generates volts in the wires, and "ribbon" where a very thin metal foil (1/50 of a human hair thick!) is placed in the field of a powerful horseshoe-shaped magnet, where there is a "north" to one edge and a "south" to the other edge of the ribbon. A tiny voltage is developed and lifted from the ends of the ribbon.

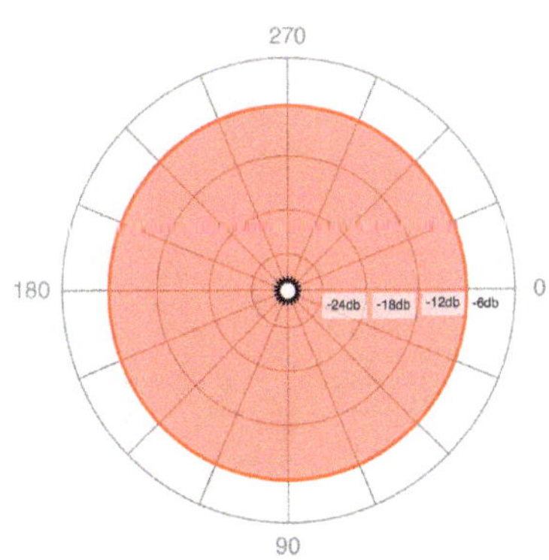

Figure 4.1 A dynamic microphone polar plot

A plain dynamic microphone can pick up sound from all directions, so it is called "omnidirectional" like the omnibus (shortened today to bus) which picks up *all* the people. Imagine putting this microphone in the middle of a round table and seating four people around it; all can be heard. Figure 4.1 shows our table (viewed from above), and the red ring, the consistent pick-up of sound from all directions.

An alternative technology is the "ribbon" microphone, which is only sensitive

DOI: 10.4324/9781003717065-4

to sound pressure that moves the flat face of the ribbon on either wide side. As you move around towards the thin edges the ribbon picks up less sound until at the sides of the microphone it picks up nothing at all. Let's take our table example of four seated people again in Figure 4.2. The two people opposite each other on the faces of the microphone can be heard, but the other two on the flanks cannot. The shape of the sensitive parts of the microphone resembles an "8"; the clever engineers christened it "figure of eight" or "bidirectional".

Each have advantages and disadvantages; the omni-directional microphone will pick up what you want, but it will also pick up all the other sounds you don't want! Also, people have different volumes of voice, so big voices will drown out quieter ones.

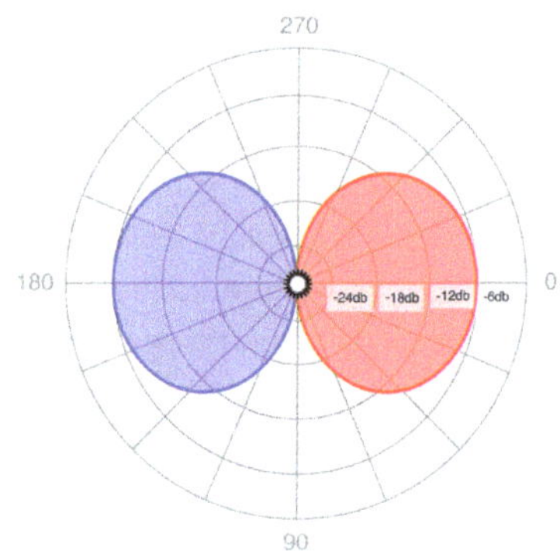

Figure 4.2 A figure of eight microphone polar plot

What about the figure of eight microphone? Well, if you just have two people it is perfect; other noises on the flanks of the microphone will be lost, and the "Studio Manager" (BBC speak for sound operator) can move the microphone across the desk between the speakers to get the best balance between the loudness of their voices. And that's what happened, literally for decades!

CARDIOID MICROPHONES

What we really wanted though was a microphone for each person. Neither omnidirectional nor figure of eight could do that. But what if we combined the two? This is the science bit. You will notice that I have coloured one side of the figure of eight diagram blue. If you listen to either side, they are identical, so what is the difference? Well, if you blow (*very* gently) on one side of the ribbon it moves away from you; it moves towards the other person. If you are on the other side and blow, it's the inverse. So, one person's blow looks to the other person like a suck. They sound the same, but depending on where you sit, if one is "in-phase", the other must be "out-of-phase".

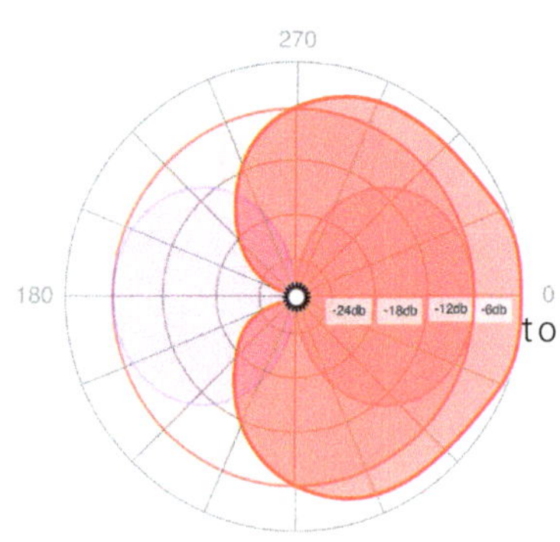

Figure 4.3 A cardioid microphone polar plot

If we add in the omnidirectional microphone on the table (very) close to the figure of eight, it will be in phase with only one side of it (red in the diagram), and so the sounds will add together make the volume louder. However, it will be out-of-phase with the other side (blue in the diagram), and so the two microphones will cancel out on the blue side.

With the doubling on one side and cancellation on the other we have a microphone that picks up sound from one direction, or "unidirectional" (Figure 4.3). Engineers, thought the shape of the pick-up looked a bit like a heart, so they called it "cardioid" – romantic these engineers.

Just to note modern cardioid directivity doesn't require two microphones to work; we achieve unidirectional pickup with just one transducer. The sound is cleverly fed back from the rear to the front (phase cancellation) to stop pick-up from the back.

So, there are three main differing pick-up directions: all around, two directions or just one direction. We call them omnidirectional, figure of eight and cardioid. *Note that today the pick-up direction no longer dictates the transducer used.* There are all variations of direction, set against different transducers. There are more directional microphones than a mere cardioid. Like a zoom lens making distant pictures close, super cardioid and hyper cardioid ("shotgun" because it has a long barrel) do the same for sound. A drawback is that none of these special highly directional cardioids has such good rejection of rear sounds, particularly at higher frequencies, and so must be used with care. Also, just a little off target with your aim and you risk losing everything!

Later in the book I will be providing usage diagrams for different recordings. To avoid confusion, I have ascribed a different small symbol for each polar pattern in Figure 4.4. The look is roughly ascribed to popular microphones in general use, the figure of eight on the BBC ribbon "coffin" microphone being a popular distraction and the cardioid design is loosely based on so many handheld vocal "mics" like the Shure SM58. Capacitor microphones (see paragraph below) are often switchable in polar pattern, and please note shape or size is no longer a certainty in transducer technology or polar pattern.

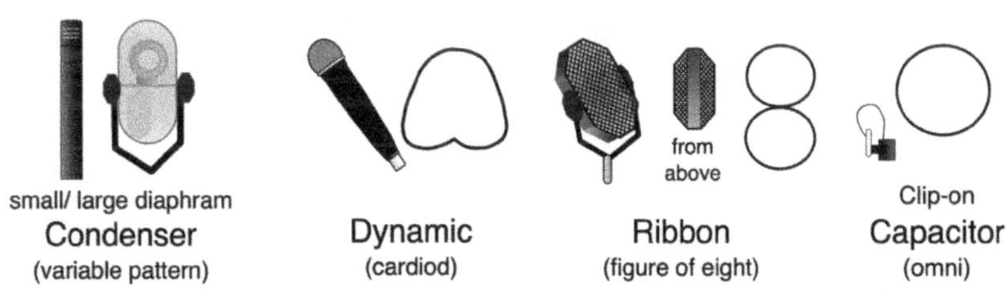

Figure 4.4 Three illustrative microphones

CAPACITOR MICROPHONE

There is one transducer technology we are yet to mention. It's not new; it was developed way back in 1916. A capacitor (old speak "condenser") is a common electronic component. Two plates of metal face each other with a small gap between them. The gap can be just air or filled with a dielectric like a gel. You can read more about capacitors in Chapter 3, "Introduction to Electronics". Here's the science bit: A simple electronic formula of $Q = CV$ shows how they work. If you place a fixed charge (Q) on one plate, direct voltages are blocked, but crucially alternating voltages (V) (like sound waves) are passed between the plates. One thinner plate is made flexible so it can pick up sound waves, the vibration (moving in and out) changes the gap between the plates, and so the capacitance (C) changes. If Q is constant, a change in C must make a change in the output volts (V).

This voltage change can only be captured by a very light touch (a very high impedance input), and so we require an amplifier inside the microphone as a buffer to the outside world (Figure 4.5). This needs to have a power supply. Also, we must have a constant "charge" across the plates, so again this requires polarising volts. So, the Capacitor microphone (or Condenser microphone in old speak) requires a power supply from the sound desk, normally 48 V (other voltages exist). A neat trick allows the powering voltage to sit unseen on the same microphone cables which deliver the sound from the microphone to the sound desk, and so it is called "phantom" power – clever these engineers! The need for a power supply was inconvenient, and the cost of production high, so for decades condenser microphones were a luxury for few.

Modern production techniques, and in particular the invention of the field effect transistor (FET) placed at the internal amplifier input, have made the cost comparable to other transducer types. The quality for the price and the ability to have several different pickup patterns selectable by switch have made them a favourite today! However, I caution that you do get what you pay for!

One further variant of the capacitor microphone is worth noting, the "*electret condenser*" or plain electret microphone. Here one plate has a permanent charge built in during manufacture. This means that a small battery just needs to power the internal amplifier, and that can work for days. Now certainly good enough to be seen in "professional" studios, they are highly popular in "semi-pro" applications like public address, churches and in mobiles, computers, etc. In fact, they are by far the most popular "sound transducer" in modern times with over a billion units being produced a year.

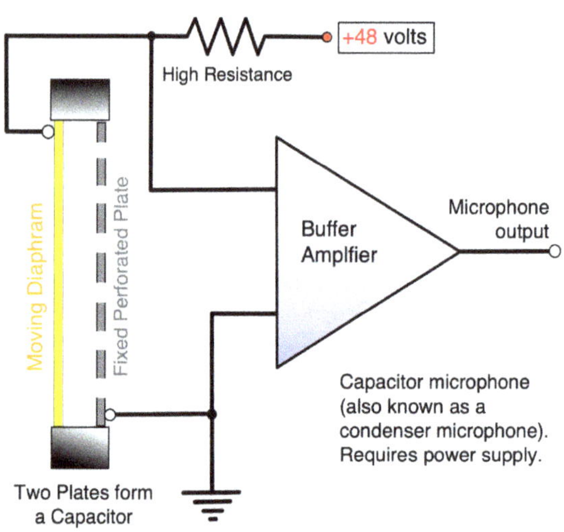

Figure 4.5 The design of a capacitor microphone

MICROPHONE PARAMETERS

We have a transducer, with a given directionality. What else is important for a microphone in a *practical* situation?

Sensitivity – a lead rock singer will want to sing loud and close creating a high sound pressure level (SPL), so a low sensitivity would work best. The opposite case would be a stereo pair of microphones slung above and behind the conductor of a large orchestra at a distance of several metres. These would need to be far more sensitive, and that sensitivity comes at a price. Ribbon and some dynamics can be less sensitive, and an inline XLR barrel preamplifier can make all the difference.

Frequency response – ideally flat through the human range of hearing, yet many specialised mics are loved because they are not flat. They enhance a particular sound they are designed for but of course would be poorly matched to produce a quality recording of a whole band or orchestra. Here are two extreme examples. A famous vocal microphone, with a much-loved presence peak, and a classic ribbon microphone with an extremely flat plot, which would suit recording an orchestra. The response is within ±1 dB between 45 Hz and 6 kHz (Figure 4.6 – see light grey marker).

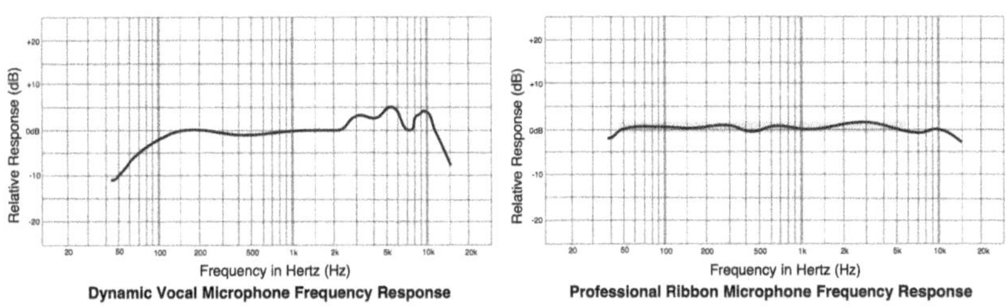

Figure 4.6 Vocal and studio microphones compared

Ruggedness – if we take it on the road will it still work after being dropped? Also, if we blast a trumpet into it can it survive? Rugged microphones, sadly, are often of poorer quality, insensitive (less output signal for the sound pressure input) and less refined at higher frequencies, but they do last!

Fidelity – this is the opposite of ruggedness (think ballet rather than break dancing). Is it capable of capturing delicate quiet sounds from a violin and delivering the high-frequency harmonics, capturing the whole of the human hearing frequency range with clarity and precision? Just don't drop it!

Close working capacity – will it sound good if used closely? Perhaps we want to drown out high background sounds happening around the sound you want. An ordinary microphone used up close and personal will sound far more "bassy", a phenomenon called "bass tip up". It may make a DJ sound "sexier", but equally many microphone types are blasted by a close loud voice and the diaphragm will be overexerted and distort sounding plain nasty! "Stocking" mesh or foam pop shield to the rescue! (Note, many "vocal" microphones have an internal pop shield.)

Price – it's not impossible to buy a microphone for US$20 or for US$20,000! Clearly at all prices there are compromises, and no single perfect solution. Experience helps sound engineers to choose satisfactory solutions for affordable investment and is a life skill worth pursuing.

It will come as no surprise that a great many differing models are made, with specific purposes in mind. Individual engineers love certain microphones for each situation and could not be swayed away from favourites, as much as changing loyalty from Man U to Man City! No answer is the right answer, and that provides lifelong debate for the audio anoraks. Me? After many failed attempts at getting a good sound for BBC Radio 1 DJs, I put an Austrian shotgun microphone in the studio. It should have sounded terrible, but it sounded fabulous and was in pride of place for many months. You never know till you try!

Despite the number of Neumann's (microphone royalty) in circulation throwing money at a problem may make you feel absolved of choice, but it is seldom the best option and often a compromise position. It's said taking a perfect portrait photograph is 90% positioning the client, the backdrop, and lights, and just 10% of the time shooting. If we were braver perhaps sound engineers could take more time to experiment with their clients to find and produce the unexpected. To take an example, let's use one of the world's most successful albums, the 1982 "Thriller" from one of the top artists, Michael Jackson. This US$750,000 album was produced by the world's hottest producer of the time, Quincy Jones, with a full orchestra, yet studio engineer Bruce Swedien used the humble Shure SM7 microphone to record Michael's epic vocals on the key tracks. The right use at the right time. And yes, he did have the guts to stand up for his choice!

MICROPHONES IN PRACTICAL SITUATIONS

Let's take several common situations. (*Note there is no perfect solution!*)

Podcasting – see Chapter 2, "Professional Podcasting".

Local band vocals – choose a relatively cheap solution in case it breaks or gets stolen, rugged for the road, you can work close into it (so low sensitivity is OK). Bass response doesn't matter, top response perhaps bright and edgy rather than boring flat to give a voice "life". Good rear rejection so good "howl round suppression" (often called feedback). *One answer*: dynamic (inherently rugged and cheap) and unidirectional (a cardioid pattern) for howl round rejection.

Studio violin – the low frequencies are not critical; however, high frequencies must be clean and pure. High sensitivity OK as there is no "blasting", as from wind instruments. Cardioid response is still a favourite to separate this instrument from adjacent ones but also figure of eight is OK if the other active side points away from other studio users. *One answer*: ribbon (whilst extended bass is often lacking on this transducer, upper frequencies are delivered flat to high frequencies with high precision). Whilst you can buy cardioid ribbon microphones, the best results would come from a traditional figure of eight given the purity of the mechanism. *Note you would not lightly take a ribbon microphone on the road as the thin ribbon will break with any abuse.*

Stage piano – a wide flat frequency response is required, low to high. As it's for stage use, a cardioid pattern is still favourite to stop howl round (and audience noise), but that must not compromise the extremes of each physical end of the piano through loss of sensitivity at the sides of the microphone. It must be able to handle quiet passages through to crescendos, so medium sensitivity is best. *One answer*: a loose cardioid "large diaphragm" capacitor microphone (large diaphragm reproduces low frequencies with more precision) set to medium sensitivity (20 dB pad in). A professional quality condenser microphone offers fidelity and a wide flat frequency response; however, there are too many internal amplifier failures among cheap ones for me to recommend them in all situations.

Kick drum – low sensitivity, rugged, high frequencies don't matter; loud sound at source means howl round is unlikely. *One answer*: use dynamic; directivity doesn't matter as the drum is overpoweringly loud, and low frequencies go in all directions. Much can be done in post-production to improve the bite and solidity. Although almost any tat would do, manufacturers do make a "specialist" product.

There are 20 more microphone positioning ideas in "Appendix 1 – Microphone Placement" at the rear of the book.

MICROPHONE IMPEDANCE

In Chapter 3, "Introduction to Electronics", we learnt that the source voltage would not be dragged down in value if the load was more than 10× the impedance of the source. This helps preserve the levels through the desk, minimising the need for amplification and so the addition of unwanted noise. It may surprise you that the microphone is a special case where the source (microphone) and destination (preamplifier) are impedance matched (the same value), so whilst the power transfer is optimised, 6 dB of useful voltage level is lost. The reason for this is the very low level out of the microphone (approximately 1/1000 of desk nominal output signal or –60 dBu). This makes it highly susceptible to local interference. We use high-quality shielded interconnects (mic cables to you and me!) which help, and these utilise another trick to keep the nasties out, the "Balanced Line". I'll discuss that below but first a reminder of what impedance is.

Impedance (Z) is a measure of the opposition to electrical flow. It is measured in ohms.

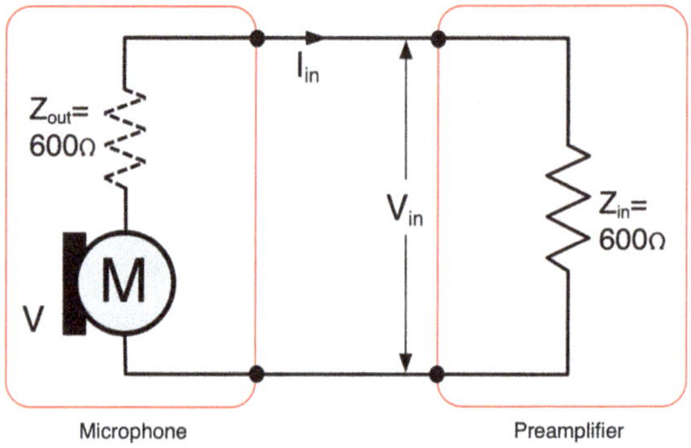

Figure 4.7
Microphone output impedance matches input impedance

In DC circuits, impedance and resistance values are identical, defined by Ohm's Law as the voltage across a device, divided by the current (R = V/I). In AC systems, "reactance" and "inductance" enter the equation due to frequency-dependent changes in overall resistance. Fortunately, in a majority of audio (therefore AC) connections the input impedance is predominantly resistive over the frequency range of our hearing; careful design ensures there is negligible phase difference between the input voltage and the input current. That's useful, as instead of getting out the bumper book of complex maths, we can use good old Ohm's Law.

A reminder that any AC generator (including a microphone) can be drawn as a voltage source with an external resistor in series. Modern microphones have a typical output impedance of between 200 Ω and 1 kΩ, and are nominally assumed to be 600 Ω when designing mixer input circuits. To meet the needs of balanced line working the mixer input will have an input impedance to match the microphone of 600 Ω (Figure 4.7 – note that transformers to create a "balanced line" are removed for simplicity).

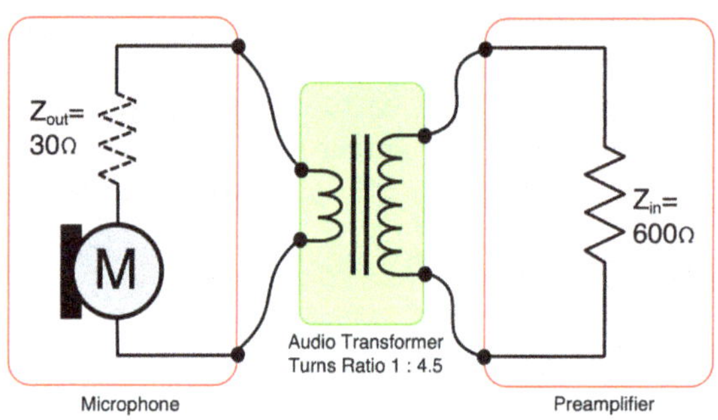

Figure 4.8
Microphone matched to input using a transformer

The assumption of a 600 Ω source impedance is a good one; the vast majority do conform. However, there are variations, particularly with vintage microphones. Ribbon microphones have a very low natural output impedance, and many vintage dynamic ones were made with a coil impedance of just 30 Ω. So, how do we deal with the mismatch? Well, we could turn up the wick on the mixer, but that brings up the noise too. We could use a purpose-designed preamp (using phantom powering), which would bring up the level with less noise than the desk, but the 30 Ω microphone is still expecting a 30 Ω destination impedance and is designed to deliver a flat frequency response into it. If it is loaded with 600 Ω, it is likely to have a poor frequency response, particularly a change in the high frequency plot. So, how do we mend (1) the very low output level and (2) the mismatch in impedance?

The answer is to insert a plug-in audio transformer in the microphone lead (Figure 4.8). We learnt in Chapter 3, "Introduction to Electronics", that a

Figure 4.9 Inline microphone amp (left) and transformer (right)
Photo by the author.

transformer has several useful features. It has voltage gain that matches the turns ratio, and impedance change that matches the *square* of the turns ratio (Figure 4.9 [right]).

A ratio of 1:3.5 gives a noiseless gain of around 10 dB, and around a 12× increase in impedance, matching a 30 Ω microphone to the 600 Ω input impedance of a mixer preamplifier. The impedance matching ensures the frequency response is as the microphone manufacturer intended it. Audio transformers are a special build and are available with different ratios for different uses. They are designed to ensure a high linearity (low distortion), and that they don't saturate at the levels they are expected to handle. That makes them an expensive option, but a highly useful one! In-line miniature amplifiers are also available and suit insensitive microphones like the Shure SM7 mentioned above. They are powered from the phantom power supply, making a neat unobtrusive addition (Figure 4.9 [left]).

UNBALANCED VERSUS BALANCED LINE WORKING

To recap, in Chapter 3, "Introduction to Electronics", we learnt that to make electricity flow we need to form a circuit. That suggests a "go" to take the signal to the other equipment, and a "return" to make the circuit complete. The enemy of low-level signals (like microphone signals) is induced noise in the cables or within the cases of the equipment, which is louder than the noise floor and so can be heard together with the audio signal. This may be from a hum loop (see "Hum Loop" below) or other induced noise from lighting dimmers, or perhaps from radio signals like your mobile phone.

We often encase audio electronics in metal boxes which we can use as a screen to shield the electronics from induced external noise. The return of the audio signal shares the same connection as the case, called the "earth". We use this "earth" to protect the send cable by wrapping it around the return cable, which should shield it from noise. Note, domestic audio is usually stereo and there is a connection for the left channel and another for the right, so two leads joined to four phono plugs. I am only showing one channel in Figure 4.10.

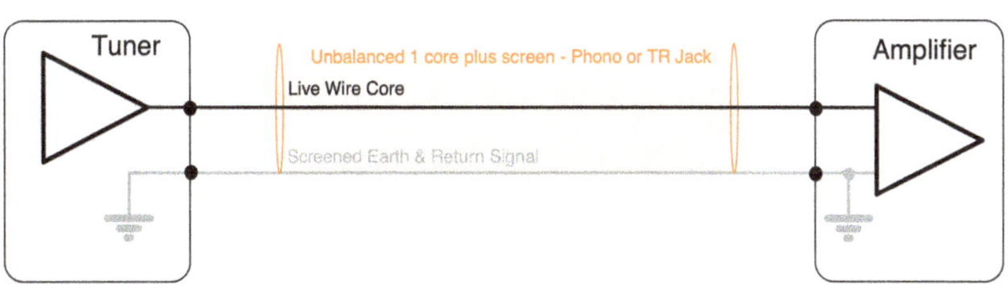

Figure 4.10 Unbalanced cable used in domestic equipment

If you use a domestic phono cable to join two devices together you are also joining their earths together by the return screen wire. This, particularly on vintage equipment, may find a path through the "earth" of the mains wiring which powers both units (no problem where one or both are battery powered) and form a large loop through the building wiring that can easily induce a low frequency hum or a buzz at either 100 Hz or 150 Hz (120 Hz or 180 Hz in the US). This is called a "Hum Loop" (Figure 4.11). If you want to hear the induced "noise" simply pull the output phono plug slightly off the socket so the outer screen connection is no longer touching. Without the "earth" the cable then becomes an antenna to such signals, and the close interference problems can clearly be heard. (*Tip: turn the amplifier volume down low, as the signal (depending on input) can be very loud!*)

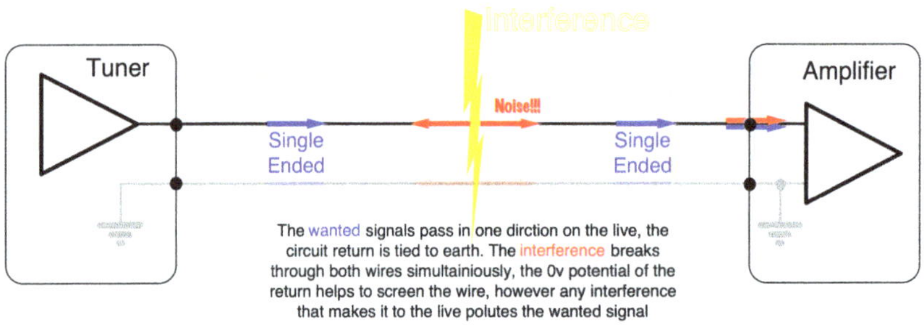

Figure 4.11 Noise is induced in the unbalanced live wire

If you remember the simplest early microphone was one where a coil of wire was moved over a magnet driven by a sound hitting a diaphragm. Traditional electronics would have it that one side of the coil goes to a common earth, and the other is the live. What if we take the two sides of the coil all the way to the amplifier; the earth remaining separate, so *both* send and return wires float above DC. Nasty buzzes on the "earth" can no longer be passed between equipment on the return wire; and because it floats like the send, it creates a genuine AC push/pull generator which is immune to interference striking the connecting wire (Figure 4.12).

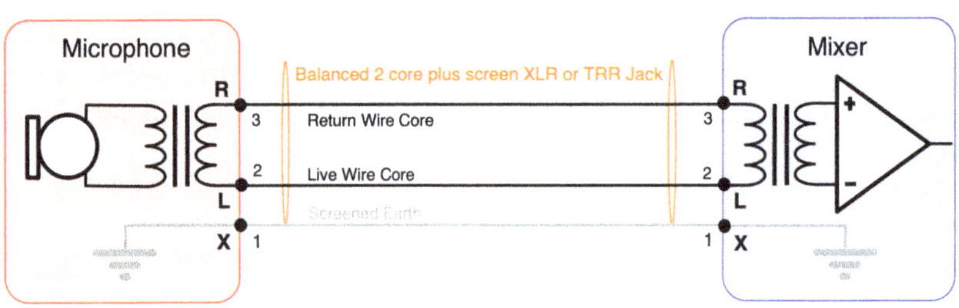

Figure 4.12 Balanced wiring provides two live wires and a screen

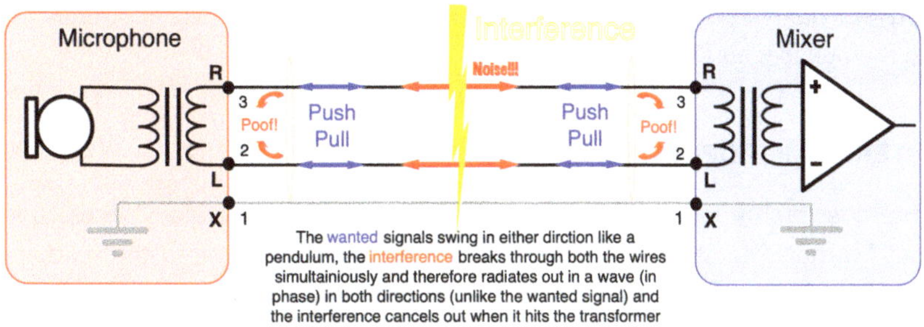

Figure 4.13 Induced interference is destroyed by phase cancellation

We can use a transformer at either end of the connection to isolate the signal from earth to achieve this. How does it work? The audio signal travels in a see-saw direction (positive/negative polarity) across both terminals. Any interference cuts both wires at the same time, so travels out on both wires away from the source. When the noise signal hits the transformer windings it is simply wiped out by phase cancelation – see Figure 4.13. (I note modern electronics can achieve similar using a differential amplifier design but best is still the transformer.)

Oh, you might say, but the earth is still joining the cases together. Yes, but if there is a problem the screen of the cable is no longer carrying the audio signal, so we can break the connection at one end with a simple switch called an "earth lift", knowing that the screen is still held at earth potential by the equipment at the other end of the cable. Frankly, it's not such a problem these days through better design; most equipment connects to the electrical supply with just two cables, and the mains earth is omitted completely.

PORTABLE VOCAL BOOTHS

At the start of the millennium a new item became available to vocal artists. An arc of acoustic foam is suspended behind a cardioid microphone, normally attached to the same microphone stand. The idea is to stop returning reflections from the hard room walls changing the acoustic and so muddying the sound. Quickly these became an essential buy. Whilst they do reduce the room reflections, it's only at frequencies 1/4λ and above the thickness of the porous membrane (see Chapter 8, "Introduction to Acoustics"). In most commercial units this means they will only be effective at around 1 kHz and above, where a noise reduction of 3 dB to a maximum of 6 dB has been observed. Further, there is evidence to show at typical air conditioning motor frequencies, or rumbling traffic, they can act as an amplifier! You might say that's OK if the colouration from the *room* is reduced. Sadly, above 1 kHz, the reflections from the proximity of the *screen*, whilst just a few milliseconds from the direct sound (so no time delay problems), create comb filtering effects, causing audible gain and loss swings of up to 6 dB in the upper frequencies of the vocal response.

Noise sources will continue to be heard from below and above (particularly ceiling reflections), the sides and of course the singer's position at the front.

Duvets, distanced 1 m or more away, hung over coat racks or washing lines, are in my opinion, a smarter solution.

STEREO MICROPHONES: X-Y MICROPHONE RECORDING

There is often a requirement to record in stereo. There are many techniques, but I'm going to showcase the oldest methods which are also the tried and tested solutions still in use today. They were developed by EMI's Alan Dower Blumlein in 1931 for the earliest patented stereo system (UK patent 394325 [1933]). The Blumlein X-Y system is simply a crossed pair of bidirectional microphones (cardioid mics had not been invented). In the example in Figure 4.14, the two microphones are clearly a left channel and a right channel and can be fed on to a left channel and a right channel of a recorder. The rear pickup of the microphones is useful to capture the atmosphere of the event, but it would be useful to tip the front of the pair down, so the rear pointed more towards the ceiling void rather than straight into an audience. Studio quality microphones were yet to exist at the time, so he created a system but could not check its success, and his work was shut down in 1934 as EMI could not find a commercial use at the time. Blumlein was also the pioneer of stereo records (same patent) and then as war broke out in 1938, he was enlisted as a sound specialist to design "Radar". While this was an amazing success, he sadly died in 1943 during aircraft tests over Wales and didn't live to see the success of his early work on stereo.

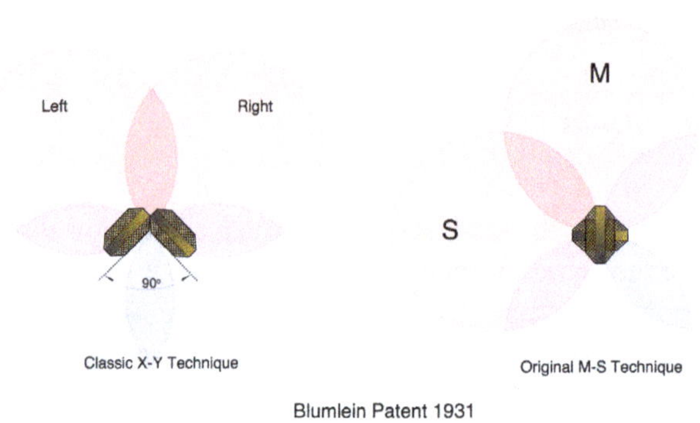

Figure 4.14 Classic Blumlein stereo placements

STEREO MICROPHONES: SUM AND DIFFERENCE RECORDING

Blumlein specified an alternative, the "sum and difference" or "mid-side" microphone array, which he said was for use in cinema sound. In this case one of the crossed pair points directly at the sound source, picking up a mono source, leaving the other to pick up the side or "difference" signals. They are often mounted one above the other. The two microphone signals can be sent to our two-track recorder as M = mono and S = difference rather than left and right. This is worth mentioning as many stereo microphones are designed to create a sum and difference output. These are particularly liked in the film world and on location as the integrated microphones can be compact, but of course they must be handled differently later in the mix. You will find a detailed guide to achieving a mid-side array using an audio mixing desk in Chapter 10, "Introduction to Sound Desks".

Forward 20 years! In the 1950s stereo long player (LP) records were being developed alongside improvements in microphone technology, and the idea of the home "stereo system" was again popular. Blumlein's stereo patent, although prolonged by the war, ran out and people across the world started a race to be first across the line with new consumer products. Blumlein's theories proved amazingly successful in practice, and with new microphone technologies and superior electronics, stereo sound was the future for vinyl records, the cinema and, not to be outdone, FM radio broadcasts.

The appearance of quality cardioid microphones led to a re-evaluation of Blumlein's principles, and a cardioid was used in a modified M-S array, as there was no concern about picking up stray unwanted sound on the rear lobe of the "M" microphone (Figure 4.15). The more sensitive cardioid needs to be attenuated in the desk so it doesn't overpower the figure of eight. Cardioid microphones also found a place in the X-Y Blumlein design. By splaying the microphones further to keep a more even response across the front (between 120 degrees and 135 degrees depending on microphones), this configuration is again more useful in noisy environments; however, many concert recording engineers prefer the original Blumlein design for the spacious atmosphere. Vintage designs of ribbon microphones such as the BBC inspired Coles 4038 still make good sales today as a result.

Figure 4.15
Modified Blumlein stereo placements

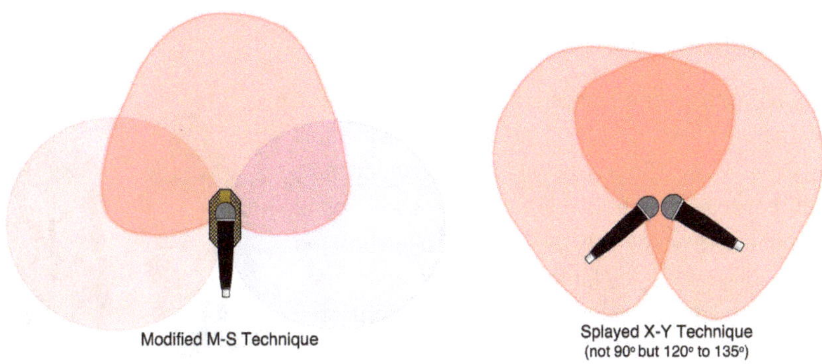

IN CONCLUSION

Differing techniques have their promoters. I have not touched yet on the simple use of two parallel microphones pointed at the stage (either cardioid or omnidirectional). These should have a sound-absorbing "baffle" placed between them (often don't!), and this technique is simply called A-B recording. In my experience, this configuration is the least compatible with mono and is the most affected by the room acoustics. Good results come from expert spacing and scientific use of the baffle on each occasion, and I leave you to experiment and find your favourite solution. Then there is the French ORTF, like the modified Blumlein X-Y but at an angle of 110 degrees and a precise 170 mm capsule spacing.

In the X-Y technique the microphone capsules are ideally mounted in the same position, which is physically not possible. The subsequent problem is some loss of high frequencies through phase cancellation between the microphones, and so some mount the microphones one above the other rather than side by side. Modern X-Y capsules are often in one housing to minimise such problems. If you know what you are doing and are prepared to experiment, the system creates superb stereo imaging, fills the sound stage fully and, if used smartly, the use of 90-degree figure of eight microphones creates wonderful spaciousness and presence. A system of great space and subtlety, and understandably a firm favourite for concert recordings. Tilt them down slightly onto the orchestra to avoid too much direct audience pickup on the rear.

The M-S (or M/S) system is best for mono compatibility, as one microphone is dedicated to that purpose. M-S microphones can be made to be very compact and so fit in a wind screening blimp on a location recording. By changing the volume between the two microphone capsules the sound appears to zoom in or out and so can be made to match a zoom lens in post-production. This makes it a perfect choice for cinema or TV location shoots. For a practical guide to set up an M-S pair on a mixer see Chapter 10, "Introduction to Sound Desks".

MICROPHONE CABLES

Figure 4.16 XLR male and female connectors
Photo by author.

Not all microphone cables are born the same, and neither are the connectors! We look at this in detail in Chapter 6, "Sound Levels and Connections", but here is an introduction. Whilst some band amplifiers may still use 1/4 inch jack plugs (phone plugs in America), by far the most popular connection for microphones is the almost indestructible three connection XLR. Unlike phono plug connections (RCA plugs) the XLR lead has pins on one end (male) and sockets on the other (female) (see Figure 4.16). An audio output (like out of the back of a microphone) is always on "pins" (male), and an input, on say a sound desk, is always on sockets (female). XLR leads are used for line level as well as microphone level, and at around 1 V line level, and less than 5 m, almost anything will do!

Microphone levels of just a few millivolts are a different proposition. We have learnt that when we amplify, we bring up the residual noise, as well as the audio we want louder. A microphone amplifier is going to need to supply a gain of 60 dB on occasion (that's amplification of 1000 times!). Any interference will grow in strength with the microphone signal. We have talked about using balanced cables, and the three connections in the XLR cable support this well: pin 1 earth, pin 2 live and pin 3 return. This alone will have a good go at cancelling the extraneous noises such as radio frequency interference (RFI).

However, other factors come into play. (1) Clearly, we want to keep any losses in the cable as low as possible, so a high purity and conductivity of copper is useful. (2) Cable flexibility is important on stage (particularly in cold conditions like winter). (3) A poorly designed cable can act as a microphone itself, picking up handling noise. (4) Robustness; the cables will be abused, particularly when packing up at the end of the evening, stretched when being pulled back, stepped on and so on. (5) Longer lengths of cheaper cable can be prone to capacitive action between the screen and the other cables, creating a low pass filter. Better cables ensure the high frequencies make it to 20 kHz.

For these reasons, I believe in purchasing high-quality cables to go between your microphone and the desk. The best of all is "Star Quad" where the balanced line system is at its most pure. The live and return are each carried on two cables, and these are carefully twisted within the outer sheath, so equally exposed to RFI. These cables are expensive and can be a little inflexible; however, they are perfect as a permanent connection between the studio patch panel and cubicle desk in a professional installation.

Last points on practical use:

We have a perfect microphone for the task at hand; what else is required?

Isolation – if the microphone has direct contact (through the stand) with the floor or a table, low frequency vibrations like footsteps, or even passing lorries, can travel up the solid stand and be picked up on the body of the microphone and so "heard" by it. These "structure borne" noises are often not noticed at the time unless the system is properly monitored and can easily be powerful enough to ruin your recording. To overcome this, either use an isolation cradle with the microphone or be smart with placement. I carry a foam block (an industrial kneeler) on location which can be placed under a mic stand for difficult venues.

Pop! If we record vocals close with a microphone certain words "blast" the microphone, particularly words beginning with "P" or "B", such as pop! (Try it, place your hand just in front of your mouth and try different words: meet, click, wish, shove – no problem! Now try *b*oom, *p*o*p*, cli*p*... Feel the pressure of your breath on your hand?) The thumping distortion heard with each "pop" ruins the recording. A sponge cover or a stretched mesh (like a stocking) between the voice and the microphone can mostly stop the pop! Not surprisingly our genius engineers call this a "pop filter", and several microphones, particularly ones designed for stage vocals, have an internal filter. A reminder, in close working, the bass frequencies are enhanced in what is known as the "proximity effect", or "bass tip-up" (Figure 4.17).

Figure 4.17 Comparison of a microphone pop shield and windjammer

Wind – the strange shape of the ear (the pinna) is thought to be for two reasons. One is so we can tell which direction a sound is coming from. The other is when we are outside to stop wind noise overcoming the sounds we want to hear. Despite many attempts we can't duplicate this simple natural solution and if we did nothing, the lightest of breezes hitting the surface of our microphone will sound like a jet passing over. Don't underestimate this problem! Fortunately, a company named "Rycote" invented a solution that works brilliantly. It looks like a bushy squirrel tail, and it simply fits over the body of the microphone (Figure 4.17). The technical term is a "windjammer", but (like hoover and vacuum cleaners) many engineers simply call it a Rycote. *A foam "pop" shield just won't do; invest in a proper furry windshield if there is any risk of working outside with your kit!*

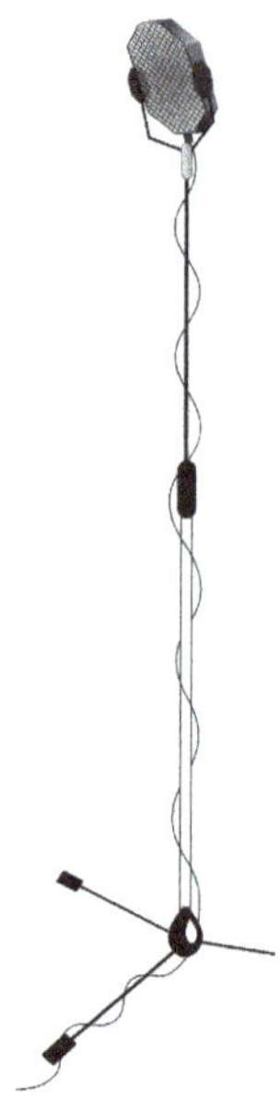

Figure 4.18 A cabled microphone stand

Stand – on a crowded stage or in a compact studio thought must be given to getting the microphone in the right place to be comfortable for the artist to perform; but also positioning the cabling to get it back safely to the wall box or multiway connector. Figure 4.18 shows the cable wrapped around the stand, and this is OK, providing you don't need to increase the distance with a higher stand or longer boom arm position. Plastic stand cable clips are great but tend to wander. You may wish to get your own (three for a US dollar – four for a GB pound), spray them an unusual colour so they are marked out as your own when you derig. Masking tape is an alternative but takes time to remove properly after the gig. The cable should be taped across the floor with 50 mm (2 inch) duct tape to avoid artists tripping. The hazard colours on duct tape (black/yellow) are great in the wings.

Power supply – it may seem obvious, but you would be surprised how often people bring a loved capacitor microphone to work with a desk with no phantom powering . Silence is not golden in this case! Note that *some* electret microphones have a small battery in the body, which negates the need for phantom power. (Don't forget to take the battery out when you are not using it; otherwise, it may rot and leak all over your precious toy.) I use fresh batteries each time in my portable kit, throwing the partially used ones in a box for later use in home needs such as remote controls.

Note when working with a professional studio or PA mixer, it's important to note that plugging in and removing microphones whilst the desk is powered and phantom powering is on, will in time damage the input semiconductors of the preamplifiers of many desks, creating an obvious intermittent blasting sound called shot noise (it sounds like a distant car being perpetually jet washed). Replacement of the damaged components is the only cure. Turn PP off on each channel before swapping out your microphones! Note that modern microphones which don't require PP will normally work OK if PP is left on; however, it is bad practice, and better to only use PP on the microphone channels where it is required.

BIBLIOGRAPHY

Blumlein, A.D. (1933) *British Patent Specification 394,325 'Improvements in and Relating to Sound-transmission, Sound-recording and Sound-reproducing Systems'*. https://adfest.pl/a/GB394325A(OCR).pdf.

Cox, T., Patil, N. and Inglis, S. (2014) How Effective Are Portable Vocal Booths. *Sound on Sound Magazine*.

Hartley Jones, M. (1988) *A Practical Introduction to Electronic Circuits*, 2nd edition. Cambridge University Press.

Hellyer, H.W. (1974) *Stereo Sound*. Hemel Hemstead, England: Fountain Press.

Robertson, A.E. (1963) *Microphones*, 2nd edition. BBC Engineering Training Manual Hayden Publishing Company.

5 Loudspeakers

It will come as no surprise to learn that a sound-creating transducer, the loudspeaker, is basically a magnet and a moving coil, just like the microphone. Whilst there are other technologies, like ribbon loudspeakers, the vast majority follow the simple design of the dynamic microphone, just in reverse. Imagine feeding a tiny audio signal into where we normally take the output of a microphone. The diaphragm will move in time to the voltage induced in the coil, and you may be able to hear a tiny sound output. The principle is the same; however the scale is vastly increased in a loudspeaker, the diaphragm (cone) movement creating a level of sound that is easily heard in a chosen environment.

A heavy-duty conductive wire coil is wrapped around a former and sits in the field of a strong static ring magnet (cutaway in Figure 5.1). A suspension system top and bottom supports the free-moving coil, the former and the rigid cone attached to the front of it. Even though there are many turns in the wire, the thickness of it means that the coil impedance is only a few ohms. Whilst the cone material is chosen to be low mass so there is less to move, it must also be rigid. Common materials are "doped" paper (strengthened with an additive) and an increasing use of specialist plastics or thin metals. I am told by leading designers that there is no superior cone material; rather it is the skill of how, when and where you choose to use it!

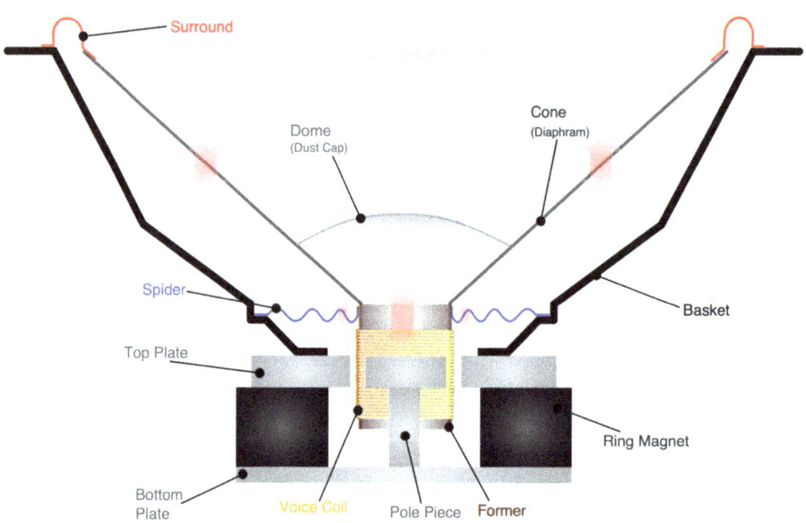

Figure 5.1 Cutaway of a loudspeaker showing major components

The two parts of the cone suspension are a flexible collar at the front called the "surround" and a corrugated "spider" at the rear. The two points ensure the cone and coil move linearly in the field of the ring magnet. A stout metal

DOI: 10.4324/9781003717065-5

chassis (basket) holds everything in place and has mounting holes in the front face so the loudspeaker can be sealed into a circular hole in the loudspeaker cabinet (often called an enclosure). When an alternating voltage is applied to the coil, the cone moves in and out in sympathy. It moves one way with positive volts, and the other with negative volts, swinging either side of a static rest position of 0 V.

THE POWER AMPLIFIER

A special sort of amplifier is required to drive the low impedance of the loudspeaker coil. It must have an even lower output impedance to control the coil accurately (see Chapter 3, "Introduction to Electronics"). This means a high *current* capability, but that's not the end of the story. The loudness comes from having a decent *voltage* gain; the higher the voltage, the higher the volume. Since we are amplifying both the current and voltage, we are amplifying the power available (V × I = P), and so the device is called a "Power Amplifier", and its output is measured in watts (W).

Power amplifiers can be sized from a few watts up to thousands of watts in large venues, and since we are talking *power* output rather than voltage, an amplifier double the size will only produce a 3 dB increase in sound output, which will be only marginally louder. So, to review, there are several types of power amplifier; they all follow the basic rule of increasing the voltage swing for loudness *and* the current capacity to create the low internal impedance at the output of a power amplifier. The amplifier will specify the impedance of the speakers it can drive; the range is normally between 2 Ω and 16 Ω; however, 4 Ω is more usual and 8 Ω the most common. In general, where the design of the power supply permits, a 50 W amplifier into 8 Ω will drive 100 W into 4 Ω (power doubles into half the impedance).

The wires used to join the power amplifier to the speaker sit in series with the load and are not lossless; they will have a resistance measured in ohms/metre (Ω/m). Thin wires may well have 0.5 Ω/m, so over a 4-metre run this would equal 2 Ω, so if we had a 2 Ω speaker, half the power will have been lost in the wiring! This is why we tend to have slightly higher impedance speakers, and of course high-quality copper cables to keep the cable resistance as low as possible. Keeping cable runs short is also desirable. Amplifier designs are beyond the scope of this book, but I will note the popular Class B or A/B amplifier design is being supplanted by Class D as it is much cheaper and lighter in construction. Many Hi-Fi enthusiasts will wish to stay with Class A/B in a high-quality home setting for some time to come, although arguably some will go earlier still with the power-hungry Class A semiconductor amp or the "enthusiast's perfection" of vintage valve power amplifiers.

It's important to talk a little more about the measurement of a power amplifier. I explained above that an amplifier is measured in watts into a known load, and for the specification sheet this load will be a pure resistance rather than a complex (and largely undefined) loudspeaker impedance. And bigger is always better, right? The professional amplifier rating is measured in "watts RMS" (root mean square) and uses a continuous pure tone test signal which is challenging to the amplifier power supply, dragging down on the supply rails. Sneaky people realised that an amplifier (any amplifier) could produce more watts on a changing music signal rather than static pure tone, or even with no load strapped across it at all! These derivations are known as watts "Music Power" or watts "Peak" and are roughly double (some will claim much more!) the pucker watt's "RMS" figure; behold our humble 50 W amplifier is reborn to the unknowing as a 100 W amplifier or higher!

Don't trust a manufacturer who quotes anything but "watts (RMS)"; this can also be phrased as "Continuous Power Output". The other sneaky tactic is to quote the output at a clearly audible level of distortion, when the poor amplifier is in permanent "clipping" of the sound peaks. A decent make of amplifier has its output wattage quoted at a distortion of around 0.05% or better at 1 kHz; if the output watts are only quoted at 1% distortion or more, spend your money elsewhere, because you can bet that won't be the only corner they have clipped! To expand on the above, I would expect a professional amplifier to present with a total harmonic distortion (THD) of 0.01% or better at the rated continuous output *across the audible frequency range*, and to have an *unweighted* signal to noise (s/n) ratio exceeding 95 dB (Ref: 1 W input).

The ability of the amplifier to manage and control its load takes us back to the explanation of impedance in the chapter on the basics of electronics (the truck towing the caravan). A figure put forward was that the driving (source) impedance should be at least ten times less than the load impedance to ensure it can control the load, without being changed itself. In rough figures a speaker with a nominal impedance of 8 Ω would require less than 1 Ω amplifier output impedance to meet this criterion, and lower is better! The ratio of speaker impedance to amplifier output impedance is known as the "Damping Factor"; I personally believe little is to be gained if this exceeds 30, as the two impedances are in series (and again cable and connection losses also sit in series, further undermining the amplifier's control). Most modern amplifiers achieve a damping factor of over 100, and I have seen 1,000 in a specification; however I don't even know how that was measured!

We've discussed harmonic distortion. Problems can also happen in poor Class B designs, at the point where the output transistor pairs switch between positive and negative power rails; it's rare now, but "crossover" distortion was a problem in the past. In addition, I feel there is one more amplifier measure worth examining. Psychoacoustic work in the 1970s pointed to the previously undiscovered inability of many power amplifier designs to be agile enough to manage transient peaks, which may well have a "rise time" of many volts in a microsecond. To put this in some context, as frequency rises the slope of the waveform becomes more vertical; the amplifier must act faster to follow the signal. You might assume that the highest frequency (and so quickest rise time) would be 20,000 Hz to match human hearing; however some have postulated that rise times need to be even quicker to accurately reproduce percussion hits. It is suggested that harmonics of a distorted signal, possibly outside of our hearing range, can beat with our wanted signal to produce a resultant sound within the range of listening. The measure is Transient Intermodulation Distortion (TID), and ideally this nasty aberration, like THD, should be less than 0.1% and hopefully 0.01% or better across the audio frequency range in a professional amplifier.

You would think, given such tight parameters for power amplifier specifications, that all high-quality power amplifiers would sound the same. The opinion of many is that there are large differences between makes, using descriptive words such as dry, bold, brash, refined, sweet, warm, forward, delicate, nuanced, light... The amplifier posse is worldwide and as eloquent as fine wine connoisseurs in their praise or condemnation of certain models or types, or of combinations of amplifiers and loudspeakers. It's a conundrum; certainly some loudspeakers can present complex (difficult) loads to the amplifiers, and some power amplifier circuit configurations are better suited to driving such loudspeakers. I have certainly heard differences, but I also must acknowledge the thousands, even tens of thousands of semiconductor junctions, leaky capacitors, dodgy leads and dirty joints between the studio microphone inputs and the mastering process, and the weight this must bear, set against the handful in a typical power amp. Psychoacoustics is alive and well, and there is always a nude king bereft of his new clothes...

LOUDSPEAKER DRIVERS

Microphones are small by nature as they take very little sound pressure to flex a light diaphragm and capture the vibrations. Loudspeakers, particularly in modern mobile phones and Bluetooth speakers, have shown that a physically small size can yield good sound; however decent bass remains elusive. We have cracked the problem with headphones and ear buds, which are physically close to the ear, but bass in a room needs a lot of energy to power the large excursions of air created by a decent width of diaphragm at low frequency.

The physical size can be reduced by acoustically loading the speaker enclosure in some manner, which we will discuss later. It is very unusual for a microphone to need to have more than one capsule; however it is normal for a dedicated loudspeaker to have at least two, if not three or even four drivers to manage the different frequencies through the range of human hearing. Bass and treble units, known affectionately as "woofers" and "tweeters", split the load of the incoming sound, each taking responsibility for a range of frequencies. Adding a "mid" unit to handle the central range of frequencies is common and there is a growing fashion of a fourth, a low-mid driver to pick up from a (sub)woofer. Just like "bass" musical instruments it takes a larger, rugged speaker to produce the low end, and as the frequencies increase the diaphragm sizes get smaller and lighter so they can move agilely to capture the flitting high frequencies. The higher frequencies are best designed using the smallest diameter units possible (for the required acoustic output) to keep the sound dispersion wide and avoid sound beaming (see Chapter 1, "Introduction to Sound").

LOUDSPEAKER PHASE

Like our microphones can be phase reversed so one sucks whilst another blows, loudspeakers in a home or studio system must be all pushing in the same direction. Therefore, a loudspeaker will always have a positive terminal (normally red) and a negative terminal (normally black) to guide connection. Where there are multiple drivers in a cabinet, each will be marked to show the positive (normally a red paint dot), and each will need to have the correct phase compared to its neighbours by connecting it correctly to the marked + or – of the frequency dividing crossover terminals. It's easy to tell if loudspeakers are phase reversed on a stereo speaker system. The bass cancels out when the out of phase sound hits our ears and so it is much reduced, and the sound stage appears to be outside of the two speakers. It sounds weird; however I have walked into the homes of music enthusiasts and the "out of phase" hits like a brick wall; sometimes people have lived with a system for years without noticing. Not for the first time I remind you to trust your ears, and your growing experience, to make personal decisions. Others may not readily agree but may well have a personal agenda of their own!

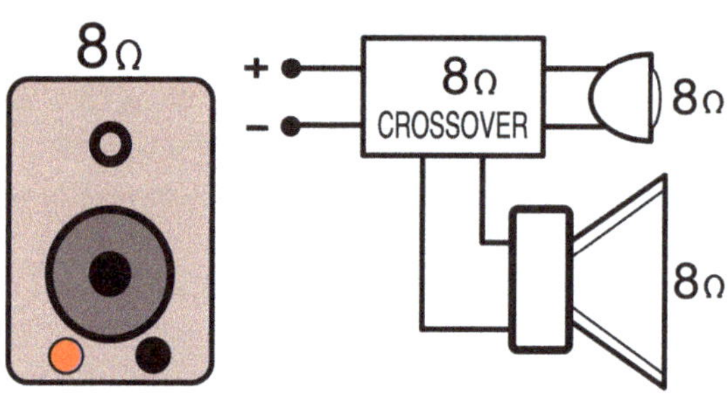

Figure 5.2 A loudspeaker shown beside a simple frequency divider

The sound to each driver is tailored to the frequency range it requires; otherwise drivers may well be damaged, and "colouration" (a distortion) of the sound will happen when certain frequencies are produced by more than one driver and heard concurrently. The network of components that separates the frequencies to the differing drivers is called the "crossover" (Figure 5.2). In Chapter 3 "Introduction to Electronics", we touched on resistors, capacitors and inductors, and these are the exact parts we will find in a "passive" crossover, although they are often physically large to cope with the greater electrical currents and voltage produced by the power amplifier which are required to drive the load of the speaker.

Let's step out of the description to understand a bit more about the measurement of manufactured loudspeakers. Ideally, the speaker should be able to reproduce sound across the range of human hearing, 20 Hz to 20,000 Hz.

The ability to achieve this is unlikely in anything either in a small package or costing less than a king's ransom; however we do need to state how near we are to the ideal. In Figure 5.3 there is a plot of a loudspeaker's output volume in decibels set against frequency; this is a "frequency response", and I would note *the left-hand dB scale can vary between manufacturers*, so a plot from one can look flatter than the plot from another, when the reverse is true! Also, a reminder this is *power* in decibels, so half or double output is 3 dB, and 10x bigger or smaller is just 10 dB! Somehow those plots start to look a lot iffier on cheap loudspeakers!

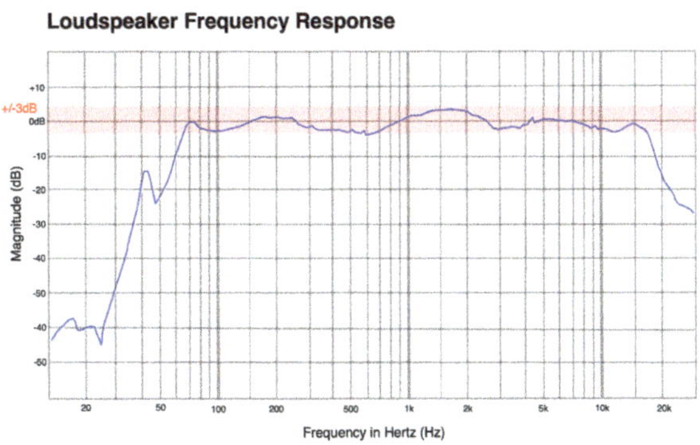

Figure 5.3 Introduction to the loudspeaker frequency response

In the past manufacturers would state a response within a certain accuracy such as ± 3 dB between a set of frequencies. Sadly, at the time of writing you must go to the small print in the brochure where you might be lucky to find a "frequency response" with defined decibel parameters. Beware of wildly fluctuating sensitivities and bands of frequencies that the speaker can only produce at useless unheard levels. In the plot of Figure 5.3 we have a decent loudspeaker, I have added an overlay that shows a useful frequency response of 65 Hz to 17 kHz; however sadly, it could also be described as 30 Hz to 30 kHz without question.

If a loudspeaker response is compared to the microphone responses earlier in the book, it is obvious that loudspeakers are far less accurate at creating a flat line frequency response. These variances, even given a margin of 3 dB or 6 dB, are audibly noticeable and provide the speaker with a unique signature. A studio engineer will pay extra to know a left and right speaker have matched drivers so the stereo image stays stable across the frequency range and, of course, achieve a flat (as possible) response that can be trusted to be faithful to the original sound. Such speakers are called "monitor" speakers, and as in all things, you get what you pay for. Often professional monitor speakers are not initially as enjoyable to listen to as the home Hi-Fi manufacturers have a lot more leeway in producing something that "sounds good" by compromising its accuracy. Terms such as "warm" or "full" can point to a hump in the upper bass frequencies to hide the lack of a real bottom end, and "crisp" or "bright", a hump in the tweeter's sensitivity which sounds superficially satisfying but trust me is wearing in time. For many professional sound engineers, working with the same monitoring day in, day out reveals its faults and features, and these are taken account of in the mix. When in doubt, carrying a decent pair of your own headphones makes a great comparison tool!

SPEAKER CROSSOVER DESIGN

Other important loudspeaker characteristics are how sensitive the speaker is given a known input. An insensitive speaker will need a bigger amplifier to drive it to the same volume. For each drop of 6 dB in sensitivity the amplifier must have twice the watts to drive to the same volume. In the past expensive speakers were often less sensitive as they used additional loss-making components in the crossover to improve the specification and to obtain a more truthful speaker. These, of course, added to the research time and the unit cost, so made complicated speakers like the LS3/5a more costly to buy. We have moved on; with improved knowledge, tools and components driving up the sensitivity of modern speakers, the speaker designer's art is to use ingenious computer-aided cabinet designs, married with smart driver choices to get it right, but keep it simple!

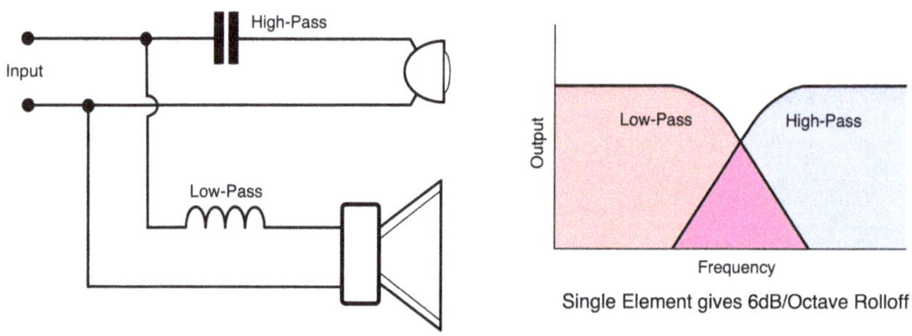

Figure 5.4 A simple single element two-way crossover

The physical constraints of a small cabinet size mean that it will be limited in the excursions it can make at the lowest frequencies, perhaps to 100 Hz. They can still sound good, and without the need for the large excursions, the driver can go much higher in frequency towards the crossover frequency required for the tweeter. Such units will not, and do not, need a third driver built in to sound great. It's only as the quest to make drivers that can approach the lower threshold of human hearing at 20 Hz that a third "middle" driver is useful to bridge the gap between the ability of the high end of the bass unit and the low end of the tweeter.

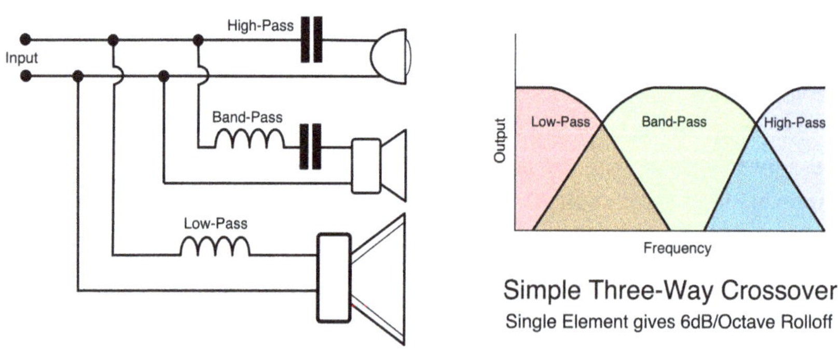

Figure 5.5 A single element three-way crossover

A simple "one element" crossover (either a single capacitor or inductor with each driver) creates a roll-off or slope of 6 dB/octave. This slope is calculated from the crossover frequency when it is at the –3 dB down point (Figures 5.4 and 5.5). Extra elements (components) can be added to give a steeper roll-off at our chosen frequency.

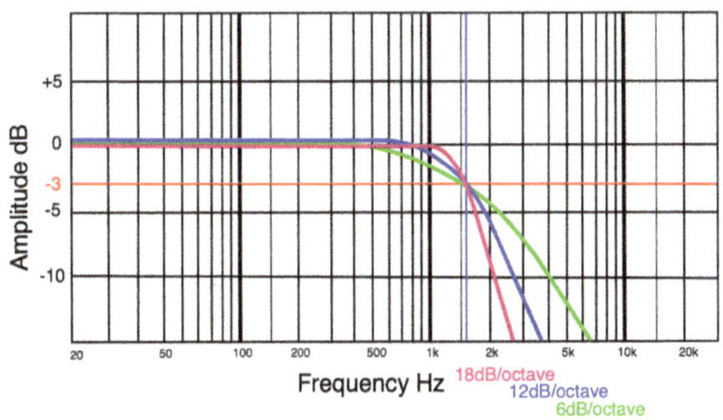

Figure 5.6 Extra components give a steeper frequency roll-off

The simple one element crossover (shown in Chapter 3, "Introduction to Electronics") is called "first order", *two* elements per driver give a slope of 12 dB/octave and is called "second order", third order uses *three* components for each driver giving 18 dB/octave slope and although mostly impractical in a passive crossover "fourth order" of 24 dB/octave is easily obtainable with a low level "active" crossover (see "Active Speakers"). A steeper slope is required in higher quality designs as the area of overlap where two frequencies crossover is reduced, and so the muddying of sound output (colouration) is also reduced as the listener is not hearing two drivers endeavouring to deliver the same sound frequencies (Figures 5.6 and 5.7).

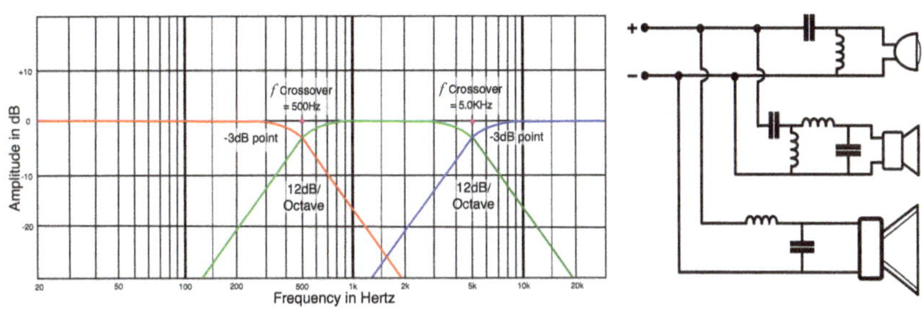

Figure 5.7 A practical 2nd order (two element) three-way crossover

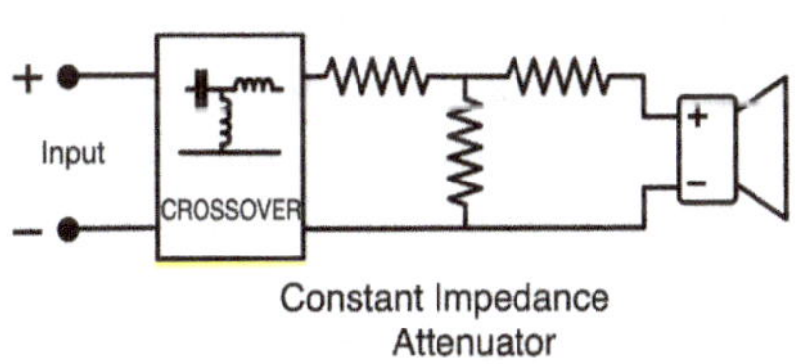

Figure 5.8 The constant impedance attenuator

It is unlikely that the sensitivity of the chosen drive units will be the same, and so normally the mid and top units need to be attenuated with a resistor network to match the output from the bass unit. We use a "constant impedance attenuator". In this simple network of resistors, shown in Figure 5.8, the parallel resistor is the value of the sum of the two series resistors.

SPEAKER PLACEMENT

Manufacturers routinely test their loudspeakers in "free field" (literally *in* a field sometimes!) They are measured whilst they are distant from the confusing reflections of hard surfaces, and whilst that is impractical in many homes, keeping loudspeakers as far as possible from walls, and particularly corners, produces a better sound. The bass boost that you encounter with a loudspeaker in a corner is unrealistic, and whilst it might suit a particular genre, it won't suit others, so your system becomes unreliable in its fidelity. Again, if your seat is pushed hard against the opposite wall, the sound you hear will be false!

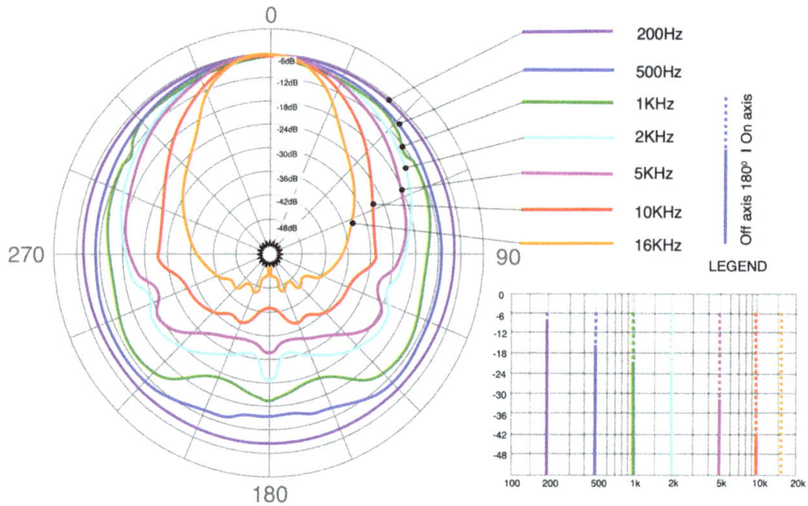

Figure 5.9 Horizontal directivity of a quality speaker (listener at 0 degrees)

As frequency rises the sound radiation from the sides and the rear of the speaker becomes more directional, so to hear "highs" accurately it's imperative that you turn the speakers to face where you sit (Figure 5.9). At 2 kHz side reflections off a typical loudspeaker can be 12 dB or more down at 90 degrees off axis, so the effects of the reflections off the walls are less significant than some might argue, but still worth acknowledging when the acoustics of the room are considered holistically. Note, many complex designs have a minimum distance away for listeners to sit to get a cohesive experience. Some loudspeakers don't work well in smaller rooms and that parameter must be part of the buyer's research.

It bears repeating that if you are listening in stereo, or multichannel, try to get the speakers at roughly equal distances to your seating position, which, if not the middle of the room, should not push back too far to the rear wall. Please read Chapter 8, "Introduction to Acoustics", to understand how the room can be further modified to create a good listening environment. People talk of the "first order" sound reflections being the worst. These are between you and the loudspeaker where the sound from the speakers glances off close objects like the room side walls, mixing with the direct speaker sound as you hear it.

This is ameliorated by turning the speakers' fronts to face you (Figure 5.10), and as previously stated, that is essential to receive the high notes properly as

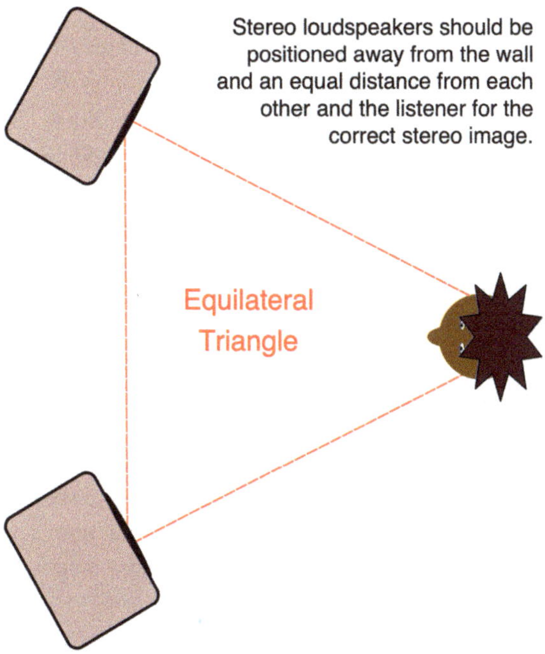

Figure 5.10 The stereo listening triangle

these become increasingly directional as the frequency rises. Further, in Chapter 8, "Introduction to Acoustics", you will learn of the Haas effect where the middle frequencies become muddy if the path length between direct and indirect sound exceeds 50 milliseconds (ms) (or 11 m), not a likely prospect in most living rooms... However, that is for a (single) point sound source, and of course, stereo has more than one; the early reflections from the walls affect the stereo image; this is particularly noticeable in the 1 kHz to 8 kHz region, and in narrow walled critical listening areas the early reflections should be damped. Please read more in Chapter 8, "Introduction to Acoustics".

Going back to the Hass effect, it is useful to break up second order reflections returning off the rear wall. I would use home-made dispersion or, for deep pockets and decent room sizes, the mathematically proven propriety diffusers, both discussed in "Appendix 3 – A Guide to Home-Built Absorbers". Porous absorbers would have to be very deep to capture the low frequencies associated with standing waves, and in any case, these work poorly when attached to the surface of the rear wall as the particle velocity is low (see "Appendix 3 – A Guide to Home-Built Absorbers").

I'm aware of "Golden Triangle" computer room placement, with online calculators to manage any room dimension. These are at least part common sense, to keep speakers away from room boundaries yet still provide a listener with a pleasing stereo image. Of course, you may want to invite the neighbours in, and then where does your golden seating position go? *Every* room is different, but a narrow room dimension is always bad, because placing speakers on the long wall makes the seating placement problematic, yet first reflections less intrusive; and on the short wall, vice versa.

I personally advocate the well-tried equilateral triangle for stereo placement, placing the listener equally away from each speaker (same in multichannel), because that's the ideal from when the studio engineer made the balance in the first place, whether she/he did it on near or far field monitors (or probably both!). It's fine to play with increasing the listeners' distance, as some manufacturers believe to be worthy; but it can't be a rule, as every speaker and every room is an acoustically different challenge. Speaker height from the ground changes the bass significantly (lower-more bass), and where ear height (when seated) is optimum for the higher frequencies, you may wish to play with differing speaker stand heights to get the best sound in your room. Finally, the obvious sitting room position of placing speakers either side of a jutting fireplace needs care! I know you will not put them into the corners but do bring the speaker fronts level with the front of the mantelpiece when you have a listening session.

Time domain effects – whilst we are about loudspeakers and time delays, one area of loudspeaker design has improved immensely in recent years. The work of Haas helped us to understand time in reverberation; however despite this some otherwise perfect speakers only sounded OK when placed in a dead acoustic. We came to understand that our ears are very sensitive to the time alignment of a natural sound source, needing this when hearing and analysing transients such as a distant twig snapping underfoot in the woods. It's a survival mechanism! We understand this basic need now, and quality speaker designers take as much trouble to align the arrival of sound at the ear from several drivers in a single enclosure as they do to get a linear frequency response. We gain a more physiologically appealing sound and the removal of listening fatigue which was experienced in some older designs.

SUBWOOFERS

We have learnt that sound gets more directional with frequency. As we go lower, the ear loses the ability to tell sound direction *at all* below around 150 Hz (see 100 Hz on the directivity diagram in Figure 5.9). This effect is utilised in a specialised bass speaker called a subwoofer and it can be placed just about anywhere in a room (but not right in the corners!). Normally only one is required or desired in a home multichannel set-up. The speaker cabinet and bass driver are tuned to produce frequencies below 100 Hz, with quality units able to extend low, flat response, bass down to 30 Hz and even 20 Hz in costly designs! The speaker often features a low pass crossover at around 80 Hz, and the conventional speakers will often have a signal path through the subwoofer with an internal high pass filter (HPF) that stops them producing anything below 80 Hz and so muddying the sound. As always unit phase matters!

The attractiveness of only needing one (large) bass speaker which can be tucked away, and the fact that others in the system can be much smaller as they don't need to produce the low bass themselves, have made this a popular solution in both home systems and quality public address in halls and theatres. Of course, where the volume of one subwoofer is not enough, others can be added, providing the phase is carefully aligned between them.

LOUDSPEAKER ENCLOSURES

Whilst the middle and top drivers are easily mounted in a sealed box, the bass can sound constricted if it is damped down by the confines of an airtight enclosure. Great designs exist, as said earlier, like the world-famous BBC LS3/5a; however this goes some way to explain the need for a decent amplifier to drive this insensitive speaker. In Figure 5.11 a sealed box is described as an "infinite baffle" as though the driver is mounted on an impossibly large board where the sound from the rear can never find its way to the front. Another sort of cabinet takes account of the natural resonance of the driver and its enclosure and places a carefully calculated hole in the box. This can be designed to be rectangular or round, in the front face or the rear. This hole will have depth (often a sleeve like in a toilet roll centre) and is tuned to enhance the bottom frequencies and extend the low frequency response of the speaker. This is a bass reflex speaker. Sometimes a poorly "ported" speaker can lose its punch and sounds flabby despite the better on-paper response, but there are some good designs too!

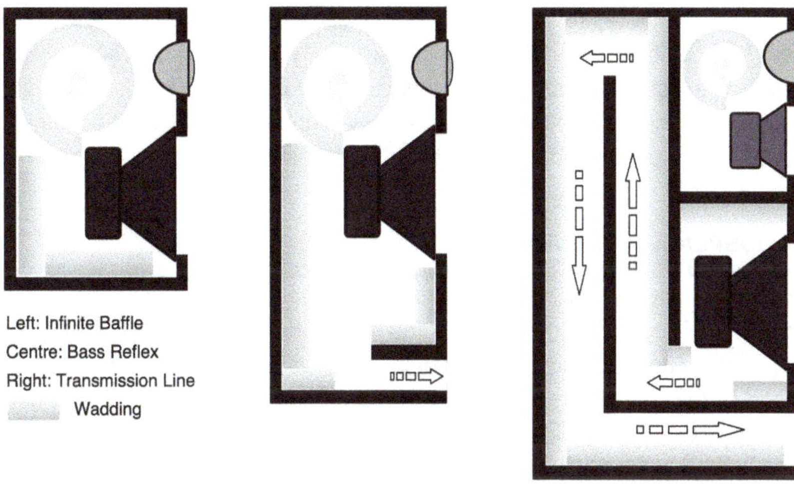

Figure 5.11 Three types of loudspeaker cabinet

Figure 5.12 Advanced transmission line pathway, photo by the author

In all designs the bigger the box, the better the opportunity for a more extended bass. The transmission line speaker in the drawing (on the right) happens to have a midrange unit added but note any of the three speaker designs could have two or three drivers. An important point is that in a quality design the midrange is acoustically sealed away from the bass speaker, so it is not affected (distorted) by the pulses of air from the larger speaker. These are just three of many designs;

however they are the ones you are most likely to currently find in a studio or a domestic setting.

The "Transmission Line" speaker shown in Figure 5.12 uses a complex cabinet that takes the low frequencies in a labyrinth around in internal walls of the box and out through a vent. The bass produced from such a speaker is natural and extended. Clearly the research and build of a complex box makes these speakers expensive, and they are at home in high-end studios or the sitting rooms of opulent music lovers. The cutaway (courtesy of PMC Ltd) is an Advanced Transmission Line (ATL) design where the loudspeaker is placed ⅓ up the length of the acoustic duct (blue path). The design behaves like a "closed pipe" (see Chapter 1, "Introduction to Sound") and so there are no even harmonics generated. The red path length cancels the 3rd (odd) harmonic, and the wadding is sufficient to absorb the 5th harmonic and above; so, the design has refined low, flat bass frequencies.

The reason for the wadding in the loudspeaker is to dampen down the series of natural resonances the enclosure cavity creates in concert with the driver. We learn in Chapter 8, "Introduction to Acoustics", that the wadding should be placed where the particle velocity is greatest (away from edges and corners); activated carbon has also been successfully used.

LOUDSPEAKER EFFICIENCY

To date we have noted that good loudspeakers have a wide and flat frequency response historically within ± 3dB through the effective range of the speaker, which we want to achieve a breadth of range as near as the human hearing range of 20 Hz to 20 kHz as possible. Whilst the top figure (or comfortably near it) is easy to achieve with decent speakers, the low end is where the money is spent. Anything "flat" down to between 40 Hz and 80 Hz is still considered HiFi, but at the time of writing, a good *passive* design will cost thousands of US dollars to be flat down to 30 Hz and tens of thousands to be flat to below 20 Hz, a prime example being the Professional Monitor Company BB5 speaker, flat to 17 Hz! We have also said we want time-aligned drivers and low distortion, although we accept it is most likely higher than the surrounding electronics within a system. A final essential requirement is to know how loud a speaker will go for a given amplifier input, the "efficiency" of the system.

We are driving the loudspeaker from a *power* amplifier; the measurement in decibels is one of power, rather than of voltage ratios we meet in a discussion of desk levels. A reminder, 3 dB (rather than 6 dB) is a doubling of power output, and 10 dB is ×10 power (as opposed to 20 dB as a voltage ratio). However, now, we want to measure the speaker's sound output. It's confusing but the output follows the inverse square law, and sound dies away at 6 dB for doubling of distance. We must choose a distance to place our measurement device, a calibrated Sound Level Meter (unweighted!), as introduced in Chapter 1, "Introduction to Sound". Typically, the meter is placed on a stand positioned *one metre* from the front face of the test speaker, in an open non-reverberant area, known as "free field" or ideally you can buy time in an anechoic chamber. We used to adjust the amplifier to output 1 watt (which can be measured as 2.83 V across a pure 8 Ω load [Ohm's law]) using an input of a 1 kHz sinewave and then replace the pure resistive load with the test loudspeaker and measure the output level in decibels on a sound level meter placed 1 m in front. This gave us a sensitivity based on dB/1 W @ 1 m. This method is now increasingly simplified. We now attach the test speaker directly to the amplifier and use an AC voltmeter to adjust amplifier output to measure 2.83 V. This favours low impedance speakers, but it is much easier and is now normal in the industry. I tell you this as you might reasonably ask, of all levels, why choose 2.83 V?

This measure of sensitivity is called the "SPL" (sound pressure level). An insensitive speaker can be as low as 84 dB for 2.83 V output measured at 1 m. Note short forms are variable, but "84 dB/2.83 V@1 m" is typical. Most decent speakers will do better; a decibel or two either side of 88 dB/2.83 V@1 m is normal, and specially designed PA speakers can achieve over 100 dB/2.83 V @ 1 m where the main aim is getting the sound out there! However, you can't get something for nothing, and in a practical design for a guitar speaker I found recently, the frequency response at 50 Hz is a full 20 dB or 10× lower than for 200 Hz. Guitar loudspeakers are a special subject on their own; like the instruments themselves, they are about bite and attitude, and frequency response is irrelevant. Got to love the Celestion Greenback though! Watch for SPL figures measured at different amplifier wattages or voltages, or at closer distances to make the speaker seem better than it is. Note, it is normal to measure larger speakers and PA speakers at 2 m or even 4 m, simply to get a complete sound front, and easy comparison can be made by adding 6 dB/12 dB back to get a 1 metre distance equivalent.

ACTIVE LOUDSPEAKERS

There is an alternative to the "passive" speaker, where the frequency determining components of the crossover are moved earlier in the signal chain to before the power amplifier and sit at low level in an electronic circuit. This means each crossover output must be amplified to drive each speaker, so there is a requirement for multiple power amplifiers rather than just one for each loudspeaker cabinet. At large outdoor events the system will be "discrete" as you can see in the separate units shown in Figure 5.13; however, in an indoor setting it makes sense to place the crossover and amplifiers within the speaker cabinet, the big giveaway being it now requires a mains power supply input on the loudspeaker unit to drive the circuits.

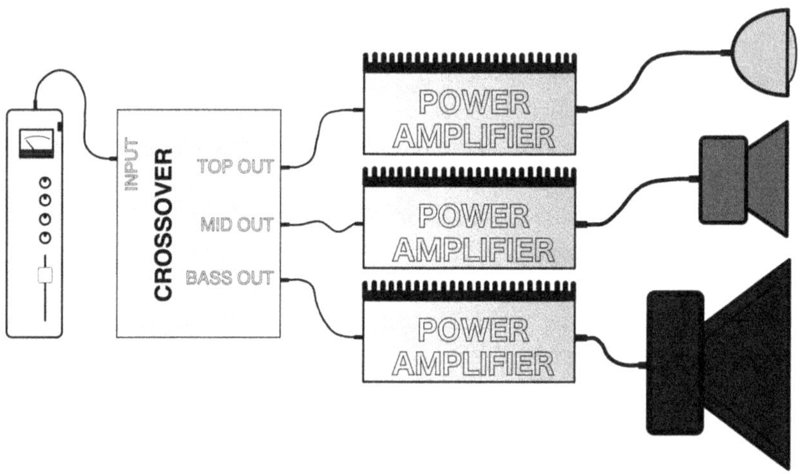

Figure 5.13 An active loudspeaker crossover

Such a speaker is called an "active" speaker and as well as the mains supply it has a single signal input straight out of a sound desk or a HiFi "pre-amplifier" at a comparatively low line level. These speakers have gained huge popularity in professional sound as the low-level crossover circuitry is cheaply tailored to suit differing drive units and by switchable presets, acoustic environments. I must give a warning that integrated active speakers are not automatically better than passive designs; economic corners are often cut on the internal amplifiers, which tend to be simple class D designs, and so as in all, use your ears to make a value judgement! An active subwoofer speaker cabinet can be designed to tackle the tricky low frequencies (nominally 20 Hz to 80 Hz), and any bumps in the response can be smoothed with tweaks made in the custom low level audio preamplifier. To give an example the active subwoofer I am listening to as I type this chapter produces down to 32 Hz as measured in situ in the study, and retails below US$1,000 (obviously managing the upper frequencies takes extra units at additional cost). Given the caveats above, active systems can give an excellent route to high quality sound on a reduced budget; however that said, there is no substitute for giving thought to the acoustic treatment of the room before placing the speakers in it...

LOUDSPEAKERS FOR VENUES AND EVENTS

In small pubs and clubs where there needs to be speakers in several rooms, we can hang more than one loudspeaker off an amplifier if we know what we are doing. It's important that we look to match the stated drive impedance of the power amplifier and not go lower as the amp will get hot and either eventually cut out or even die... Fortunately, we can treat speakers just like resistors (Figure 5.14) and add them together in series, parallel or both!

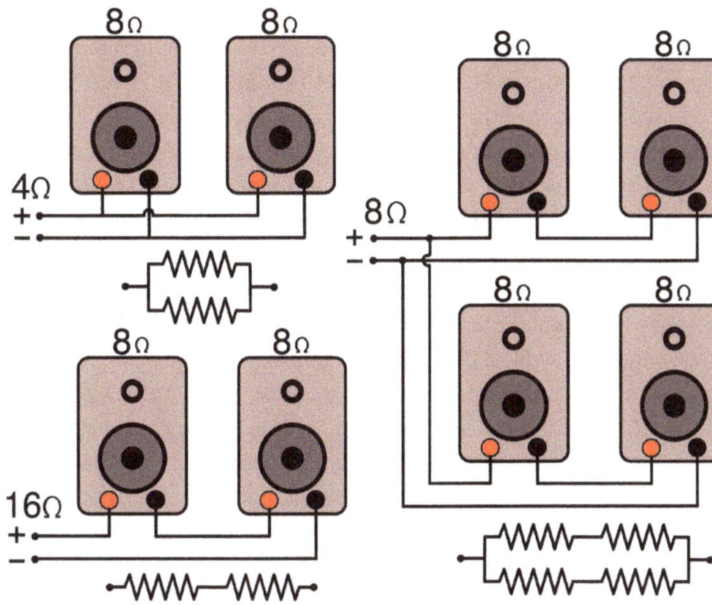

Figure 5.14 Multiple speakers connected to a single amplifier

If we daisy chain two speakers (series connection) the impedance will double, and we will have less load. If we parallel the leads we will halve the impedance (double the load). A useful configuration is series/parallel where four speakers can present the same impedance as one and be fed from a single amp. This can be scaled up to larger numbers, 9 and 16, but we are running out of reliability and common sense. Just remember with these longer lead runs to use quality cable so the losses in the wires are kept low.

VENUE AND ARENA SOUND

Bigger venues, anything over 500 W, are better managed with active speakers, where the active crossover unit can sit as a separate unit to feed many amplifiers and their respective speakers throughout a large auditorium. The saving grace here is the power amplifier stacks can be placed near the speakers to cut losses in the wiring runs; however, a detriment is that three times as many line-level XLR lines must be run out. In high-end solutions, a digital interface is used, perhaps with radio links.

There is an old-fashioned way to move amplified sound around a large site with less losses. Such a system is called "70-Volt" working in America and "100-Volt Line" in Europe, but little separates them. A transformer is attached to the amplifier output to step the voltage up to a high level (this transformer can be internal to the power amplifier casing creating a "100 V Line Amplifier", or external). The output impedance steps up as a square of the voltage (see Chapter 3, "Introduction to Electronics") and so is a *much* higher impedance source. Any resistance in the speaker wiring sits in series with the source impedance but this is now likely to be negligible in comparison, and so long runs can be made without noticeable loss. At the speaker end the voltage needs to be stepped down again in another transformer to provide the current to drive the loudspeaker, and this too can sit as a separate unit or form part of the loudspeaker chassis. Some loudspeakers have several taps off the transformer coil so the load (and so the volume) can be chosen to suit the positioning.

In the first chapter where we introduced the subject of sound we touched on the difference between a "point source" and a "line source". Theoretically a point source sound dies away at 6 dB for each doubling of the distance (inverse square law) and a line source decays at half that, just 3 dB for a doubling of distance (Figure 5.15). Another way to think about it is to see the point source as a radiating sphere, the sound travelling off in all directions. The line source can be thought of as a cylinder, with sound radiating from the sides, but not the ends. What if we could design a speaker that followed the line source principle and could shove the sound twice as far?

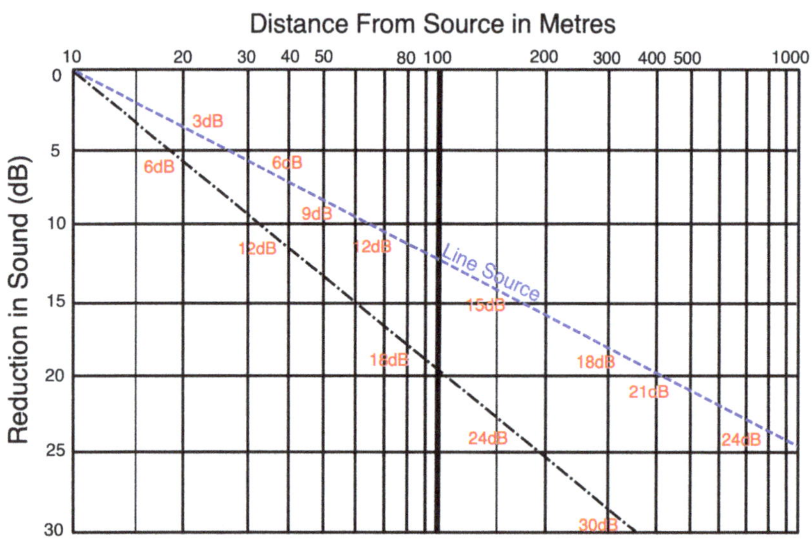

Figure 5.15 Sound decay of a line source

This is the theory behind the column speaker (Figure 5.16) which places several speakers in a vertical line to create the line source effect. Many designs use elliptical speakers (oval rather than round) to create an even slimmer, taller cabinet. These speakers have always been popular in medium sized buildings such as churches, and for good reason! Aesthetically the tall thin structure is less cumbersome on the eye, and the extra reach of the design is very

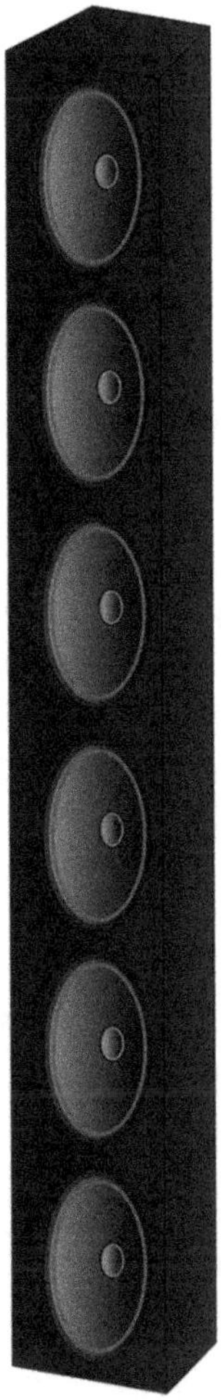

Figure 5.16 The column speaker

welcome. The sound is beamed to the audience off the "sides of the cylinder" and so is more directional, meaning reflections off the unforgiving hard walls of the church are less likely to send the system into feedback. Further, a line source is also a "steerable source"; using digital synthesis, with multi-band FIR and IIR equalisationi, the sound can be steered where it's needed, and importantly steered away from reflective destinations which will encourage system howl round.

So why aren't all systems like this? Bass sounds by their very nature are always going to be "point source" so the advantages of the design are lost at low frequencies. Further as mid frequencies turn to high, the ear becomes better at pinpointing the direction of the sound and is confused if there are multiple sources, even if they are in the same cabinet, and so high frequency sounds can appear muddled. So, the column speaker is efficient, directional, avoids early feedback problems and blends in better *but* it is only effective for speech frequencies, and so loves a long sermon from the pulpit...

I find it amazing that way back in 1933 a member of Bell Labs in America, Stokowski, predicted multichannel sound in "Giant Recreation Parks" using "High Towers of Loudspeakers". It is understandable that supergroups will want to go on the road with the minimum kit to get the job done. How many pantechnicons will be to individual choice!

There are bespoke cabinet designs for high volume scenarios, and this is a truly fascinating science, but the detail is not for this book. These designs focus the sound where it is needed and often make use of the "horn effect". A trombone mouthpiece can make a fun sound, but once it's connected to the trombone brass horn it can be deafening! Loaded horn enclosures can have efficiencies of many times the domestic speaker and can be created in low, medium and high frequency enclosures. As you would expect, low frequency units are complex in construction, very large and heavy and so have wheels (or castors) built into the box to help move them around a stage.

Midrange/top speakers are often slung high in a vertical curving line, resembling (to me at least) elephant tusks. Like the column speaker above, these Optimied Line Arrays are "steerable" using digital signal processing (DSP) and can be focused on sections of the audience, avoiding annoying spill into the surrounding neighbourhood. DSP is integrated with the active crossovers, splitting frequencies to different banks of speakers, and providing time alignment of the multiple arrays. In passing, I note DSP technology is recently finding its way into the studio as well, with active studio monitors beginning to come with similar tools to match each speaker to the others, and to its acoustic environment. Venue speakers are robust for transport, and of course water resistant. Other than that, it's a matter of scale, and a balance between the frequencies and directivity to keep the sound where it's supposed to be. Sound Designers are well paid for the job, and so they should be!

A word about timing. Previously it has been pointed out that if there is a path length more than 11 metres between two identical sounds reaching a listener the resultant sound will be muddy and confusing. At large events the sound from the PA stacks is delayed by DSP to meet the sound coming from the stage, so they share the same timing. In large concert halls where speakers are placed up the sides of the hall, multiple time delays (greater delay the further from the stage) will be generated so the sound from all the loudspeakers reaches the listener at the same time. These systems are normally house systems which stay in place year in, year out; the setup, as you can imagine, is complex and time consuming.

SPEAKER POSITIONING IN VENUES

Whether the installation is just for the evening or the life of the building, the same simple rules apply. Don't point any speakers directly at the rear wall as that will encourage standing waves and howl round (see Chapter 8, "Introduction to Acoustics"). Do point speakers inwards from a wide stage up a narrow hall, or outwards (closer together) in a wide venue. Get the speakers high so the sound passes well over the heads of the audience; this is particularly important for high frequencies as "grazing" over the crowd absorbs them. Where several speakers are used remember the 50 mS (or 11 metres) rule and watch that people do not receive more than one strong path of sound from differing speakers. This is particularly important where there is sound reinforcement that is a gentle support to a natural voice such as a pastor in a church. Placing a speaker above the pulpit is a good tip; using one speaker carefully angled for each section is another. The popular placement of speakers at points up the side walls (of say a church) to feed an audience or congregation asks for confusion between the original voice and multiple clones and needs careful evaluation to have any chance of success.

STAGE FOLDBACK

In the studio a musician can hear her/himself sing and play by listening on quality headphones. This is vital, particularly for vocalists to be able to pitch correctly. This essential was ignored in live performance on stage and in theatres until the rise of amplified instruments, such as the electric guitar. Speakers had existed on stage before then, as personal amplifiers, to bring the vocalist up to a level to match a band made up of acoustic drums, horn section, piano, double bass and acoustic guitar. In the early 1960s, the emergence of the "beat combo" using largely electronic instruments on stage, such as Hammond organ, electric piano, rhythm and bass electric guitars, took up the venue volume, on and off the stage, by several orders. For a while, musicians still carried their own amplifiers, which would be stacked behind them as they played, so they could hear themselves.

As "pop groups" grew in stature, audience numbers dictated larger venues than cinemas and theatres, and a new approach to sound reinforcement was required. There was a time when the personal amplifiers remained, and a microphone was put in front of each cabinet to capture the sound for the main PA. Crucially, this meant each player could still hear their personal contribution. Unfortunately, more microphones resulted in increased, complex, howl round problems. In time, the main speakers were moved out from behind the band to the wings to retain a vestige of the performers' failing hearing. The various instruments were captured by plugging into "direct inject" boxes (DI box) and fed into inputs on a PA mixer. For the first time the drums needed miking, as well, of course, as the vocals. It was now like a studio recording; a balance engineer mixed the band's entire sound and returned it through power amplifiers to large stacks of speakers.

The elephant in the room was now that the performers failed to hear themselves, and each other. At first ordinary speakers were placed either side of the stage pointing back at the artists, but the distance from them (and so increased sound levels) created colouration and feedback problems with the main sound rig. Then they hit upon the idea of placing low profile speakers in front of the artists, so the view was not compromised, and the distance close, so lower colouration. Also, this speaker positioning, just in front of the artist was no longer problematic due to the growing use of new cardioid microphones like the Shure SM 58, which ensured howl round was contained. The favourite low profile was the wedge-shaped speaker, pointing up to the artist. They quickly learnt to use strong mesh in front of the speaker as they frequently got stomped on! The system was variously called floor monitors, stage monitors, wedges and, of course, "foldback" (Figure 5.17).

Figure 5.17 A floor monitor or foldback speaker

In time, the wedge has remained a staple of onstage monitoring, with many similar designs of differing sizes and qualities. At the turn of the millennium there was a huge growth of robust *active* speakers for stage use, with cost and weight savings from using class D amplifiers. These utilised tough plastic moulded cases in place of traditional ply and medium-density fibreboard (MDF). Available in several sizes and, of course, qualities, these had an unusual case design, which has become a very common sight today. The speakers tapper down from the front to the back, meaning they can stand on feet as main PA on the stage or be laid down as a wedge. As in all active speakers, the units are individually mains powered and receive the audio signal from the desk at line level, normally on an XLR.

The main PA was retitled Front of House (FOH) to distinguish it from the stage monitoring. The monitors in smaller gigs could be fed from the mixer "auxiliary outputs" (some of these were even retitled "foldback" mixes, see Chapter 12, "Outboard Devices"), which could favour aspects of the mix by the amount of level fed from each channel to support vocals and beat to keep the band together. In larger gigs, the roles became separated into FOH and Foldback mixers, where the monitor mixer started to create mixes for different band members to favour their participation in the band. The introduction of stage monitors provided a solution that has lasted for decades, however clearly an imperfect one! The artist is tied to their monitor to get a useful sound, so mobility on stage is compromised, the stage monitors erode the margin to a howl round, and it's more kit to lug around and install at each venue.

The rise of first analogue, then digital radio, mics created microchip transmitter/receiver pairs which are tiny, cheap and high quality. It was a simple matter of time before these chips would be utilised in a return path to the artist. The wish for mobility on stage is now complete, with artists freed to move around with radio microphones and instrument interfaces, and a separate wireless return path back to the band members with their in-ear monitors, supplied from the foldback balancer with a personal mix (Figure 5.18).

Figure 5.18 In-ear stage foldback Photo by Christoph Stowasser.

THE NON-SPEAKER LOUDSPEAKER

What do you do when the loudspeaker is not loud enough? The hard of hearing have a right to hear your brilliance too! At the beginning of this book the "Cocktail Effect Problem" was described, where older minds in a public space cannot separate the wanted sound from the general background buzz. It is also the grey army who are more likely to own hearing aids. Although increasingly smart, at its heart, the hearing aid is simply a personal PA system. A miniature microphone on each ear, feeding a battery-powered amplifier to a tiny speaker, mostly sealed into the ear canal by pressure, just as modern audio ear buds.

Fortunately, there is a technique to pump sound directly to the hearing aids using an "Inductance Loop". For some reason this very simple technique appears mysterious to many, so let's try and bury some myths. The Induction Loop is not some sort of radio transmitter; it's nothing more than a coil of one or more turns of wire placed around the perimeter of the venue.

The coil is fed by the output of a power amplifier, just like the loudspeakers; however, it may be specialised with current feedback rather than voltage, and it may have a transformer to better match the output impedance to a known coil resistance in a particular venue. This is also managed by choosing the gauge of wire (and so the resistance/metre) used in the loop. If you read through Chapter 3, "Introduction to Electronics", or look at balanced line working you will understand that a coil of wire will induce a current in another coil of wire magnetically, simply by sitting in a parallel aspect to it. The hearing aid wearer traditionally switched from the normal microphone to a coil in the hearing aid marked with a "T" for "Telephone" (its original design use). Modern aids will now switch mode automatically.

The listener will get the best experience if the loop is placed at ear level, although decent floor or ceiling-mounted loops achieve their purpose too. Also, the hearing aid coil has a vertical aspect, so expect a drop in level if the wearer ducks their head or lies down. It's important that you understand this if you choose a career as a sound provider, as inductance loops are mandatory under discrimination legislation in many countries "wherever information is verbally provided", so expect to see them and to need to keep them working!

Note that a coil connected to the output of an ordinary power amplifier in place of a loudspeaker can cause it to go unstable and oscillate; the electrical version of a howl round, and although the oscillation might be inaudible, it will heat the amplifier up and can cause failure! A second point is, just as a power amp can clip a (too) loud audio signal causing harmonic distortion, a poorly sized induction-loop power amplifier will distort the signal into the loop. Induction loops are a skill on their own, and although it is straightforward physics, I advise "would be" designers to make use of the great material available on the internet.

USING MONITORS IN THE STUDIO

This paragraph is not about the acoustics in a monitoring area; we assume the direct sound from the main speakers dominates the reflections and that they are distanced to provide a coherent sound from the drivers at the operational engineer's seat. Appendices 2 and 3 provide a detailed guide on monitoring room design; please give them a look! The copper coils in our quality speaker drivers provide accurate displacement of the sound waves because of the high electrical conductivity. Unfortunately, copper is also a brilliant conductor of heat, and during a working day, the coils will heat up, in some cases to over 200 degrees Centigrade (>400 degrees Farenheit). Our speakers are literally cooking! Manufacturers are aware of the problem, using ferrofluid, special adhesives and large metal heatsinks (often the driver chassis itself) to transport the heat away from the coil. The reason this is so important is that the resistance of copper rises with temperature, often to double, meaning the power input to the driver has halved.

Whilst that could be seen as a useful protection against abuse for the driver, this thermal limiting is different for each driver, and so the expected sound of the speaker changes. In particular, the midrange normally remains in better shape than either the bass or top-end drivers; however, it still goes through a natural compression of the sound in relation to the input signal (just like an audio compressor). Even this is not linear, as the top end of the driver (each driver) becomes more inductive near the crossover point and so is less affected by the resistance change than the lower frequencies. Put this all together, and a well-used monitor speaker becomes increasingly harsh in the upper-mid range as the day progresses. I am indebted to audio legend Ben Duncan for his insights here. A natural protection of the ear also acts at high volumes; the higher frequencies being temporally suppressed to save your hearing from damage. What to do? The obvious answer is to check balances early in the day, or even to just balance the final mix first thing after the equipment has warmed up. Breaks are important for recovery of both your ears and the speakers. Knowing is to plan!

Nearfield monitors started life as a cheap low-quality check to give an idea how a track would sound on a tinny transistor radio. In time people realised that the stereo image was much better in the nearfield, so engineers started improving the performance of these speakers to give a second high-quality check. Whilst the nearfield is refreshingly untainted by the room acoustic, with certain speaker types (horn loaded being the worst), the speaker is simply too close to give a coherent performance, as the sound from the speaker drivers has not had time to coalesce into a stable image. Conventional cabinets with a maximum bass driver size of between 4 and 8 inches are the most likely to give a decent nearfield experience. Note that the sweet spot is compromised to just the engineer, or someone standing close behind.

LOUDSPEAKER CONNECTIONS

The modest subject of power amplifier to loudspeaker connections attracts a lot of discussion, perhaps more than is outwardly appropriate; yet a whole industry hangs off the apparent quality of "connects" and a justifiable price for specialised products. There is a lot going on, yet the very low impedance of the system renders the many problems that happen at higher frequencies redundant. The loudspeaker delivers a complex load; the speakers themselves are large inductors, and we have learnt sizable capacitors and additional inductors sit in the crossover to divide the audio range into the correct frequency ranges to suit the selected drivers. Further, resistors make more changes mainly equalising the volume of the drivers. Complex indeed! The power amplifier output needs to be a very low impedance source (damping factor ≅ 30 so 0.3 Ω for 8 Ω), yet it is normal to add final components to make the loudspeaker appear less inductive as a load. This addition, shown in Figure 5.19, is known as a Zobel network.

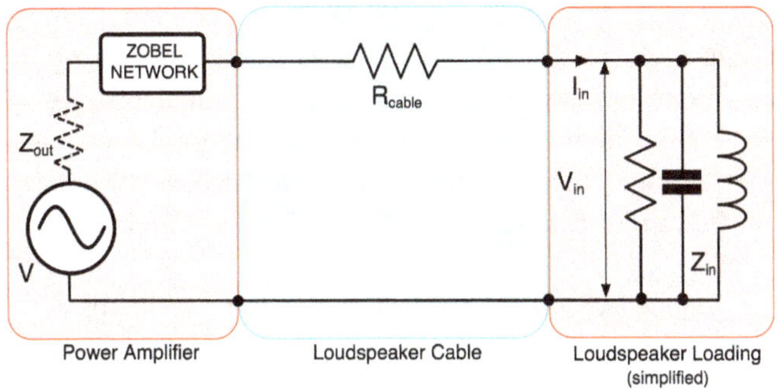

Figure 5.19 Loudspeaker cable effects

With the ends complete, we need to string them together, the loudspeaker cable itself. In the evolution of sound, the work of Zobel allowed better transmission of analogue audio, and Boucherot improved high power electricity distribution using power factor correction. Good high current output power amplifier design draws on both research philosophies. Further, whilst loudspeaker connecting cables have capacitance and induction values in theory, at audio frequencies theory also says we can confidently ignore them and concentrate on resistive degradation alone (that is unless an esoteric manufacturer has purposely built capacitance in… Don't go there!). To overcome resistance, we need high conductivity which can be simply assessed by (1) the purity of the conductor, (2) length of the connection and (3) an adequate cable cross-sectional dimension (i.e. current carrying capacity).

Taking point (1), the purity of the conductor is difficult for the end buyer to assess, and even a slight impurity of 0.05% can hamper conductivity by 15%, so I would always suggest buying a known brand that has facilities to test, and a reputation to uphold. Some enthusiasts advocate using copper mains wire; I note this is allowed a small level of impurity. UK ringmain socket wiring is 2.5 mm^2 (24 amp) and standards allow resistance of 18 mΩ/m for solid core and 16 mΩ/m for multi-strand flex. To take point (2) a reasonable length of this 20 A cable, say 5 m to connect to each speaker, is still under 0.1 Ω. (3) Thicker purpose-built connects will do even better.

Ordinary 2.5 mm^2 home "mains" cable (UK) is manufactured to work conservatively at 2,000 W. If your speakers are 100 W or under, and you are using this (or a similar area of copper strands), you would only be utilising a tiny fraction of the capacity available, yet we must also consider the damping factor of the amplifier. With inferior connects this will be compromised by the cable resistance sitting in series with the power amplifier output impedance. However, moving up to 4 mm cross sectional area (CSA) of "mains" flex brings down the resistance of a 5 m length to 0.05 Ω, and again decent proprietary brands should offer more purity and therefore lower resistance/metre. Tests show solid and flex copper connections are virtually equal, the flex being just slightly better.

At the time of writing there is misdirection to avoid. I've quoted cable size by CSA (cross-sectional area) in mm^2. I could equally have quoted by "gauge" meaning the US Standard of American Wire Gauge (AWG) or cable diameter in millimetres (mm). *That tiny "2" (the "square" of CSA) is important but is often conveniently left out.* At 16 AWG cables are virtually identical (1.3 mm), but as they get thicker, it makes a big difference. To take the examples above, 10 AWG would be the equivalent of 2.5 mm *diameter* cable, but the CSA figure is now 5.3 mm^2. 4 mm *diameter* cable (my go to) has an AWG of 6 and whopping CSA of 13.3 mm^2. Beware of confusion where the CSA figure appears to be the copper diameter.

The best high conductive copper available on the commercial market is "Oxygen-Free Electronic" (OFE) copper, a purity of 99.99%. It has the lowest (common) level of volatile impurities. It's made using an induction melting process with a copper cathode covered in a granulated graphite bath to protect it from the atmosphere, which contains nasties including non-oxidizing elements. "Oxygen-Free" (OF) copper is the next standard down and is 99.95% pure. Electrolytic-touch-pitch copper (ETP) is the copper used in everyday electrical applications; it is 99.9% pure. The question is, does being "oxygen-free" improve conductivity? I can find no evidence to suggest it does. The required copper standards (IACS) are near identical for OF and ETP, and the much more expensive OFE is just 1% better, and that sort of difference is completely overwhelmed by surrounding factors. So, buy your copper cables from a manufacturer based in a country where you can be confident the international standards have been truthfully applied, and from a supplier who has a reputation to uphold. In my view, the product of Copper Clad Aluminium (known as CCA) should be avoided in loudspeaker cables; its conductivity is much lower, and it can corrode over time, creating problems at the terminals.

With the cable quality sorted, just a word to point out that audio levels to the speaker are frequently below a volt in quiet passages of music. Tarnished or dirty connections and connectors at the ends of the cables will have increased resistance and may have joint rectification issues at particularly low volumes. Keep the cable ends clean and shiny! Also braided ends straight into the speaker terminal will settle and need an occasional gentle tighten to maintain a great connection.

ZOBEL NETWORK

A resistor damped inductor and the RC series circuit to ground on the output of a power amplifier are collectively called a "Zobel" network after the 1923 findings of Otto Zobel of Bell Laboratories, who flattened and widened the frequency response of copper phone lines so transmissions could travel further. The circuit is used to isolate the feedback network of the amplifier from the inductive load of the loudspeaker coil, making the speaker look more like a resistive load (Figure 5.20).

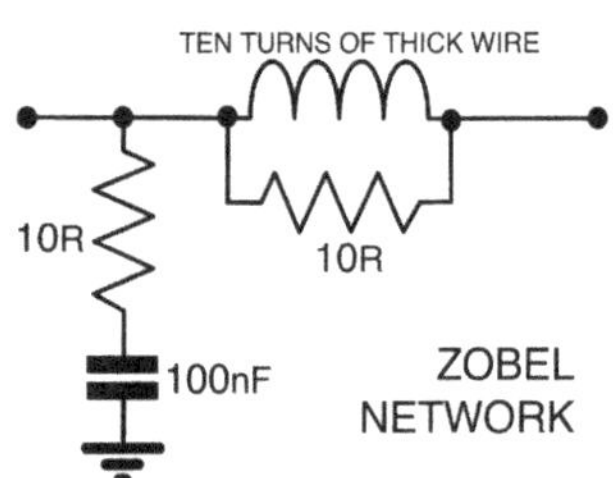

Figure 5.20 The Zobel network

This is necessary to prevent the Power Amplifier from turning into an oscillator at very high frequencies and to ensure it is more stable driving into a wider range of inductive unknowns in esoteric speakers. Certain manufacturers omit them on the claim of purity of sound. That claim may or may not be significant, but in my experience amplifiers without Zobel networks can make some complex speakers sound surprisingly disappointing. In truth, we really don't know why some cables sound better than others in certain situations. It would be a bold claim for a manufacturer to assert their cable would be better in all situations.

WHY POWER AMPLIFIERS ARE TREATED DIFFERENTLY BY AUDIO ENGINEERS

Modern quality audio gear, whether it be analogue or digital, uses comparatively low voltages at low currents. Because of the small power needs, they don't require much thermal control, so the metalwork can be insulated from the casework, and only a 2 pin (live and neutral) mains connection is necessary. That is why a 2 pin IEC is often used in domestic audio equipment, even on low power amplifiers (under 100 W/channel). As the power requirements to feed your monoliths of speakers increase, so does the need to think carefully about your installation.

Class A amplifiers will always be power hungry and hot; so, these and the larger class B or A/B amplifiers require decent ventilation otherwise they will overheat. There are likely to be external heatsinks, which are obviously metal, so expect there to be the need to safety ground these with a third mains pin of an earth. It's important to check and use a decent (10 A or 13 A) three pin mains to IEC cable to connect your beast. A quality powerful amplifier will be pulling both large and variable currents, so it is a bad idea to use a series Power Line Conditioner in the feed to the amp. To be clear, the amplifier designer will ensure the power supply has decent supply interference suppression. It's hard to improve over that, and placing in a "conditioner" may ensure the tail wags the dog, and the required current will not be available to meet the surge needs of the amplifier on louder passages. My view is to plug the power amp directly into a decent wall socket using a quality lead of the right length (not fed through an extension cable).

As a lifetime broadcaster, who lives and breathes "balanced line" working, I am here to tell you that on short cable runs balanced design inputs do not improve sound quality and sometimes are noisier. Balanced line is suitable for long cable runs in hostile circumstances and that is unlikely to be the case in a semi-detached bungalow in Chipping Sodbury. You will learn in the next chapter, balanced and unbalanced use differing interface levels, and so it is unwise to frig a link between them. At a level of several hundred mV input sensitivity and with a typical modern low impedance send, an unbalanced connection is perfect up to around five metres in length. If you have a run nearer to 50 m (some lounge!) and have a welding shop next door, you will love balanced line! Use the shortest link you can. If the amp and preamp sit together, buy a quality short connect, rather than have a long one draped over the power supply extension sockets and the local area network (LAN) cable.

Earlier in this chapter we went through the thorny subject of loudspeaker connects, and I'm not going to repeat myself. Whether bi-wired or not, keep cable runs as short as possible and use decent high current connectors, whether they be banana, spade or terminal. I will remind you that if you are using the bare braid of the cable, tighten by hand as well as you can into a clean clamp (contact cleaner is a good toolkit option) and revisit a week later to retighten the "settled" braid in its terminal.

As to amplifier wattage size, this is a thorny subject. It is bad practice to undersize an amplifier, as an amplifier driven into clipping is nasty to hear, but worse, harmful to your speakers. Simply, a clipped audio signal approximates a square wave, and that is a speaker killer! A match of speaker wattage (at a given impedance) to the speaker would seem safe and optimum. My preference, though, is to oversize the amplifier, so it runs cool and can handle transients without effort. If you are a "foot to the floor" buddy this is not a sensible option, but for music with a wide dynamic range, and/or complex nature, sizing up and using volume economically is the way to go. A sensible precaution, particularly with a larger amp, is to make sure any plugging and switching of your system is done with the volume right down, or better with the amp turned off!

BIBLIOGRAPHY

Brown, C. *Introduction to Hi-Fi* (Editor Hi-Fi Sound Magazine).

Duncan, B. (1983) *Control Room Monitoring Part 2.* Home and Studio Recording-November 83. Music Maker Publications (UK), Future Publishing.

Hall, A. (1991) *High Power Loudspeaker Enclosure Design and Construction.* Germany: Neu-Anspach.

Hellyer, H.W. (1974) *Stereo Sound.* Hemel Hempstead, England: Fountain Press.

Mackenzie, G.W. (1964) *Acoustics.* Bungay, Suffolk: Focal Press Limited, Richard Clay and Company LTD.

Pannett, W. (1962) *High Fidelity Pocket Book,* 1st edition. London and Edinburgh: Morrison and Gibb.

Parkin, P.H., Humphreys, H.R. and Cowell, J.R. (1979) *Acoustics, Noise and Buildings,* 4th edition. London and Boston: Faber and Faber.

Pops, H. (1997) *The Metallurgy of Copper Wire.* https://www.copper.org/publications/newsletters/innovations/1997/12/wiremetallurgy.html.

Watkinson, J. (n.d.) *Time Domain Accuracy in Human Hearing and Loudspeaker Design.* https://www.thebroadcastbridge.com/content/entry/7066/loudspeaker-technology-part-1.

6 Sound Levels and Connections

ABOUT METERING

In the chapter on Loudspeakers, we define speakers used for professional listening as "Monitor Speakers". The idea is that by satisfying several goals, such as extended bass, a flat frequency response, time alignment of the drivers and fast transient response, we gain an accurate reference that is repeatable in other locations with satisfactory acoustics. I guess the outcomes of such a strategy are questionable; sadly, speakers that should sound neutral and revealing often fall short of the claim, as successful monitoring speaker concepts can also initially sound dry and dull compared to many commercial models. Also, the speakers are tested in "free field", and once placed in a room, the characteristics will change from the ideal.

We have done better with metering, with two world standards dominating sound level visual measurement for almost 100 years. Say what you like about the antique British, black-faced PPM meter (and its many European variants), and the American Volume Unit (VU) Meter, the vital point is that wherever these meters are professionally built, fitted and properly aligned, the results they create are completely repeatable; they provide a reference an experienced sound engineer can trust. Others may well look great, they may create their own standards which gain a local foothold, may call themselves "PPM" or "VU" and may, or may not, copy the tenets of the old standards. Frankly it's a bad idea if they do, as both old meters fail to accurately capture audio peaks, show true loudness or even give a clue as to the clipping level. The point is seasoned audio engineers *know* their limitations, and work around them.

False prophets of either metering ilk may be technically fabulous, particularly when they don't rely on a mechanical movement. However, engineers must invest time to know the limitations of each one, in each location, to be able to trust what their eyes tell them. That is why I feature the PPM and VU meter in this chapter, which historically require a reference wattage into a precision loading at each location they are installed (0 dBm and 0 VU). We have had to wait around 100 years for a new global reference to be created, and to find wide acceptance from many professional bodies: the "Loudness Meter", standardised in ITU-R BS 1770. I'll take you through that as well.

MORE ON THE DECIBEL

Decibels are 1/10 of a "bel", the unit of sound intensity. Like using a ruler to measure an ant is not ideal, the *deci*bel better suits the range of measurement

DOI: 10.4324/9781003717065-6

we require in sound. The unit of the decibel (dB) is a similar invention to the octave; whilst frequency is linear (measured as hertz or cycles per second), the ear hears on a logarithmic basis. The octave is a linear way of expressing a logarithmic function. Each octave is simply a doubling (or halving) of frequency. Equally, the ear hears intensity (loudness) on a logarithmic basis. Again, decibels give a linear approach to a logarithmic problem.

If we talk about the weather, we can be subjective or absolute. Hotter or colder by five degrees is a subjective move of temperature, without boundaries. Saying +30 degrees Centigrade, assumes an absolute measure against our freezing point of 0 degrees Centigrade; get the sunglasses! So it is with decibels; we can go louder or softer by say 10 dB, and a sound engineer will know roughly what to expect and achieve, or we can measure the sound level against a reference such as "0 dBm".

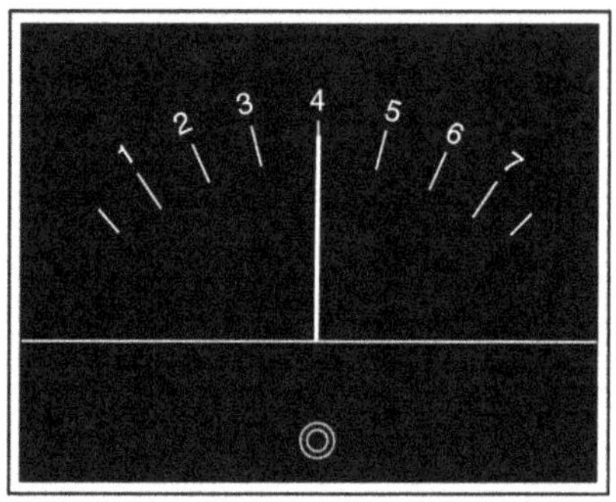

Figure 6.1 The PPM meter invented by the BBC

"0 dBm" is the original old (British) General Post Office term set against a standard value of 0.775 V (1 mW into 600 Ω) and is the "nominal" (think normal) level of line-up tone at the output of the mixing desk and also bang in the middle of the familiar, black-faced BBC designed PPM at level "4" (Figure 6.1). So 0 dBm is *roughly* 1 V of signal. (See other terms and uses at the end of this chapter.) Note it is not the peak signal we can read on this meter; that is 8 dB higher at PPM 6.

Let's try and put some practicality around this. The slightest change a typical ear can hear in volume is around 1 dB, but that's on a pure tone. On music or speech, it can be more like 3 dB. A change of 6 dB is a literal doubling (or halving!) of level in volts, but it is not perceived as that big by the ear. A 20 dB change in level is another useful benchmark. That is a full ten times louder (or ten times quieter). 10 dB is a less useful measure at a messy ratio of $\sqrt{10} = 3.162$ voltage multiplier change; however, it is still a useful approximation.

Here's the bit where some get lost at first. By going between voltage levels and decibels we must treat them differently. We simply ***add*** decibels, where we must ***multiply*** voltage ratios. Note that we only need to get in the right ballpark; *precision is not necessary* for the studio operator as the desk has a wide tolerance of voltage. If we add amplification to make a sound louder it is called "*gain*"; if we place in an impedance to lower a sound signal it is called "*loss*". Table 6.1 contains a simple dB ratio table you can use day to day; however, I don't think it's needed in the adjacent three "count on the fingers" examples.

Table 6.1 Examples of dB/voltage conversions

Loss/Gain in dB	*Multiplier (within 1%)*
–60	0.001
–40	0.01
–20	0.1
–14	0.2
–10	0.316
–8	0.4
–6	0.5
–4	0.63
–3	0.708
–2	0.79
0	1
+2	1.26
+3	1.41
+4	1.59
+6	2.0
+8	2.5
+10	3.16
+14	5
+20	10
+40	100
+60	1000

Example 1. A microphone nominally produces 1 mV (1/1000 of a volt). What amplifier gain do we require to match with a desk output at around 1 V (we must amplify it by 1000×).

We know +20 dB is an increase in level of 10 times.

10 × 10 × 10 = 1000 voltage change.

20 + 20 + 20 = 60 dB change.

Answer: we need 60 dB of gain in our amplifier.

Example 2. A guitar nominally produces 50 mV. What amplifier is required to match it to the desk?

We need approximately 1 V (1000 mV) output, so we need 1000/50 = 20 times increase.

2 × 10 = 20 times voltage change.

6 + 20 = 26 dBm (6 dB is ×2 and 20 dB is ×10).

Answer: we need 26 dB of gain in our amplifier.

Example 3. A keyboard produces 230 mV. What pad (attenuator) is required so we can plug it into a microphone input on the desk? (Microphone input nominally 1 mV level.)

1/230 = a change (reduction) of *around* 200×.

10 × 10 × 2 = 200 (roughly near enough in volts!!!).

20 + 20 + 6 = 46 dB.

Answer: we need a pad to produce a loss of *around* 50 dB.

In the last example we were content to find the nearest simple multiple that puts us in the right ballpark as *you will struggle to hear the difference*, and it's well within the tolerances of the desk. Also, whilst a 50 dB pad is an absolute value, it can be placed in the circuit at a subjective level. Whatever the level at the input, it will be 50 dB lower at the output of the pad, i.e. input –10 dB, output of 50 dB pad is –60 dB (–10 – 50 = –60). It's the same for an amplifier, but obviously we are *adding* decibels of gain.

Good sound engineers think about decibels every day to place sounds at the best level to pass through each stage of the mixing desk, including passing the signal to "outboard" external boxes of tricks (see Chapter 12, "Outboard Devices"). Too little gain at the wrong point and we risk dragging up the background noise of the mixer. Too much gain and we risk driving the desk outside of its ability to faithfully produce the sound, clipping off the extremes of the signal. This sounds awful and is called harmonic distortion. The margin for safe working above our nominal output of a desk is called "headroom". (Think of a doorway in a house, a shorter person will have lots of "headroom" to avoid hitting their head on the jamb (top of the frame), a giant will have little, perhaps none, and **ouch!**) We need a nominal person...

Note "equalisation", (if you like, the "tone controls" of a mixer) introduce gain or loss at specific frequencies; these count in the thinking and must be at the back of the mind to avoid going outside of the headroom at *any* frequency. The term **dBm** specifies a known system *source* impedance of 600 Ω loaded by an identical 600 Ω *load* impedance, where 1 mW will give a voltage of 0.775 V. Practically, that's not as relevant these days as we tend to place a load on an output that is a much higher impedance, and if the load is 10× or more of the source impedance, it will not affect the output level. We have new terms to better explain what is going on.

dBu – this is very similar to dBm; however, "u" is for unspecified loading impedance. We are only interested in measuring the voltage. Note that 0.775 V remains as the reference and is known as "0 dBu". Some have used a little v (dBv) for the same measure, not to be confused with dBV below. Outside of a workshop line-up it doesn't matter, and dBu is now normally used in place of dBm on common chains of equipment.

dBV – the domestic Hi-Fi industry didn't like the fact that 0.775 V is such an odd figure. They rounded it up to a 1.0 V reference (0 dBV) and made the standard for domestic Hi-Fi 10 dB lower at –10 dBV or 316 mV which is also an odd figure, and so quite confusing. Luckily, we have a lot of leeway in our levels.

dB (power) – a power amplifier for a loudspeaker is measured in decibels using *power* in *watts*, not volts. A halving/doubling is *3 dB* and a 10× change is just 10 dB. A 200 W power amplifier is ten times greater than a 20 W amplifier, yet for all the money, to the ear it sounds moderately louder at 10 dB more. It's worth being aware of, but don't let it confuse the needs and requirements of the "everyday" operational engineer.

To recap, unless you are lining up a single source sound system in a laboratory, decibel levels are about being in the same ballpark, and understanding what sort of loss or gain will match one item of equipment to another. That said, *where there are two channels or more, it's important to match the line-up levels (1 kHz tone) to within a single dB to ensure the soundstage will be accurate*!

DYNAMIC RANGE

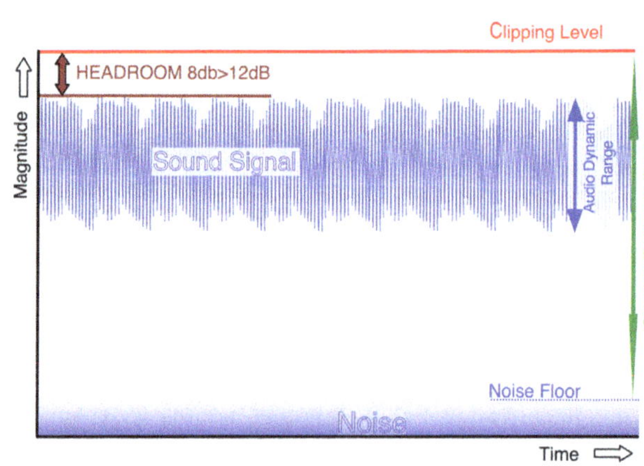

Figure 6.2 Audio system dynamic range

Several times I have referred to the need to get the levels right, and I guess that is predominantly the skill of the sound engineer, whether it is across one mono level or 64 channels across a major sound project. Input the sound signal too low, and the noise of the system starts to invade the sound we want; we call the residual level of noise in a sound system the "noise floor" and it is measured in decibels relative to the highest undistorted signal. The difference between the two extremes is called the "Dynamic Range" of the system, and the larger the number of dB the better the system (Figure 6.2 and Table 6.2). Note we measure dynamic range in analogue and digital systems in a similar manner.

Table 6.2 Common terms applied to an audio signal and its environment

Audio Signal	Required AC voltage that we want to recover as sound at the end of the process
Noise Floor	Unwanted AC voltage created through imperfections in the audio circuits such as shot and thermal noise in semiconductors and the induction of local wire-borne and airborne interference.
Clipping	The audio signal swings between the DC voltage rails in the amplifier, getting larger as the gain increases. There comes a point where a loud signal meets the power rails, and the audio peak is sliced cleanly off creating distortion of the waveform that cannot be undone.
Headroom	To avoid clipping, the maximum permitted level (or "nominal level") must be pegged at a lower point typically indicated as the peak on the metering. The difference between this chosen peak level and the clipping point is known as the system headroom.
Dynamic Range (System)	The amount, measured in dBs, between the system noise floor and the clipping level. The bigger, the better the system.
Dynamic Range (Audio)	The difference between the loudest and the quietest levels, measured in dBs of a *given piece of audio.* Limiting the dynamic range to a smaller amount (amplifying the quieter sounds to be nearer the louder sounds) creates subjectively more loudness.

If a sound signal is too high, we come dangerously close to the DC power rails and the extremes of the signal get chopped off, squaring off the tops of the peaks. The result of this "peak distortion" is that the level of bad harmonics goes through the roof, and the sound is strained and nasty. We add this to other imperfections of a system, to create a measure of the "THD" and it's measured in percentage terms. The mechanical "transducers" at either end (the input of a microphone and output through a loudspeaker) are still likely to be the highest distorting components in a modern well-engineered system where the electronic circuit is capable of 0.01% distortion or better, whilst it will be a very good loudspeaker to get better than 1.0%. Where there are other measures of sound distortion, they are in general used in the designing of equipment which is beyond the scope of this book.

We have gone to some depth on the dangers of taking signals too high, yet there is danger in being too careful. The noise floor of each analogue circuit we pass through adds to the last, building on the noise level as it goes. Where a quiet source needs a lot of amplification, say 60 dB (×1000) for a microphone, the *noise is also amplified by 60 dB*. The noise floor tracks the sound levels through the desk, and so it pays to keep the wanted signal as far as possible from the noise by keeping the levels as high as we comfortably can (often called the nominal level) at all points. It's worth saying that outside recordings will have additional unwanted environmental noises such as wind, passing trucks, playground chatter and so on, and we will talk more about that later. Often system noise becomes irrelevant against these odds, and so the sound engineer should choose to have a larger headroom to avoid clipping by unexpected sound peaks, over the worries about additional noise from the equipment, which will be masked by the surrounding environment.

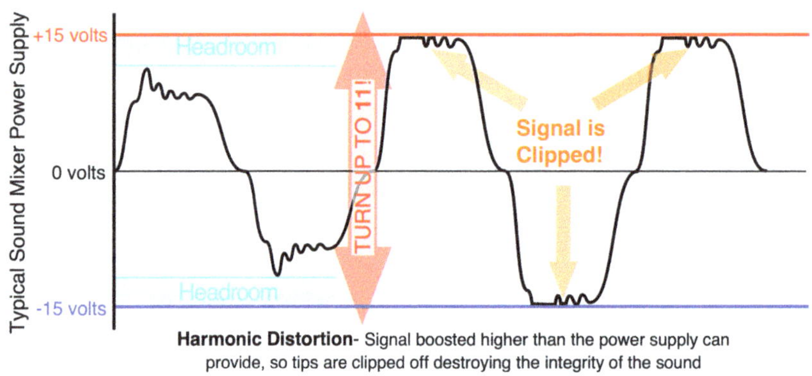

Figure 6.3 Audio signal driven to clipping

Figure 6.3 shows an audio signal, swinging between the positive and negative of a typical sound mixer DC power supply. This AC audio signal is "peak to peak" and is clipped off as it approaches the DC power supply lines. The actual point at which the signal is distorted will depend on several factors, the main one being what sort of semiconductors are used and the configuration they are laid out in the circuit. Whilst some modern "IC" operational amplifiers can swing all the way to the DC power lines, it is normal in both discrete and integrated designs that clipping starts earlier. Factors other than circuit design play a part; for instance the amount of gain in the circuit and the load impedance of the following equipment can alter the parameters.

As an example, in Figure 6.4 I have shown one design where a single silicon junction in each leg drops the voltage by 0.6 volts. However, the ultimate headroom of the system is not this 28.8 volts peak to peak. We measure audio levels on meters showing *RMS* volts (V_{RMS}). So, the actual headroom is 10.2 V or around +12 dB above the BBC peak PPM level (PPM 6).

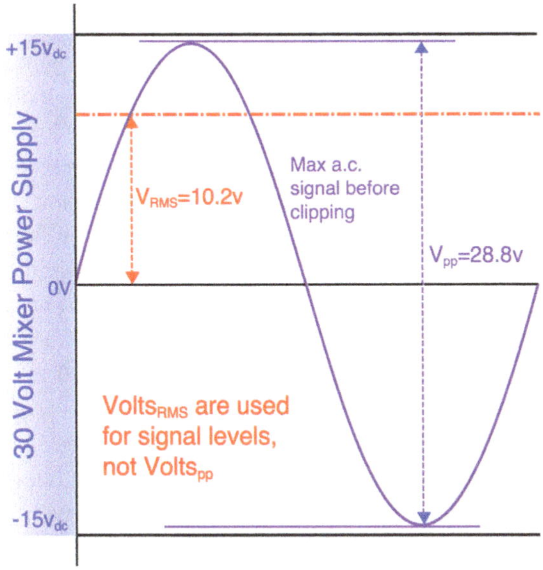

Figure 6.4 We measure audio level voltages in volts RMS

Using equipment correctly is central to this book; let's take some time out to understand what is happening here. We will investigate how equipment is joined together, both in the home and the studio. *There are two major levels of sound we can plug into a professional sound desk, and so we fudge anything else to fit this model. The two are "Microphone" level and "Line" level.*

MICROPHONE DESK INPUT

The sound output from a microphone will depend on its construction, and so the sound desk will have to cope with a variance here. We traditionally consider a typical microphone to have an output of 1 millivolt which would need to be amplified by 1,000 times (or 60 dB) to be at the level of one to two volts we need at the desk output. The second variable is how close the microphone is to the source. The same microphone may be an inch from an enthusiastic rock singer's tongue or strapped a metre above a serene solo violin in an orchestra. Remember, sound dies away to half the level (6 dB) every doubling of distance (inverse square law), so the output of the microphone will vary tremendously. Professional microphones often have a switch to take down the sensitivity called a "pad" which can help. These are typically 20 dB, which takes the sensitivity to one tenth of normal and can protect the microphone amplifier (where fitted) and the input of the sound desk from being overloaded. But whether the microphone has a switched pad or not, the sound engineer must decide and adjust the microphone channel gain to ensure it is not too low or high in the desk. A microphone input, in professional applications, is always a "balanced" input. Please read this up in Chapter 4, "Microphones".

LINE DESK INPUT

"Line" level sources are plug-in sources that already have amplifiers in them to bring them up to a professional sound level of *around* 2 volts (currently +4 dBu = 1.228 V is favourite). Line inputs might be one or two channel inputs (mono or stereo): a keyboard, a professional CD player or playback from some sort of sound recorder. Also, we have effects units which also plug in to a mixer at line level like limiters or echo; however, these are not plugged directly at the input; they are "outboard" devices which we talk about in Chapter 12, "Outboard Devices". Unfortunately, domestic equipment (like a home CD player) has a different (lower) standard of levels to professional sound equipment ("Pro" CD Player). You might think that an industry that has been running for 100 years would have "absolute" standards of what levels should be; unfortunately, we don't. There is also sadly confusion between "Line Up" levels, "nominal" levels (maximum normal working levels) and "peak" levels, and to what peak level is: is it the "maximum" metered levels, or the point where the audio will clip and distort? A Line Input is often "unbalanced" on prosumer or domestic equipment and normally "balanced" on professional equipment.

As touched on earlier, we call the difference between "peak" system level (as shown on the meter) and the point where actual signal clipping occurs, the "headroom" of the system (measured in dB). The headroom is the protection zone between the engineer missing the peaks (or poor metering failing to register them) and calamity. Whilst 8 dB headroom is enough for most audio situations, I would always allow (unless well-rehearsed) for 12 dB and maybe more on location using a digital recorder. Yes, the digital noise floor will be slightly more apparent, but at least you will have something you can use! Read more about digital systems and recording in Chapter 9, "Digital Audio".

Let's start with an analogue professional system, and I choose the BBC standard, as it's international (but not universal). *Just to note the PPM is still in use though most of Europe but scales (but not dB levels) vary. In several countries, the centre number "4" is replaced by "Test" and "8" is replaced by a variety of terms, 8 dB to 9 dB above the centre reference level of the meter.* In Chapter 3, "Introduction to Electronics", we learnt that the BBC line-up level is 0.775 V. which is "4" (PPM 4) on the PPM. However, "Peak" Programme is "6", and given that there is 4 dB between "4" and "5", and another 4 dB between "5" and "6", BBC peak programme is 8 dB higher than line up, which is close to 2 V (1.946 volts actually). Also known as "+8 dBu" or "0 dBr". Then we have the system headroom on top of that, which I've found to be between 6 dB and 15 dB higher before the clipping point; that's a healthy 4 V_{RMS} and 10 V_{RMS} respectively.

Figure 6.5 The PPM compared to the VU meter

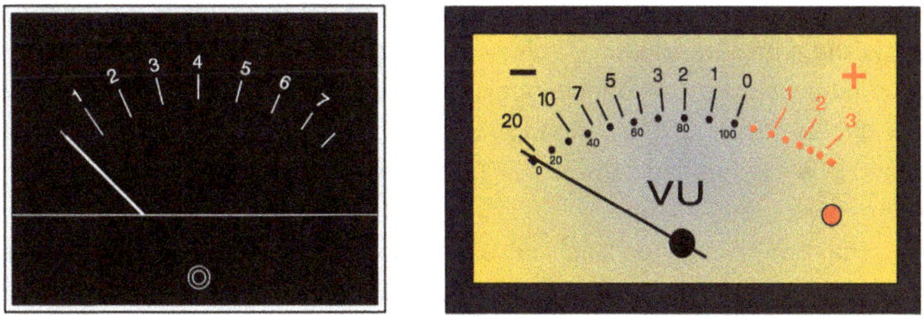

The BBC developed PPM and the American VU meter. Line up level on the PPM is 0 dBu = "4" and peak programme is +8 dBu = "6". Line up on the VU meter is +4 dBu = "0" (used to be +8 dBu in early broadcasting) and peak programme on the VU is *anyone's guess, but don't let the little red light flash a lot!*

You should know that both the PPM and the official VU meter (many VUs are cheap substitutes for largely cosmetic effect) are highly specified expensive equipment, so results are repeatable wherever they are installed and professionally aligned (Figure 6.5). The VU meter is *not designed to show peaks*; the attack time favours "loudness" over true peaks, and the specification *requires* a further 10 dB of volume margin *above the scale the meter shows* and on top of that a further 6 dB as a headroom against clipping of the signal!

Audio equipment manufacturers like a simple reference to a nice rounded 1 V, (around +2.22 dBu); this is known as a reference "0 dBV" and is generally ignored. However, they reference domestic equipment at *a nominal working level of –10 dBV (316 mV)*, which *is* an internationally recognised standard. So, commercial practice is to equate –10 dBV as a domestic standard against a professional level obtained during the sine wave line up of a VU – the +4 dBu standard. To move between domestic and professional systems we require a box with gain or loss in it (depending on which way we are going), and these have been available to buy in the past. These days people often wing it, relying on enough adjustment in the surrounding gear to get away with it, but sometimes they fall flat on their faces. We will come back to specific "digital" system metering later in this chapter (see "The Digital Loudness Meter (LUFS – Loudness Unit, Referenced to Full Scale)" below), and please see Chapter 9, "Digital Audio".

Comparing Sound Level References

In recent months, I've seen a lot of false information regarding the movement between professional and domestic analogue sound levels, some going in truly bizarre directions to explain a relatively easy situation. Whilst it is true, we are stuck between two different reference systems, dBV and dBu, they are both RMS voltage levels expressed as dB. It does not help that semi-professional equipment often gives level switching marked "–10" and "+4" without noting they are on different reference scales. In Figure 6.6 I've created a diagram to show the complete situation; I chose the linear spaced PPM meter face as a base, showing the sound logarithmic progression as it is much easier to follow through. The European Broadcasting Union (EBU) PPM shows specific dB sound levels (reference 0 dBu = TEST), and this is then compared to American sourced VU meter levels.

The slim "12 dB" red label above the black EBU PPM shows the true level difference (11.784 dB to be more precise) between –10 dBV and +4 dBu, and so the *official* amplifier/pad value of 12 dB to go between domestic and professional equipment. Please note that in general, European equipment has peaks of at least an additional 4 dB (nominal peaks of +8 dBu), and I will remind you that the VU meter standard expects peaks of level well beyond the meter scale, *so expect with vintage professional equipment such as tape machines to need to have at least 16 dB gain in hand, if not 20 dB to match to domestic standards.* The difference between +8 dBu and +6 dBV is under 1/4 dB, just a needle's width! If only dBV had referenced to two volts rather than one volt everything else would have fallen into place. Ho hum!

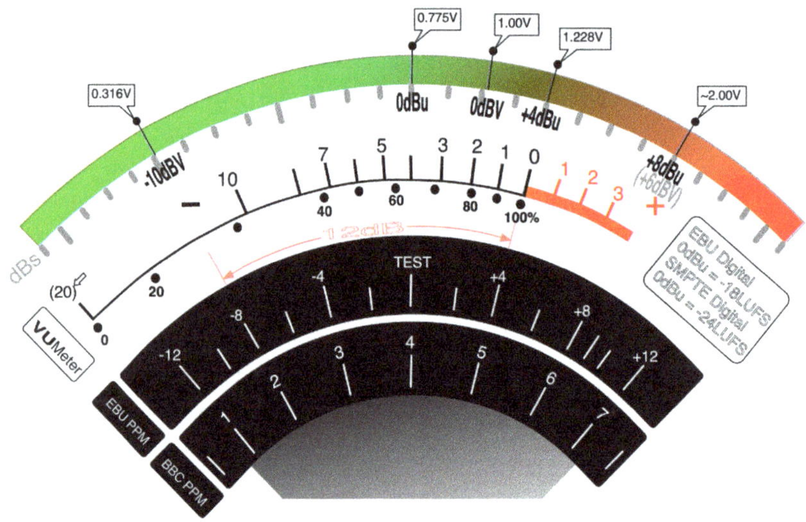

Figure 6.6 Common metering scales compared

When I started in audio "DIN" plugs and sockets were still around in domestic settings, and these tended to be lower level than the other (enduring) system, the "RCA" or "phono" plug. Levels of 30 mV to 120 mV (0.1 V approx.) were common for DIN, and 250 mV for phono plug systems. No one mentioned whether this was a nominal or peak level, and frankly I don't think people, in my experience, gave it any thought! DIN audio has largely disappeared, so "old fashioned" domestic levels are typically eight times lower than the professional levels, which equates approximately to the 16 dB difference mentioned above. 12 dB stubbornly remains a modern reference.

Fortunately, we have a clue as to whether something is going to have "domestic" or "professional" levels, from the leads we use to join equipment together. Where phono plugs are the normal domestic connector, the professionals use "XLR" connectors. Developed in the 1950s they are robust (you can jump on them), electrically and mechanically secure, and if a decent lead is bought, able to manage delicate microphone levels as well as robust line levels. In the pictures of popular connectors (below) it can be clearly seen the phono plug has two contacts where the "XLR" has three. You can find out more about this often-misunderstood topic, "Unbalanced versus Balanced Line Working" in Chapter 4, "Microphones".

ANALOGUE EQUIPMENT CONNECTIONS

The XLR plug is the only connector that tells you how it is wired! Each 3 pin plug or socket has the terminals inscribed with 1, 2 and 3. The earth (X = shield wire) is connected to 1, the Live wire is connected to 2 and the Return is connected to 3. Three other facts about the XLR: (1) unlike the phono or jack, the XLR lead has a socket on one end and a plug on the other. An output is always on "pins" (male) and an input is always on sockets (female). (2) The "X" earth connection is engineered slightly longer than L or R so the earth connects first to ensure there is no chance of buzzy-hum problems when plugging or unplugging (unlike the jack or phono!). (3) Both line and panel versions feature a locking tab on the male which is released by a catch on the female.

We have not yet touched on Jack plugs as they are, well... complicated! The first thing to note is that there are two popular sizes, normal quarter inch (6.35 mm) and miniature (3.5 mm). (*a 2.5 mm jack standard exists at the time of writing but is disappearing. This book will not cover P.O.316 style jacks and jack fields in detail; they are used in plugging bays and patch panels and are in decline. However, note they have the same crude dimensions so fit each other's sockets, unreliably!*)

There are also two wire and three wire versions making a standard for the 1/4inch "mono" (2 wire) "TS" jack (tip, sleeve connections) to be used in band PA as unbalanced (microphone or guitar to amp) connections or in "stereo" (three wire) "TRS" (tip, ring, sleeve) as an alternative to the XLR in balanced connections in professional or semi-professional equipment. In this case the tip is "live", the ring is "return" and the sleeve is the "earth". These are mostly used at line level, and seldom at microphone level, but I have seen it. Prosumer kit goes further to allow a TRRS 3.5 jack, (tip, ring, ring, sleeve) which allows a microphone feed as well as a stereo headset.

To add to the confusion many Americans refer to "jacks" as "phone plugs", which sounds really close to "phono" connectors, and coincidentally, jacks of both sizes are used as stereo headphone connectors in both domestic

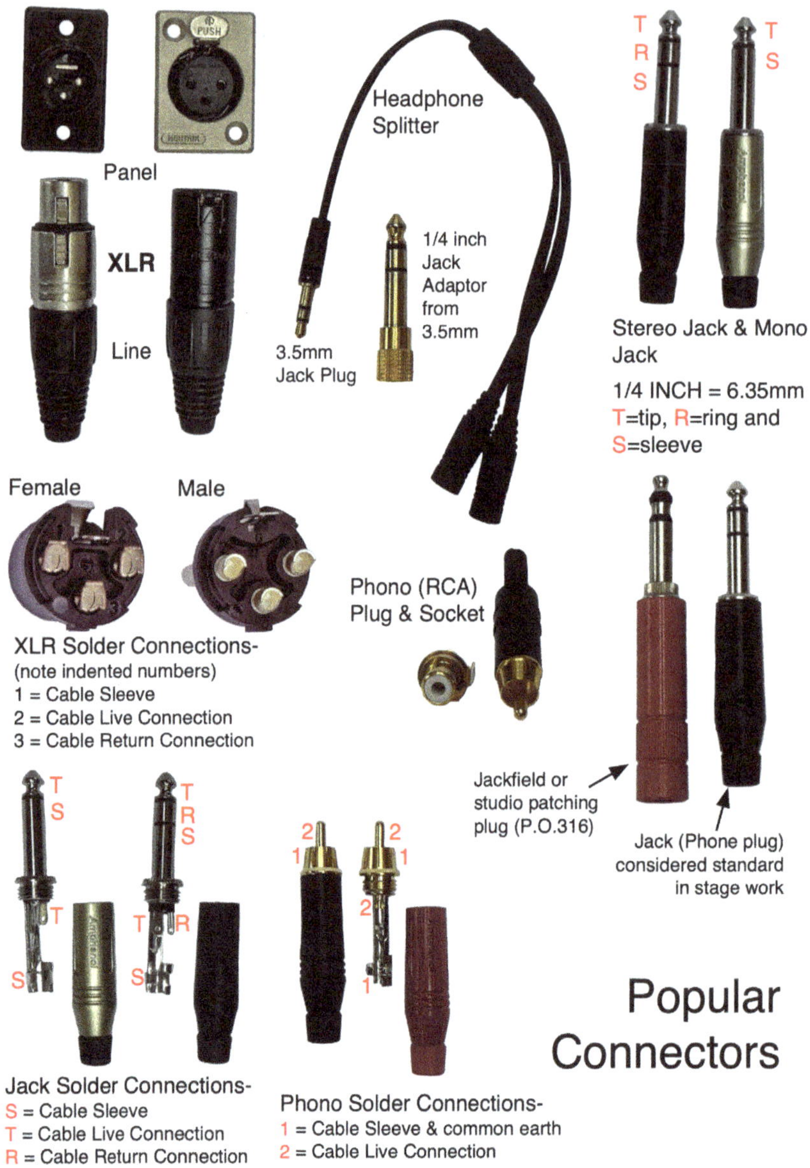

Figure 6.7 Audio connectors in common use
Photos by the author.

and professional equipment. This time the tip is left channel, the ring is right channel and the sleeve is a common return. Phono or "RCA" plugs do make an appearance on semi-pro mixing desks (sometimes called "Prosumer") as they take up less space than either XLRs or jacks, but it's anyone's guess what audio levels they will conform to! I've brought together some popular connectors in Figure 6.7.

Back to the humble coil of the microphone. Years ago, manufacturers were poor at agreeing what side of the coil was live and which was live return. If I take you back to the description of the blow and suck of a "figure of eight" microphone, if the microphone coil was not wired to match the other microphones in the box, it would suck when the others blew. It is "out-of-phase". Although a wrongly wired microphone is rare these days, it can cause havoc! The check is to place the suspected rogue alongside a "good" microphone with both made "live" and panned centrally; if you have a rogue the bass frequencies will cancel out when you bring them together, making the sound thin and nasty.

Many sound desks have a "phase reverse" button to solve the ancient problem of rogue microphones, but I think *the bigger problem is people not understanding why the button is there and using it inappropriately*, causing chaos in the session! Knowing it can happen will make you cautious, but you will only ever find a rogue microphone by the test I have advised above. It really is straightforward simple physics! When attending a session arrive early, work methodically and choose your microphones well; don't be over-awed by the clients, stay calm and all will be amazing!

Over the years manufacturers have made musical instruments and equipment which, by simple physics, have an output somewhere between microphone and line levels. To cope with these odd levels, we need a box (an interface) to change the signal to either one or the other of the expected levels. A guitar pick-up is a prime example of something that does not suit a sound desk and needs to go through an interface called a "Direct Inject" or DI box, which *then* connects to the sound desk. Cheap DI boxes are simple attenuators (networks of resistors and perhaps transformers) which reduce the level to microphone level, and others are amplifiers to add gain (so they must have batteries or power supplies), which lifts the signal high enough to be plugged in as a "line level" input. Obviously misunderstanding where to correctly plug the output of the DI box will result in tears.

A traditional and formally essential role of the DI box was to "uncouple" the guitar from the sound equipment to ensure there is no possibility of a mains electric shock passing back through to the instrument. Tragically, several prominent guitarists lost their lives on wet outdoor stages when valve equipment was king. That risk with modern electronics is vanishingly small. We leave it to the music critics now.

MORE ABOUT NOISE

You will not miss out hugely if you jump this section; it is for the curious who like to experiment with their equipment. However, I would state that much of this applies to digital too, and that this largely unavoidable noise is the major limitation stopping all types of audio circuits improving past the current generation. I have stressed the need to keep the noise floor as controlled as possible, but what is it? Noise is an undesired audio signal that conflicts with the audio we do want. Each time we copy an analogue signal, noise (and distortion) from the source is added to noise (and distortion) in the destination electronics, so the noise floor grows increasingly intrusive. We measure noise degradation with the Noise Rating (NR), simply how much more noise is present at the output compared to the input, after taking the gain (or loss) into account. A decent result is <3 dB. Copy on copy creates further degradation until the audio file is no longer usable. There are several differing sources that combine (conspire?) to produce an unwanted signal in every audio application. Some sources are partly avoidable by good design, and others are just part of the cosmos we live in.

Johnson noise (aka thermal noise) – if I held up any resistor to you in mid-air, made of any construction, it would still produce measurable noise across its terminals! Listen and you would be literally hearing the universe, the thermal agitation of atoms. The level of this "open circuit" noise voltage is dependent on (1) the value of the resistor, (2) on the ambient temperature [cooler is lower noise] and (3) on the bandwidth of the measured frequency range (10 kHz is appropriate for most audio tests). Nyquist (more from him in Digital Audio) created a formula to predict the amount of noise in any situation. Johnson noise has a flat frequency response when measured against a *linear* frequency response (remember – in audio engineering we measure frequency on a *logarithmic* scale, as it then sounds "linear" to our ears, as octaves do). When listening the sound gets more obtrusive as the frequency rises, becoming 3 dB/octave louder. How does it sound? Like an FM radio tuned off station, or a nearby minor waterfall. This sound is called "white noise". It is literally all the frequencies sounding at once! Whilst it is unavoidable, using a lower range of resistor values in our circuit design (where possible) yields lower Johnson noise.

Shot noise – our resistor is now placed in a circuit and has current running through it. There are small but relevant unwanted currents produced in the flow, eddies if you like. Wiring and pure resistance produce little noise, but semiconductor junctions in devices like diodes and transistors create a lot! An analogy is a hosepipe. A garden hose carrying water is silent, but turbulence in a spray head or a partially open tap can easily be heard. Shot noise grows more intrusive as lower current levels are used. The sound, like Johnson noise, is white noise. The shot noise level can be reduced in electronic circuits by use of "feedback" (which reduces the gain but usefully lowers the distortion of a circuit). Sadly, shot noise can be easily heard in an audio mixer fault where the input transistor junctions are damaged beyond repair (commonly by voltage spikes from plugging microphones into desks with the phantom power switched on). It then is a spluttering, intermittent sound, like a distant neighbour jet washing their car.

Flicker noise (aka 1/f noise) – in both Johnson and Shot noise, the level is set by the resistor *value*; the noise is the same however the resistor is made. Real resistors, and by implication transistor junctions, build noise because of impurities in their construction, and these are proportional to the current flowing through them. Unlike Johnson or Shot noise, Flicker noise has a flat frequency response across a *logarithmic* scale; it sounds flat to the ear and is linearly the same level across the octaves. This is called "pink noise" and sounds like persistent rain drumming on a tin roof, or a large busy motorway in the distance. The major source of flicker noise used to be cheap carbon composition resistors, but this has been reduced by a factor of ten in modern circuits with more use of wire wound (no flicker noise but often impractical in compact audio circuits), metal film (best composition for noise), metal oxide or carbon film devices. It is still a concern in the junctions of both semiconductor and valve components where it is the largest contributor to noise in bipolar transistor circuits at low to medium frequencies.

Distortion – crossover, harmonic and transient intermodulation distortions of the sound signal are not technically noise; however, as a non-linear product of the sound system, they sit as an unwelcome addition to the unwanted total NR of the system. I'd like to think that on modern, well-designed systems, the audio distortion is low enough to be masked by the traditional noise sources, lost beneath a blanket of white noise if you like; however, it does grow as the signal passes through each stage of the audio chain, and again each time there is another generation of analogue copying, so it cannot be discounted.

Interference – there is an important final ingredient in our soup of noise. Unlike "true" noise sources this irritant is not predictable by sound spectrum or amplitude. Try as we might, with metal shielding, balanced lines and careful circuit layout, some extraneous interference will get in and be amplified along with the signal we want. Sources may be connected to our equipment, such as the power supply or control or data leads, or it can be airborne RFI such as data cables running close to our sound cables, radio transmitters, motors, lighting (especially dimmers) and sometimes the weird phenomenon of cables themselves becoming "microphonic" and picking up physical vibrations.

The cures are many (several detailed in Chapter 16, "Real World Problems with Audio Systems") and the stuff of audio legends; however, you can't go far wrong if you buy quality balanced audio cables and keep them well away from data cables. Where they must cross do it at 90 degrees to each other. Further don't fall into the myth of extending your audio usage *well* above 20 kHz to get a better sound from including the ultrasonic frequencies in your chosen equipment chain. There is a lot of garbage lurking up there! To sum up, after 50 years of design improvements, interference, distortion, 1/f noise and shot noise have been largely tamed in modern quality designs and components. That leaves the background thermally induced noise, Johnson, the presence of which invades all, setting the cosmic baseline. Lowering levels by low impedance designs can only go so far. A reminder, Johnson noise is "white noise" and is often filtered by 3 dB/octave to create the useful flat frequency response "pink noise" used in sound equipment analysis and room acoustic testing.

LINE UP AND LEVELS

In general, to line up equipment levels we use a single mid-band tone oscillator producing a sine wave (normally in the vicinity of 1 kHz) plugged into a mixer, either the line level or attenuated to be plugged into microphone level to line up each channel to create the correct expected level at the output, as shown on the meters.

> Whilst 0 dB**m** is traditional (1 mW into 600 Ω = 0.775 V_{RMS}), the use of 0.775 V 0 dB**u** (Hi Z so not considering load impedance) is now normal. A reminder, the BBC works at 0 dBu which is *PPM 4*, and 8 dB below the nominal or peak level of +8 dBu or "PPM 6". The venerable US-inspired VU meter is set up to VU "*0 dB*", which is +4 dBu (on the testbench 2.5 mW into 600 Ω ≈ 1.228 V_{RMS}).

BBC studios and outside broadcasts traditionally have sent line-up tone to the network continuity prior to going on air. The tone (1) shows the source is there, (2) allows for the "outside source" channel "pfl" (pre-fader listen) to be lined up to the correct level [and balance if stereo], (3) checks stereo by testing which is A leg (left channel) and B leg (right channel), (4) allows a phase check between the stereo pairs and (5) iif the tone is heard to be distorted or crackly this will lead to further checks to ensure there are no faults (particularly common are dirty connections!). Outboard devices used to have line-up sources, tape players a 1 kHz line-up tape and a 10 kHz azimuth band and vintage grams a series of tone tracks on a vinyl disc. Every tape was lined up each time it was transmitted with level and balance (between stereo channels) on a 30 second burst of 1 kHz tone at the front of the tape reel prior to broadcast. Life is a lot simpler in the digital world.

So, talking of the "Digital World" when recordings need to be made, how is the alignment made? You will recall that whilst PPM 6 is the traditional maximum level recorded, the actual point of clipping of the audio is close to the DC supply rails. A typical middle-of-the-road analogue mixer has +15/0/–15 DC voltage rails, a swing of 30 volts peak to peak, which is approximately 10 V_{RMS}. This is a decent headroom above the maximum of PPM 6 (approx. 2 V_{RMS}) and provides a huge margin for operator error. Equally, on a professional VU meter, the standard *demands* there is at least 16 dB above 0 dB, and there is likely to be more! It's all a bit vague really. As will be explained in Chapter 9, "Digital Audio", the point of *digital* "clipping" is when all the available bits are at "1" which has no correlation to the voltages and impedances of our analogue world. It is the "True Peak"; there can be nothing above this maximum value. You have hit the ceiling, there is nowhere to go without causing damage. While using "True Peak" gives no headroom, no margin for error, it is an absolute value and so a valid reference point.

Further, it works across all digital platforms, regardless of other parameters such as bit rate or sampling levels. Since you can't go up in level, every reference to "True Peak" must be a negative one, and the negative numbers must grow as the audio gets quieter. On our meter "True Peak" is naturally the top value; the meter is at "Full Scale". The units of the meter can still be in decibels, just like our analogue meters. Stepping back from the edge of this digital abyss, I might choose to have a safety headroom of 16 dB, therefore I would peak my audio signal to –16 dBFS (decibels Full Scale). Note, a reminder that a peak signal value is not in itself a value of sound "loudness"; that is more to do with the audio dynamic range of the source.

These digital peak values are way above the maximum analogue mod of PPM 6 and further traditionally, line up at has been 8 dB below PPM 6 at PPM 4 (0 dBu). So, if we were to equate the two, our traditional tone (0.775 V) would sit below dBFS at between –16 dBFS and –20 dBFS on the digital desk. *It does vary between users* but to take two, the European standard for 0 dBm alignment is –18 dBFS, and the American standard is –20 dBFS. As 24-bit recording becomes widespread the reference level may well fall further to take account of the increased dynamic range, allowing the opportunity of sharing lower noise levels with *the peaks of increased dynamic range.*

In conclusion, a professional asks the customer what is required before the recording and notes reference levels clearly in the documentation so there is no confusion up the way. I note that on many digital desks the bargraph meter turns from green to yellow at these moderate dBFS levels, leading to the digital desk operator maxim "yellow is the new red". Follow this, and you may be wrong, but not so far wrong, and you know you will have a usable recording at the end of the day.

LOUDNESS AND THE ANALOGUE METER

So far, we have talked about meters in a steady state, PPM 4 for line-up and peaks to PPM 6. Let's join this up to the Dynamic Range Diagram at the start of the chapter. We look at the "System Dynamic Range" as a measure (in dB) of the quality of a system. The greater the number, the better. The range is quoted from the noise floor to the clipping level and is a manufacturer's specification. Arguably a more practical measure for the sound tech would be the dynamic range open to us, which is from the noise floor to the peak meter reading; the so-called "nominal" level, or the start of the system headroom.

However, back to the script, if a sound is peaking on the meters at the top (PPM 6, or a VU meter in the red), clearly it shows it is as "loud" as the system will allow; however, there is another measure of the loudness. The "Audio Dynamic Range" gives us the difference in dB between the loudest and the quietest sound we are currently playing. The *less* this range, the louder the sound. If we looked on the PPM (to take one meter type), and the audio was very occasionally hitting 6, and often at PPM 2, we could surmise a wide audio dynamic range. If the meter hardly went below PPM 5, hitting PPM 6 all the time, that would point to a narrow dynamic range, and so a subjectively much louder sound (for the same audio peaks). Traditionally the BBC has used the PPM to show loudness both in peaks and dynamic range (Figure 6.8). The active area of the meter spans 20 dB; from PPM 1 to PPM 6 is 10× the level. Let me give a few practical historic examples.

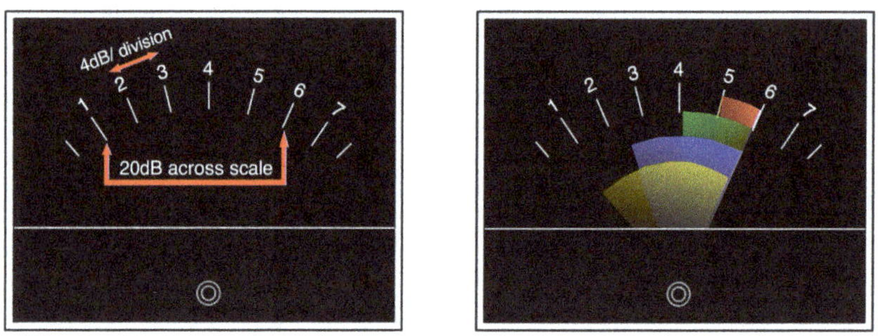

Figure 6.8 The PPM as a guide to loudness

The traditional model for the BBC National networks followed differing guidelines:

Radio 3 (Classical): 20 dB+ dynamic range (yellow) to allow for quiet passages in classical music to remain quiet, and crescendos to blow the roof off! Often listened to in solitude on a decent radio. Announcer peaks at a low PPM 3 to 4 to match in with the quieter mean volume.

Radio 4 (Talk Radio): 12 dB+ dynamic range (blue). Mainly speech-based, but also a good number of dramas. Dramatic effect requires from a whisper to a roar; however, the radio is often on in a kitchen, so it needs to compete with a busy domestic background. Announcer PPM 4 to 5.

Radio 2 (Adult Radio): 8 dB in the daytime, more in the evening (green). The dynamic range needs to suit a mixture of programming, but mainly music-based. Various genres (jazz, country, band, blues, vintage) need a bit of space to breathe! However, the highly produced daily commercial music will have a small dynamic range to cut through to the consumers on the school run and the factory floor, so the presenter's voice will need to be made louder with an external device to match the loudness of the music. Presenter PPM 6 with slight compression, say 4 dB of soft limiting.

Radio 1 (Youth Radio): 4 dB dynamic range (red). Music in the morning bathroom, on the move, in the gym! Modern highly produced and processed music allowed to peak to PPM 6. This means the presenter's microphone will need to be made *much* louder to sound the same level as the music. Presenter PPM 6, strong compression of 8 dB to 12 dB. The amount of dynamic range is easily picked off the meter simply by looking at how far the needle falls back. If it seems stuck to PPM 6, that's Radio 1!

Of course, the same is true for bargraph meters and to an extent VU meters, and you must bear this in mind when mixing different sources; it's just that the explanation makes more sense on the PPM, as it was designed for just this purpose. For interest, the presenters' voices are made louder with an extreme version of an "Audio Compressor", called a "Limiter". It's well worth learning about this device, as it is a powerful and effective "must have" for professional voice work; please read Chapter 12, "Outboard Devices", to learn more.

LOUDNESS AND METERING

The PPM has existed over my lifetime in sound, and I have used it with highly diverse audio material from classical through speech to heavy rock. I understand whilst the design is over 90 years old, it is still a great tool! The PPM is better at programme peaks than many devices, but we have always known it is not perfect; transients of a few decibels do get through. The VU meter is better at showing perceived loudness, but peaks are completely lost on it. Modern bargraph meters (whether physical or implemented in software) offer the opportunity of faster rise times to catch peaks and of smart metrics to capture the "loudness" of the sound; however, they are not created along universal specifications, so are often an unknown when compared to traditional metering. The quest for "loudness" has been hinted at across the chapters in the book and is currently (and rightly) a hot topic in the industry.

Chapter 12, "Outboard Devices", introduces equipment that makes your masterpiece stand out from others; the quest to sound louder and so more obvious is an old one. From the mid 1950s, targeted at the new phenomenon of affluent "teenagers", commercial recordings were available to hear either on a jukebox in the local bar or over AM radio. The crowds in the bar, or the sheer loudness of transmission hash on amplitude modulation, meant that all production means were used to make your commercial recording stand out from the noisy background. One popular exponent of creating loudness was 1960s producer Phil Spector with his "Wall of Sound", but they were all at it! Restriction of the dynamic range was the most popular tool, using a compressor to bring up the volume of quieter sounds to that of the louder ones. The engineer's secret was to retain the dynamics of the sound, as unwise compression tended to flatten the life out of the music. Motown Records were brilliant at this!

Despite high-quality FM radio, the growth of HiFi in the home and the huge diversity of modern music available, the quest to be louder (perhaps seen as better) has remained, with digital effects units overtaking analogue ones, mathematical tricks transcending the previous "possible" in the analogue world. However, the tide is turning as music streaming sites become the major means of consumers' daily access to music. These sites carry out "Loudness Normalisation" as part of their upload process of new music. The over-compressed tracks are rendered dull, and the more dynamic ones prosper. However, this new age has not hit the TV advertisers who want their soap brand to be more in your face than the competitors. This has led to the adverts sticking out from the normal programme material, shrill and annoying as people across the world leap for the remote at each TV commercial break.

THE DIGITAL LOUDNESS METER (LUFS – LOUDNESS UNIT, REFERENCED TO FULL SCALE)

The hero of the hour is US Congresswoman Anna Eshoo, who was astonished when her proposals in 2010 to limit differences in TV loudness went through unopposed by colleagues. The Commercial Advertisement Loudness Mitigation (CALM) Act led to a revaluation of metering, requiring for the first time that "loudness" be quantified and measured by "Loudness Range" and "Maximum True Peak Level" (dBTP) (drawn from standards EBU R128 and ITU-R BS.1770). At first the purchase of a standalone Loudness Meter was expensive and tended to be used only where the law required it, but thankfully they are easily simulated in software and are now common in Digital Audio Workstation (DAW) programmes. Affordable LUFS accredited metering (Loudness K-Weighted relative to Full Scale is an earlier but identical term) is on its way to becoming obligatory in streaming services as well as broadcast; and LUFS, already crucial in multichannel mastering, will overtake traditional meters before they reach their 100th birthday, so look out for them!

HOW DOES LOUDNESS METERING WORK?

Traditional Metering is designed around a flat frequency response and gives an instantaneous display which is used, second to second, to adjust levels on the fly. As soon as we consider "loudness" we have to concede that the ear does not hear all frequencies with equal intensity.

In Chapter 1, "Introduction to Sound", we looked at the concept of "Equal Loudness Contours" and it was shown that the ear is at its most sensitive between 2 kHz and 5 kHz.

If you remember, the units are still the same as the dB, but the measure is called phons because the level varies with frequency. LUFS units are also in

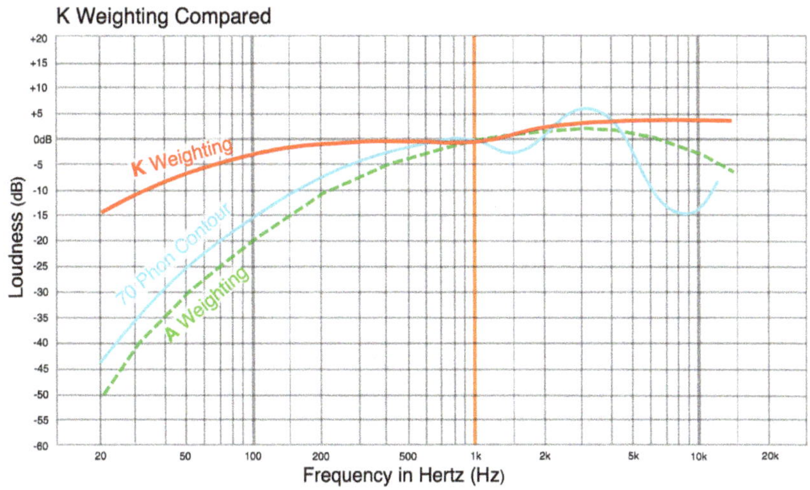

Figure 6.9 New K weighting compared to traditional curves

decibels but are not called that, again because of the modifications made in measuring. The 70 phon curve has been chosen as a model of normal household "living room" volume, and it's plotted on the graph in Figure 6.9 showing an insensitivity to bass frequencies and a sharp sensitivity to the upper mid-range. (Note the diagram is inverted from the traditional Equal Loudness Contours graph in Chapter 1, "Introduction to Sound", to show "most sensitive" at the top, which we are more used to). At the end of Chapter 1 we learnt about the idea of "weighting" sound on a sound level meter to measure "loudness" from a personal perception. Whilst there are several, the popular "A weighting" curve which is used for assessing background noise nuisance is shown here as a simple graphical comparison. A newly developed curve is shown, *K weighting*, which is designed to mirror the perception of loudness discerned whilst hearing a broadcast in the home and it is used as a vital modification in the new meters.

Here's the science bit: in our Loudness Meter we have "weighted" the sound, now we compare and average the channel values (be it stereo or multichannel) to find a common instantaneous summed value for the programme channels (surround channels are +1.5 dB to account for being nearer the listener). We digitally sample this value of sound ten times a second, aggregated into 400 mS blocks. The output of this averaging feeds a traditional bargraph display to give an instant and recognisable value of "Loudness", which is perfect for managing transient peaks to keep below the critical digital sound level of "True Peak" (0 dBTP). A second output is created from the same 400 mS blocks, this time averaging sound samples over 3 seconds. This is called the "Short Term Loudness" and whilst this near instantaneous output is visually easier to watch, it is likely to miss transient peaks and so needs using with care by following local guidelines on maximum permitted levels. Your Loudness Meter may feature, one, the other or both these "real time" modes.

Crucially, the 400 mS blocks are also fed on to create an aggregated "loudness" level for the whole programme, regardless of its length. In addition, we need to understand when we have silence and when we have programme. A BS.1770 algorithm gates any "blocks" under −70 LUFS and they are marked down as silence and not measured. A second gate is called the Foreground Element Relative Gate and addresses loudness concerns during dialogue in quiet passages discovered in early testing of the meter in multichannel sound. The blocks successfully jumping both gates are passed to a final integrator which continually plots the programme until its end to create a value of overall "Integrated Loudness".

This final output is a "Loudness Number" for the *complete programme* to compare to the network (and the other contributions to that network). World broadcasting bodies, the EBU and SMPTE agree that averaged volume shall not exceed a loudness value of −23 LUFS. If a programme is found to be too loud, say −21 LUFS, it could simply be digitally attenuated by 2 dB to make a legal −23 LUFS. That simple! A peak limiter is employed in the programme chain to ensure that peaks do not exceed −1 dBTP to protect the digital stream; *the two measures together create a compliant programme.*

Note that at the time of writing streamed services can use a louder output of −20 LUFS or even −16 LUFS as lower volumes have been found unable to sufficiently drive the headphone outputs of mobile phones or tablets in current devices. This is at odds with the requirements to protect people's hearing from damage when some listen to a loud volume for long periods, and international work is currently under way to integrate loudness with safety in mobile devices.

TESTING EQUIPMENT FREQUENCY RESPONSE

Back to analogue and our traditional line-up tone oscillator. It can be used for other technical reasons, to check for distortion by inputting differing levels, or by checking at different frequencies of tone to see what the frequency response of the equipment or chain of equipment is. Short cuts to check frequency response can be an automated stepping-tone or a smooth glide tone across the human frequency range.

There is another way; what if we could have a signal that is all tones at the same time? This "real-time" analysis could check and allow the adjustment of equipment, or particularly see how sound behaves in a room and adjust the acoustic material to get a neutral listening environment. So how could this be done?

White light (daylight) is an equal mix of all the colours of the rainbow. "White noise" takes its name from this phenomenon and is an equal mix of all frequencies of sound. That is to say, it has an equal power spectral density, regardless of frequency. This complex sound approximates the sound of a waterfall or what is heard by tuning an FM radio off station. Earlier in this chapter, we looked in detail at the "noise floor". Thermal (Johnson) noise is developed inside all our audio circuits, and whilst it is linear in terms of frequency, it grows in level in logarithmic frequency terms at a steady 3 dB/octave, so it is louder to our ears at high frequencies than low frequencies. If we want to amplify and use this "noise" as a sound source, we call it "white noise".

Our ears perceive noise energy (rather than power density) as linear across each octave of sound. Clearly, to measure an audio device, or perhaps a room, we want the measurement tool to have a flat frequency response across the audio range, and we can achieve this by modifying the white noise with a –3 dB/octave filter. This useful new tool is shown in Figure 6.10 and is called a "pink noise" generator or sometimes "1/f" noise (note, native flicker noise isn't a generating source).

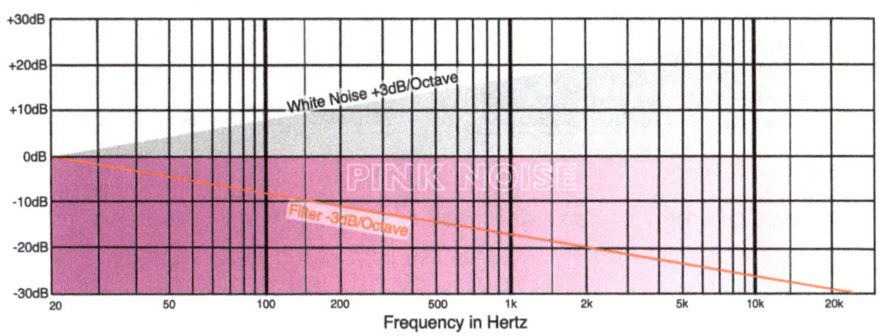

Figure 6.10 Pink noise made from white noise

If we output pink noise from equipment (most often a monitor loudspeaker in a room) we can capture the changes to the sound made by the room acoustics using a calibrated receiving microphone. We hook its output to an audio spectrum analyser which gives a real-time display of the levels of many frequency channels across the audio range – very useful! It's worth noting that the energy in pink (or white) noise is logically much more than a single tone; it's a big fat 18 dB louder, so be sure to expect the large jump up in level when using test equipment or find a new set of speakers; or worse, ears!

In time, engineers have produced other "colour" sounds to create specialist uses out of the phenomenon of white noise, but these are not currently useful to the aspiring audio engineer and can be ignored in our context.

STEREO METERING

When we have the two channels of analogue stereo sound, we need a different solution to monitoring. Classic strip or bar meter monitoring is easy; two columns can just be placed next to each other for easy visual comparison. Less so for VU meters; however, two can sit next to each other to get a good idea. But does a simple left and right meter tell the whole story? What it doesn't say is how much the stereo sound field is concentrated in the middle of the sound stage, and how much it is out to the edges. If sound sits centre stage (a mono signal, say like line-up tone) the left and right signals are "in-phase". Fully out of phase would be a fault as the sound would appear to be outside of the speaker sound stage. Stereo sound is likely to sit between the two, and phase information is useful to a smart engineer as we want to know how "coherent" a signal is. Unless for special effect, "M" should mostly lead "S". The same problem often exists for PPM meters, but here we have a choice, it's getting rarer now, but you may well see it.

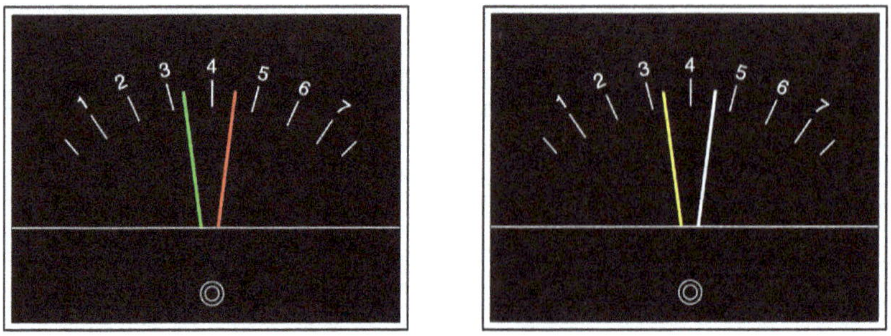

Figure 6.11 A pair of stereo PPMs

In studio nomenclature the left channel is called "A leg" and the right channel is called "B leg". At first, it seems like a conjuring trick; a Stereo PPM has two needles! There are two flavours, White and Yellow needles, and Red and Green. White = A + B (sum [mono]) Yellow = A – B (difference), Red = A (left) and Green = B (right).

It's not unusual to see both meters sit next to each other to give the whole picture, as seen in Figure 6.11. I note that a coherent (in-phase) sound peaking "4" on left and "4" on right should mean the white needle should sit at double the level, that's 6 dB higher at PPM 5.5; however, a 6 dB pad is inserted in both the sum and difference inputs so the "M" is also lined up to PPM 4 (known as A = B = M).

This chapter has been a tough read in places, and I sincerely don't expect it's all committed to memory on the first pass. It's here so when it is relevant, you can check back on the detail to meet a new situation you find yourself in. Whilst it's detailed for completeness, a reasoned overview will get you 90% of the way!

BIBLIOGRAPHY

Hartley Jones, M. (1988) *A Practical Introduction to Electronic Circuits*, 2nd edition. Cambridge University Press.

Mackenzie, G.W. (1964) *Acoustics*. Focal Press Limited, Richard Clay and Company LTD.

The Audio Engineering Society. *K-weighting and other* Frequency Weighting *Schemes*. https://aes2.org/resources/audio-topics/loudness-project/learn-more/#true-peaks.

7 Analogue Sound Storage

This book would not be complete without a journey through the history of the capture of sound before the digital era. The most compelling reason for this is that enthusiasts refuse to relinquish ancient technologies despite the purity and availability of modern digital devices at a fraction of the cost. Famously King Henry VIII, he of the six wives, made his poor romantic decisions based on the only visual medium of the day, a painted portrait. Much as the photograph should have killed off portraiture, and the digital photograph should have killed off film, these ancient mediums are not only with us but gaining a new ascendancy. They bring an alluring and artistic perspective to the easy and simple capture of an image. So it is for sound; digital sound capture, like visual capture, is capable of levels of refinement and enhancements previously impossible in the analogue world. Not only is it available, but the cost savings, storage space and reliability are simply breathtaking over previous analogue iterations. Yet vinyl and cassette sales are booming; why is that?

"Let's start at the very beginning, it's a very good place to start!" In passing, the film of *The Sound of Music*, made in 1965, was artistically and technically unopposed in its own time!

Sound recording began over 100 years before the Austrian-based musical by Rogers and Hammerstein burst onto the stage in 1959. The invention of the phonautograph by Frenchman Léon Scott in 1857 gave us the first method of capturing sound waves as they travelled through the air. His device traced sound pressure changes onto smoke-blackened paper wound around a revolving cylinder. He used a small bristle to create a pattern on the paper. This was a laboratory tool, designed to increase our knowledge of acoustics, and it took a further 20 years or so for the idea of using these "phonautogram" graphs to reproduce the captured sound. However, we had to wait until 2008, using advanced optical scanning and computer processing to recreate the sounds committed to paper over 150 years before!

RECORDING ON DISCS

The phonograph, later to be called the gramophone, record player and now in the 21st century the turntable, started life as a cylinder, rather than a flat disc. Thomas Alva Edison's first job was as a telegraph operator, receiving messages and news sent on cables by the first railway tracks in Michigan. By 1877, he had become an inventor working on a device to transcribe frequently required telegraph messages by embossing them on to a paper strip. Alexander Graham Bell had already invented the telephone, turning speech to

DOI: 10.4324/9781003717065-7

electricity, and Edison thought maybe he could record sound in a similar fashion to the telegraph messages.

His first sound recorder did use paper; however, within months he moved to a metal cylinder with tin foil wrapped around it. A steel needle moved slowly down the length, creating a fine spiral of indentation. His detailed drawing was made into a prototype by his engineer John Krusei in just 30 hours, and to Edison's amazement it worked the first time he tried it, with his famous quote, "Mary had a little lamb".

The early machines needed expert operation, and the tin foil quickly wore out. In the 1880s Bell's "Volta Laboratories" changed the needle from moving up and down (hill and dale) to side-to-side (lateral) and started using wax-covered cardboard cylinders. By 1901, whilst foil was still used for recording, reproduction cylinders were made of a hard black wax, and vitally, copies could be produced using moulding techniques ramping up from a modest production of 120 to 150 cylinders a day. This was no longer a novelty for the rich, or at best a dictating machine; now people could hear recorded music in the home, although it was for a playing time of around two minutes a cylinder.

At the turn of the 20th century, Emile Berliner introduced the "Gramophone Record"; this flat disc could take a four-minute recording, but crucially it could be turned over to hear another recording, which gave eight minutes in total. The commercial 10-inch discs ran at 78 revolutions per minute (rpm), and the steel "needles" which tracked the lateral groove wore quickly, so were sold in small tins of 200. The needle had to be replaced every couple of plays to avoid the record being damaged.

Whilst the "Phonograph" was Edison's machine, the public applied the term to any sound reproducing equipment. All these machines were purely mechanical; the amplification of the audio was by a horn, from small ones like a trumpet mouth to opulent logarithmic flares 25 inches (625 mm) across at the end. The Victor Talking Machine Company cunningly hid their horns inside a compact cabinet, calling this portable player a Victrola. There was no volume control; if you wanted the sound quieter, you would stuff a suitable piece of cloth down the throat of the horn to muffle the sound. This led to the expression, "Put a sock in it!". At the start of the 1920s electronic valves enabled the production of Marconi's radio receivers, which were often placed in a large wooden cabinet with a gramophone unit and were popularly called a "Radiogram". A valve power amplifier was shared between the two sources. Eventually Berliner won through with his disc as it was easier and quicker than a cylinder to copy in bulk, stamping out copies on a production line.

THE AGE OF ELECTRONICS

Valves, of course, were also the kickstart of the professional sound industry, with each microphone having an amplifier the size of a modern fridge and sound technicians passively mixing these outputs before a valve line sending amplifier placed the combined output on its way to the transmitter. Programme recordings were rare in the early days of radio, but in the BBC, they were always made in a specialist area by cutting a gramophone "disk" made from cellulose-coated aluminium. This could be played back from the same machine or carried through to a Disk Reproducing Desk, such as the DRD1. The BBC stocked two sizes of disk, 12 inch and 17 1/4 inch, and used the turntable speeds of 78 rpm and 33 1/3 rpm. Longer programmes featured live changeovers from disk to disk to gain the duration, and news bulletins featured operators jumping from track to track to edit what was said live on air! Chinagraph wax pencils were used to mark the out and in points to guide the disk playing operator. When you consider the desk was faded down to make the jump, and then faded back up again, you must admire the speed, accuracy and coordination of the craft! Interestingly, the rotary faders used on the type B desks of the time could be snapped open and shut quicker than either the later Painton quadrant or linear Penny and Giles faders we see today.

Cinema sound is an entire subject on its own, and I would strongly encourage further reading on this complex and intriguing history. In short, in 1927 discs were used to create the first soundtrack for "sound on film" movies or "talkies" (as opposed to silent film) using large discs at a slow speed (so less quality) to gain a running time of 15–20 minutes, which sadly, quickly ran out of sync with the pictures, and so was hated by film producers as it undermined the integrity of their pictures. As far back as 1902, Duddell had found a way of recording sound by means of the audio being converted to an optical waveform sitting beside the photographic film. Sadly, he could not play it back! Reproduction of his system was not viable before the invention of valves; however, with the latest electronics in America, Lee De Forest created photographically married sound, and in Europe, Miller backed by the Philips organisation perfected a track-cutting stylus for celluloid film, which created a transparent track that was read in the cinema by shining a light through the film onto a photocell. This was called an Optical Soundtrack and this visual storage of sound next to the picture ran through some 30 years, including the move to stereo cinema sound where, simply, two transparent tracks ran side by side.

The sound was restricted in bandwidth and sounded hollow and distant, as though you were hearing it down a tunnel, but it was the best available! In the 1960s four ferrous tracks (magnetic tape recorder technology) started being pressed into the side of feature films for a surround sound experience grand enough to pull people from a fast-emerging new competitor, "Television". An acceptable signal to noise figure could now be achieved with the new "Dolby A" compression. This required serious investment, so it was deployed in the major city cinemas. In the 1970s, Ray Dolby created "Dolby Stereo™", which explosively made the 1977 *Star Wars* a milestone in immersive sound. Here was a system which would work in the provincial theatres as well as the cities. Ray's pioneering work on surround sound went on into the digital age with Dolby Digital™ and Dolby Atmos™, but that's for another chapter.

Back to records: the 10-inch 78-rpm commercial "record" had been traditionally made from shellac (a resin secreted onto trees by the female lac bug from India – yuck!). These discs were brittle, and would easily snap if badly stored, or shatter if dropped. Clearly, no good to send to a war zone, so the army experimented with a new substance made principally from oil and salt, a flexible plastic called polyvinyl chloride (PVC), or "vinyl" for short. The American troops received audible news from home on a handy new format of a neat 7-inch disc. It worked well, and after the war, vinyl gradually started to appear in general manufacturing, including the record industry where it replaced the frail 10-inch record with a 7-inch 45-rpm "microgroove" record. This could easily contain one quality music track a side and so became known as a "single". Simultaneously the concept was introduced of a 12-inch long player "LP" record containing a sequence of tracks, like photos stored in an "Album". Both names quickly became synonymous with the new 12-inch format played at a speed of 33 1/3 rpm. This record would happily run to 20 minutes a side; however, if you were prepared to turn down the wick to a lower volume, and compromise on the bass end, over 30 minutes per side was possible.

Cutting a disc was always a complex specialist role, constraining the possible wide groove exertions of bass that were managed by attenuating it to create more time on the disc (this hit-and-miss process was standardised in the mid 1950s with the Recording Industry Association of America (RIAA) Equalisation Curve which is still in use today – see "The Record Turntable"). However the commercial pressure remained to make your record louder, meatier and brighter than other 1960s productions to sound good on the new tiny tinny transistor radios. In the mid-1960s stereo records started to appear, adding to the complications, although fortunately, stereo records would play well on mono turntables. The stereo disc continued the idea of lateral stylus movement but additionally reintroduced vertical pickup to provide two opposing 45-degree channels from the stereo pickup cartridge. These dual signals were fed on to a new two-channel "stereo" amplifier. Later in the decade of the Concept Album, the 1970s, savvy Record Cutters became an essential cog in the commercial process. Outboard devices, previously more at home in a full professional studio, became part of the Disc Cutter's armoury. She/he had to manage these technology leaps, and the record time duration was directly opposed to the loudness on the disc, particularly if there was fair bass content. Smart people! One of the early pioneers was the BBC's Reg Pidsley.

Reg was a quiet, unassuming "C Engineer" who travelled with BBC wartime journalist Lewis John Winford Vaughn-Thomas on his seminal excursions to report progress in World War II. Pidsley once found himself strapped into the bomb bay of a Lancaster so Vaughn-Thomas could peer out of the open bomb doors and report firsthand the explosions of bombs falling on Berlin. It's extraordinarily scary stuff as they were under intense fire from the ground and from fighter aircraft above with no protection. Listen yourself by searching "BBC Vaughn-Thomas Lancaster". To add to the danger, Reg was required to make a "disk" recording, cutting a groove in a cellulose disc with a sensitive cutting arm (think record player arm) on a "portable" disk recorder (it weighed over 50 lbs!). How he kept it steady is unbelievable! You may ask why he wasn't using a decent tape recorder, so much more stable and reliable in difficult situations? Because at the time we didn't know it existed! Read on...

THE PROFESSIONAL TAPE RECORDER

As early as 1899 Poulson invented the "Telegraphone", which used a changing magnetic field to represent sound in both amplitude and frequency. He used the system to transmit Morse code at high speed, so the time spent on the telegraph wire was cut back. Whilst he reasoned high-quality voice and music would also be possible on his system, it was impossible with the technology of his day. The appearance of (you've guessed it!) electronic valves made people re-evaluate his work, and before World War II in the late 1930s, a new breed of recorder was making an appearance on several continents. Wire or perhaps steel tape moved from one spool to another, passing over a transducer which varied the magnetic flux applied to the wire in concert with the applied sound. This metal tape was dangerous, especially if it broke whilst spooling and went whipping around the room, and frankly the sound quality off these early machines was disappointing.

During World War II, the German company AEG made two brilliant improvements; the first was the substitution of wire for a plastic tape invented by Fritz Pfleumer. This had millions of tiny beads of black iron oxide impregnated on it. The first commercial "tape recorder" was called the Magnetophon. The magnetic plastic tape was much safer; it was kinder to the tape heads, reducing wear, and the quality was better. It could also be physically cut and pasted to remove items of sound which were not wanted, and so "sound editing" became a new skill. The second improvement created a machine that was arguably capable of better sound reproduction than its rival, the gramophone disc (the argument still rages today!). The Magnetophon added a high frequency current of 100 kHz to the audio signal, and this "AC bias" cut the inherent distortion to levels where it could no longer be heard.

The allies did not know about these technological advances in Germany during World War II and were highly puzzled how Hitler could be in one place and heard broadcasting from another. After the end of the war in 1946, America and in England the BBC "acquired" a few Magnetophon tape machines from Frankfurt and from Radio Luxembourg for experimentation. At first the BBC Research Department could see no use for them when comparing these early tape recorders to the refined high art of the precious BBC disk cutter.

The tape recorders sent to America were copied by Ampex and had the advantage of a ground-breaking Fe_2O_4 (brown) oxide coated PVC tape developed by BASF. In England EMI produced the "British Tape Recorder One" or BTR/1, and then the BTR/2, which added in the distortion-busting bias current, whilst Swiss engineer Willi Studer, with cultural precision, quietly created the world's finest machines including the seminal C37, and later the gorgeous A80. I personally worked with the C37 and A80, as well as the mono EMI BTR/2 and its in-house conversion to stereo by the BBC, jokingly called the BTR/2 GT. It was the same Reg Pidsley, of Lancaster Bomber fame, who was put in charge of the BTR/2 conversions.

Improvements in the quality of magnetic tape continued in America, England and of course the home of the tape recorder, Germany for more than a quarter of a century. More changes occurred in the construction of the backing plastic than in the oxide formula, making strides to cut down the phenomenon of "print-through" where, in storage, adjacent layers of tape would magnetise each other, the louder passages winning and causing a ghost of the sound to appear after the main sound. Many early important recordings were damaged in this way. Being able to edit tape changed the way programmes were constructed. The idea of post-production, where mistakes were easily removed and sound effects added, meant less time in the studio.

Comedy was highly popular at the time; producers started scripting a recording for an extra five minutes and then would lose the jokes that bombed or add more laughter where they could. Interviews could be recorded prior to the live programme, edited to hone the pace and delivery and then delivered as though the person was live in the studio. Short, sharp trails and promotions found their own place in the network's day as the presenters could record a short speech insert in the rehearsal of their show, which, when combined with others and music in a separate suite in the network production offices, made a strong pitch!

In music production everyday sounds were speeded up, slowed down, reversed, spooled... All to create effects to run alongside the music. The Beatles and Pink Floyd were huge fans of these tape machine-based effects. I was recently reminded of phasing. We used to have great fun playing the same recording on two machines in sync and putting a finger on a spool to minutely slow it to produce some wonderful phasing effects. The drums on "Itchycoo Park" by the British band the Small Faces are an early great use of this effect. Another great effect was to loop the tape around, pulling off sound from the replay head and feeding it back to the record head. This could produce reverb and echo, depending on tape speed.

Professional tape machines came in three speeds, 30 inches per second (ips), 15 ips and 7 1/2 ips. America, and many top music recording studios in the 1980s favoured the top quality 30 ips; however, the BBC only used the two slower speeds for mastering: 15 ips for music and 7 1/2 ips for speech. Sadly, in smaller outposts and during the occasional times of austerity, we were forced to use 7 1/2 ips for music. Why was this a problem? The higher tape speed gave both a lower noise floor and an increased high frequency response; the difference was not only noticeable but hurt if you listened back after recording the 84-piece BBC Radio Orchestra. We could pay for around 90 people to attend a music session in Maida Vale, but we could not afford the extra pound or two for more tape... Politics! Also, all our music was recorded in stereo, and the use of the standard 45-degree angled tape block to edit the music created a wobble in the sound image as the music played through. A nightmare! I cut a 70-degree slot in my personal editing block with a junior hacksaw to solve this problem; this was much admired by colleagues and eventually copied by the manufacturer.

However, back to the plot! Tape reel sizes were 10 1/2 inches in diameter with the addition of a "NAB Adaptor" to fit the larger hub (30 minutes at 15 ips, 1 hour at 7 1/2 ips), 7 inch and 5 inch for short programmes and inserts using the much smaller 1/4-inch hub. In addition, special 5-inch spools with a large hub were used where speed was needed at the start like snappy news inserts.

The point about the need for news spools highlights a serious problem with tape machines. The drive to push the tape snugly against the erase, record and replay heads came from a steel rotating shaft called a "capstan". It only worked when the tape was pressed onto it by a freewheeling soft "pinch wheel". The tape machine tried to keep the tape tension by gently pulling the spools in both directions, taking up small variations of slack tape with tensioners either side; however, at the extremes of a fully loaded spool on one side and an empty one on the other, unfortunate effects can happen, including a painfully slow start. The larger "NAB" centre to the short inserts of a "News" 5-inch reel practically cured the problem.

THE TAPE PATH

All tape machines, whatever the size, share the same tape path. The tape is fed from the spool on the left to the spool on the right. On the Revox B77 (Figure 7.1) there is a single tape tensioner on the feed (many have two, a second one on the take-up). Tape guides sited at either end, and again by the capstan, position the tape to run accurately across the three heads. The first head uses a feed of the bias oscillator to scramble the oxide particles on the tape to effectively wipe the tape clean. This is the erase head. The second head uses the same bias oscillator to modulate the analogue sound signal, placing a copy of the audio signal on the tape by patterning the oxide particles. This is the record head. The third is used to pick up the audio signal from the tape and reproduce the magnetic pattern as audio again. This is the replay head. Clearly this is required to playback tapes made in the past, but it has a brilliant second use! Positioned after the record head, it picks off the audio that has just been recorded and returns it to the output, where the sound tech can listen to check the quality as the recording is being made (note, cheaper tape and cassette machines combined the record and replay heads as one). The tape is continually tensioned by a soft backwards pull by the feed spool and a soft forward pull by the take-up spool. The tape tensioner(s) softens abrupt changes, such as a start or stop. The accurate speed of the tape is set by the rotation of the sintered capstan.

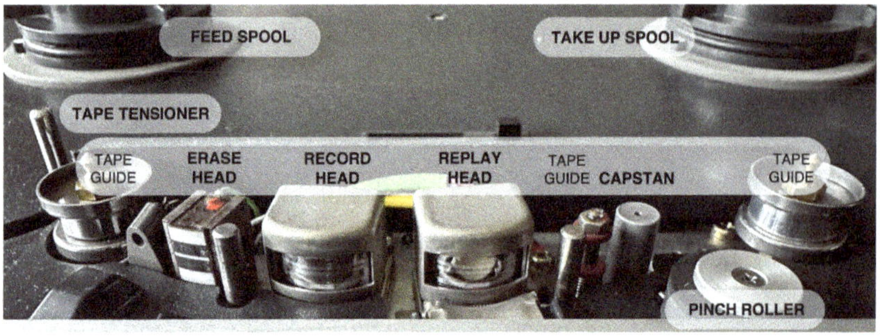

Figure 7.1 Head block of a tape recorder
Photo by the author.

The capstan rotates all the time the machine is powered, but the tape is stationary until "Play" is pressed and the freely rotating rubber pinch roller snaps onto the capstan. The capstan drives the roller, which in turn pulls through the tape. When the machine is stopped or running in forward or reverse wind two of the movable tape guides (one to the left of the record head, the other to the right of the replay head) keep the tape from contacting the heads. This is to mute the sound whilst spooling and to protect the heads from wear due to the abrasive tape passing over them at high speed.

TAPE RECORDERS IN THE HOME

In the late 1950s valve tape recorders started to appear in the home market, and those with deep enough pockets could record themselves, or more often, programmes off the radio. Rather quaintly, there was no provision for leads to connect the two directly, so the tape recorder microphone would be placed in front of the radio speaker, and everyone in the house had to shut up for the length of the show! Many is the time that someone burst in at just the wrong moment! Most home recorders used a slower 3 3/4 ips speed, but like the professional machines, took the 1/4-inch wide tape. In time, home recorders were capable of recording two channels for stereo, called two-track machines, and some were even four-track machines, where the user could record a stereo source (tracks 1 and 3) and then turn the tape over and record on the remaining blank tape strips (tracks 2 and 4). This idea went forward to be used in the Compact Cassette. Sadly, every halving of tape speed led to twice the noise (tape hiss) and a loss of high frequencies, and every halving in available tape width, (single, dual and four-track formats) also doubled the tape hiss.

America was quick to encapsulate magnetic tapes in plastic containers in the early 1950s, so they were more robust, manageable, reliable and didn't roll off on the floor under the nearest desk when dropped! The first enclosed tape format was for Talking Books for the Blind, followed by the Fidelipac, a cartridge introduced at the NAB convention in 1959, for use by broadcasters to play in relatively short duration tapes for station idents (jingles) and radio commercials. This tape cartridge (called a "cart") had one spool and a single loop of lubricated tape. The feed to the replay head was pulled from the centre of the spool (hence needing to be slippery) and the return was wrapped around the outside. The tape ran at 7 1/2 ips, so was decent quality, and early mono machines were quickly updated to stereo heads for the FM Radio Stations. The "Cart" stayed in use for 40 years until the turn of the millennium!

William Powell Lear wanted music in his luxury planes. Gramophone records had not worked out as the stylus would skip during turbulence or manoeuvres. From the beginning he saw the tape cartridge format would work in cars as well, so he got Ampex, Ford, General Motors, Motorola and RCA records around a table. The first eight-track cartridge music players appeared as an option on new Ford cars from 1965. This provided endless play of four stereo channels (hence eight tracks) and could not be jolted in a car – or even in your personal jet! All eight tracks could be accessed without turning the tape over, and it would click through all four stereo channels before starting again after (up to) 80 minutes. The casing and mechanism were of a similar size to the Fidelipac jingle cartridge and improved on the precision plastic housing, the guides and the encased 1/4-inch tape inside it. When the tape completed a loop it switched tracks, and so on until the fourth track finished when it started again from the top.

Whilst the eight track did run at a decent speed for domestic sound reproduction, splitting the 1/4-inch tape into eight separate tracks took its toll on the quality; tape hiss was a problem, but luckily this was largely masked by the sound intensity of the popular music format. The narrow ribbon of tape reproducing sound in the eight track, however, did suffer from weaving effects as it crossed the heads, making the upper frequencies waver.

In general, the pioneering encapsulated tape machines were replay only (no erase head, record head and the associated electronics) for cheapness and reliability. They played centrally produced material preloaded into the plastic case for hire or sale like vinyl or CDs. Whilst recording cart (jingle) machines were available, better quality was obtained by recording on reel-to-reel and then inserting them into the Fidelipac case. This was my job for a season as a junior in the BBC; the most difficult were creating the never-ending loops of music for the Radio 1 DJs made by cutting the start to the tail, removing the stop pulse. It was great when it worked! Later, at the turn of the millennium, looping was at the centre of the skill of scratch DJs, creating new works from those that had gone before.

In 1963 the massive European technology company Philips invented a tape format that was to change the world for the remainder of the millennium. The Compact Cassette was first created for office dictation to replace wire recorders. It was of so-so quality as it ran at a very slow speed of 1 7/8 ips and was monophonic (stereo was still in its infancy). At only 1/8-inch wide the tape did not allow much area for quality, and you could turn over the tape and play the other side, so the width available for audio was less than half! Stereo came later and halved each channel width again... Therefore, I find it an amazing journey from the formal launch of a naff home recorder in 1968, the Philips EL3300 (though it seemed like a miracle of engineering at the time!) through to the finest cassette recorder (in my opinion) the 1979 diminutive Sony Walkman Professional WM-D6C, which was good enough to be used by the BBC for broadcast journalist purposes.

That Philips, already a leader in tape recording, decided to encapsulate miniature reels in a plastic shell, using a tiny 1/8-inch width tape, a slow tape speed of 1 7/8 ips and four-track recording was an amazing technology leap. The compact cassette shortened quickly to just "cassette" should have bombed on quality, but people loved the easy portability (tapes would fit beautifully in a top shirt pocket), and players were eventually added to all new car radios. Pre-recorded music cassettes fought through to commercially rival LP records in the mid-seventies, mainly because they would play well in the car. Amazing gains in quality came through two endeavours: (1) dramatic improvements in tape formulations using CrO_2 (chromium dioxide) from Advent Corporation in 1971 (and later cobalt), which in turn would have worn the tape heads away if it wasn't for the development of much harder alloys in the head construction and (2) utilising a modified version of Ray Dolby's compression system (Dolby A™) to compress the high frequencies during the record process, and decompress them on playback which took down the invasive background noise level (hiss) contributed by the tape by as much as 20 dB. The use of Dolby B and C systems was financially possible due to Dolby creating integrated circuit chip versions of their compression system, taking both the cost and size to a fraction of former levels.

The humble Cassette Recorder became a 1980s shining star, a HiFi stack essential and held off the medium of Digital Audio Tape (DAT) cassettes, making DAT (formally Rotary Digital Audio Tape – R-DAT) a professional format instead of the proposed mass consumer market. I see that just as Vinyl albums have risen in popularity and value in the last ten years, the pre-recorded music cassette tape has also become a nostalgia fashion favourite. It appears analogue still reigns supreme for many enthusiasts.

MODERN REPRODUCTION

Much of the information in the foregoing chapters has been dual skilled, equally suitable in the professional studio or at home. I'm now going to take a bit of time to explain what is lost to the professional workplace, however, is common in the history of home listening, and is key to engineers and enthusiasts alike who want to revive or "remaster" old recordings for use with a new admiring generation.

If I had written this book ten years earlier, including the subject of record turntables as a music source would have been seen as bizarre! But it didn't stop there; people at the time of writing are actively collecting tape cassettes, and even reel-to-reel tape has a recent resurgence. All these analogue sources have weaknesses that digital audio has sought to conquer, and on paper at least, it has! In Chapter 9 on "Digital Audio", I point to poor mastering practices in the production of early CDs which undermined the introduction of this first digital home system, and whilst the engineer in me knows digital on paper is capable of far higher fidelity than previous analogue home systems, I am easily swayed by the sheer musicality of a well-engineered turntable playing a carefully crafted vinyl record.

Let's get "digital" out of the way. The early digital consortiums like the Motion Pictures Experts Group (MPEG), like generations before them, looked to place all the complexity in the digital encoding process, so the consumer electronics in the home could be as simple as possible, and so cheap to produce in huge numbers for a mass market. I'm not going to step through the same points made later in the chapter "Digital Audio"; we know that many differing lossy and lossless formats deliver the same popular music material in endless versions, and comparison is difficult, as the online providers want to give as little costly bandwidth over to storage and distribution as they can. Ever more clever algorithms are created, but it seems precious quality is doomed to be sacrificed by using excessive bandwidth to stop the piracy of material by a few chancers making illegal copies. Uncompressed digital files of some concerts are sold at high prices, as are popular lossless formats such as Free Lossless Audio Codec (FLAC).

Whilst some digital to analogue converter (DACs) are sold for the price of a car, similar technology can be produced for the price of a can of coke. At the time of writing, I feel many in the mass market are failing the home user; fidelity is suspect in all available routes, and newly remastered "old classics" on vinyl apparently recreated "by loving hands" often disappoint when compared with a decent pressing of a 1960s or 1970s LP original. I worry there is a pervasive industry fear that if the best quality is available in the home, it will stop future rebranding of the same material for the next generation. It seems there always must be a path to "improvement" to get people to put their money on the table. How sad, and how different from the birth of popular music, where through to the 1980s truly jaw-dropping fidelity was available for the home enthusiast, sanctioned by the pervading pride of professional engineers in the industry. I think the piles of CDs making their way to charity shops at present will reverse as people realise this is the only cheap way to get uncompressed digital quality. Like unscarred LPs, prices may well rocket on mint CDs of the late 1980s to 2000s featuring the bands' material before vinyl LPs made a comeback to compete against them.

As a young boy I was aware of wind up "Gramophones" gathering dust in the corners of living rooms or moved into attics or basements where they are still found today. It is hard to explain to a generation who are five-button-clicks away or even a single spoken phrase from a favourite music track how extraordinary it was to have the ability to hear a concert in the living room or even out on a picnic. It was so unbelievable that Conan Doyle wrote a Sherlock Holmes mystery about the sleuth fooling violent visitors at Baker Street by stepping into the next room to "play the violin". The possibility didn't register that it could be a recording, and of course he got his men. With wire "tape recorders" and the Edison Phonograph wax cylinder, early copies had to be recorded in real time. The jump in recording quality in the 1920s came with the new "electrical system" of capturing sound with a microphone and amplifying with electrically powered thermionic valves. The later flat 10-inch 78-rpm gramophone record could be pressed and so took seconds to produce each one. Easy to store and use, all other systems took a back seat.

In America, the first 12-inch microgroove record hit the market in 1948, and the large centre 7-inch "45" came in the following year. In Europe we used the same small quarter inch hole on the 45 as the LP, using plastic adaptors on import singles. The two sizes have both been available in 33 1/3 rpm and 45 rpm, but the larger is usually the slower speed and the smaller the higher. To recap, the 12-inch was created for multiple tracks put together as an "album" where the 7-inch was created for a "single" track which led to its name.

THE RECORD TURNTABLE

A vital point is that a standard turntable outputs its signal directly from the coils of the pickup cartridge. *Not only is the level nominally 40 dB below other inputs at 1 kHz, but it is also modified by backing out fixed equalisation in the cutting process to create more time on the record; there is a massive 20 dB gain at the bass end, falling to 20 dB loss at the extreme top frequencies.* This modification is called the RIAA equalisation curve (Figure 7.2), and a "phono" preamplifier is required to recreate a flat response at a nominal domestic line level input of around −10 dBV. These circuits were routinely installed in old-fashioned HiFi preamplifiers, but in the age of digital are often left out for a cost saving.

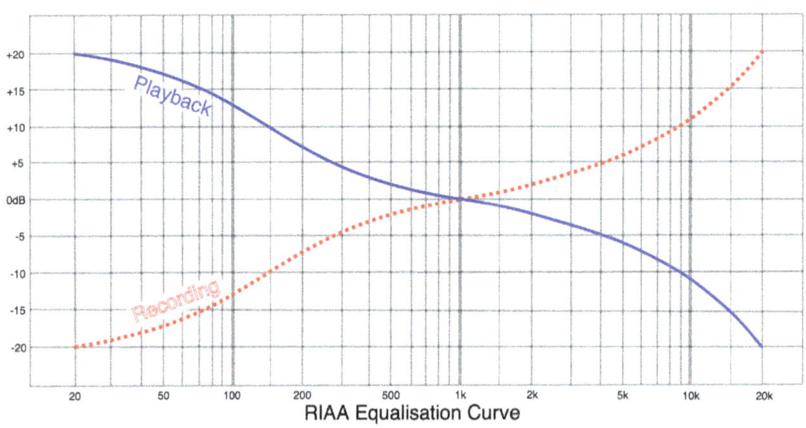

Figure 7.2 RIAA Equalisation Curve

A low noise amplifier is required, with precision resistors and capacitors in a feedback loop to ensure the response curve is correct, and is matched across the two stereo channels, so the image doesn't wander. External pickup preamplifiers are available to buy, and at the time of writing some turntables are beginning to provide a line output (i.e. internal RIAA preamp), and some have a digital output on USB too. Frankly, the quality of these is suspect, as I said above good RIAA amplification costs, and decent DACs are a serious premium! Note that pickup cartridges, like microphones, are available with differing transducer modes. Whilst most preamps can handle the sensitivity of "Moving Magnet" designs, the lower output of most "Moving Coil" pickups requires specialist pre-amplifiers to give their best.

The turntable has a rotating "platter" which, at the minimum, must be able to run at both 33⅓ and 45 rpm, but more than that, it needs to be governed so the speed is true and does not vary. Whilst at one time this used the cycles of mains electricity as a strobe guide, these days reasonable turntables are locked with a quartz crystal oscillator (similar idea to a modern watch). The platter must also not waver as it rotates; low-speed axial aberrations (up to 20 Hz) are known as "wow" and those above 20 Hz as "flutter". A motor will normally drive the turntable through a geared link, and this also allows for a speed change. Earlier units used a freely rotating rubber wheel to join the motor drive to the turntable rim, called an "idler wheel", and this eventually wore flats with frequent starts, leading to wow and flutter. Good turntable platters were made heavy in the past to smooth out imperfections through the flywheel effect. Later turntables used rubber drive bands, and with the introduction of the professional Technics SP10, the motor directly drove the turntable (Later this technology was placed in the iconic Technics SL1200 disco turntable which is still made). This technology has a high motor torque, and so made for "instant start" to cue records in at the right time for both broadcast and disco DJs. ("Instant" is in reality about a third of a turn, so the DJ needed to talk to the gap as part of his/her skillset, as a prelude to a "voiceover" of the introduction [like it or hate it].)

Our record is smoothly rotating at the correct speed on the turntable; we need to "pick up" the sound off it, and so engineers developed an arm to swing over the diameter of the record to capture the audio off the groove. They imaginatively called this the pick-up arm (or "tonearm" in many markets). The arm holds a pick-up cartridge which has a stylus at its front to track in the groove. The cartridge, which can cost from dollars to thousands of dollars, converts the lateral and vertical wiggling of the needle in the record groove into two electrical signals which travel up fine flexible wires, connected to tougher wires at the base of the arm which pass out of the back of the turntable casing as a stereo pair to the system amplifier (if it has the appropriate "phono" input). The stylus (old name "needle") encounters a great deal of friction, so is tipped with a tough precious stone, normally a tiny diamond. This was traditionally spherical in shape, but more modern designs have tried to mimic the shape of the disc cutter head to follow the groove more accurately. The most popular is "elliptical", but there are others, with more esoteric names, for obviously higher prices.

The record arm has several parameters to get the positioning right for the correct operation of the stylus in the groove. The most obvious is the tracking weight, simply how heavily the stylus rests in the record groove. Weights range between 1 and 6 grams, with better cartridges showing more compliance (flexibility), and able to track at lighter weights. Lighter tracking also means less record wear, however, any debris in the record groove will be exacerbated, as the stylus can't plough through it! Modern cartridges should track at 3.5 grams or lighter (1.5–2 grams is most common), the heavier groove strippers of early broadcast turntables now rightly consigned to the dustbin. As the turntable rotates, there is a centrifugal force (think playground roundabout) that looks to push the stylus outwards off the record. Pick-up arms have an adjustment called "anti-skate" (weights or spring pressure), which applies an inward pressure (a centripetal force) on the arm to counterbalance the outward force and keep the stylus tracking uniformly. This is normally set to the same value as the tracking weight.

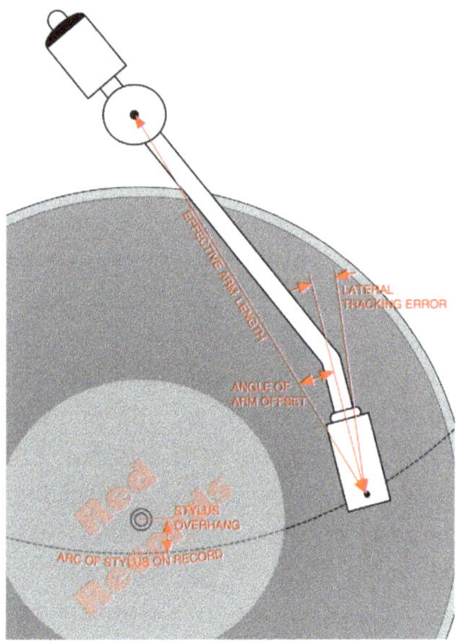

Figure 7.3 Tonearm tracking

When the master recording is made, the cutting head of the disc cutter moves tangentially across the disc in a straight line. Most pick-up arms pivot from a single point, and so the cartridge is adjusted (normally at the cartridge head-shell) to make the most accurate tangential arc possible (Figure 7.3). There are simple protractors, most of cardboard, to adjust cartridge "overhang" at the time of fitting. The final adjustment is the height at the pick-up arm pivot point, so the stylus sits at the correct angle to the record. This rarely needs to be changed on modern equipment.

Well set up systems are still sadly problematic. The RIAA 20 dB boost of the bass can expose rumbly turntable problems and often makes the cartridge/platter combination slightly microphonic; if you are unlucky this can coincide with the fundamental frequencies of the room standing waves and move towards a howl round. Decent turntables are set on isolation feet (see Chapter 8, "Introduction to Acoustics"), but it's worth noting the effect and listening out for it. Records gather dirt and dust, which can be greasy, particularly from fingerprints. This grinds away in the base of the groove and creates hash, pops and clicks. Cigarette ash was the biggest problem in the heyday of vinyl; luckily, it is much rarer now. A dry disc, fresh from its clean sleeve, can be a static magnet for small particles, and static electricity can jump to ground through your body, creating tiny welts which we hear as clicks and pops on the record. Wiping the disc before playing it, using a slightly damp proprietary cloth, pad or carbon fibre brush (moistened with diluted record cleaning fluid or iodised water) will cut static and so the attraction of dust and static discharges. The stylus will build debris as it plays and should be gently cleaned after each use with a stylus brush (back to front, gentle wiping); otherwise, it will ride up the groove on its own dust hydrofoil, and it will sound like the socks in the horn! The precious stone can chip in use, and whilst sounding "OK", the stylus will chafe the groove as it passes. The next play will sound a little fuzzy (one channel worse than the other), and each successive one worse, before you realise there is a problem and investigate.

In conclusion, using a turntable is a tactile delight; the performance of choosing and preparing the disc and then gently setting the tonearm off on its journey across an inky vinyl lake is poetry. The West's version of a tea ceremony! The open and airy sound is a breath of fresh air after a clinical digital rendition, and an LP's artwork can be as expressive as its music. The use of a turntable isn't just sound; it's *ART* and all the more emotionally complicated because of it!

TAPE PLAYERS TODAY

I think it is less likely readers will own a tape recorder, as digital sound recorders are cheap, effective and cost next to nothing to use. There is, however, something quite hypnotic in watching the spools turn... All sizes and types of analogue tape recorder work along similar lines and have similar characteristics. I will assume the device you are using is working and so producing sound; however, like turntables, there is a lot to know. Above I explained one-, two- and four-track machines, and in general, a tape needs to be played back on a machine sporting the same number of tracks as the one that recorded it. You can get lucky; I lifted a treasured programme off a four-track tape, using a two-track playback head, as the tape had not been turned over to record on the other side. In general, a tape will playback well on the machine that recorded it; however, there are several problems if the tape is taken for playback on another machine. Obvious ones, like track placing mentioned before, but also speed accuracy and "Azimuth". Whilst it sounds like a heavy rock band, this is simply adjusting the tape heads (record and replay) to be exactly 90 degrees to the passing tape. If the replay head is a little out, the record head will have been adjusted to match it, and so all will sound OK... Take the tape elsewhere, and the maladjustment will kill the high frequencies. This is what happened on occasion with creating sub-masters from the studio original analogue tape for either record or CD replication plants.

As well as adjusting the azimuth, the AC "bias" needed to be optimised for the tape before making a recording. Not to worry here, as the recorder will be set up for stock tape, either in initial manufacture or when serviced. However, if you put your reel-to-reel machine in for service, supply a blank sample of your favourite tape for the service engineer to use in the adjustments. Other checks include speed accuracy, wow and flutter; like a turntable "wow" is a low frequency irregularity (less than 20 Hz), often from irregular spools where the spokes could rub on the tape; and "flutter" (higher than 20 Hz) often from flats on the pinch roller or worn tape guides. Recorders, particularly early cassette recorders, would build a coating of ferrous material stripped from the tape to contaminate the tape heads, which would need to be routinely cleaned off with diluted isopropyl alcohol, or the high frequencies would suffer. The heads, in time, would also become magnetised, producing a warble effect on the recording. This required a "demagnetizer" with a strong AC field to put it right. One other simple operational point is the elephant in the room for cassette tape users. Was Dolby encoding employed in the recording (almost certainly yes)? If so "B" or "C"? The number of times I hear the encoding left without being uncompressed (Dolby "off" on playback), as people like the zingy crisp top! Is it correct, or is it what you want to hear?

The last point I would put forward is the more subtle, and often ignored. Levels and frequency corrections occur within tape recorders, and all but one will cancel out in a well-maintained machine to create a flat frequency response. However, recording tape has never been perfect as a medium; the high frequencies became saturated with earlier formulations of tape (and still do to a lesser extent). This was managed in early American Ampex machines by tweaking the replay amplifier. Note this is not on the same basis as RIAA, and whilst a set amount of record pre-emphasis may be used, the replay EQ changes are mainly due to (1) the replay head's native frequency response rises at 6 dB/octave and (2) the tape's response (flux density) drops off at 6 dB/octave from around 3–4 kHz. Below this frequency correction of the rising response is made by a 6 dB/octave filter, and above it, there is a flat response to counter the tape losses (e.g. the rising frequency response of the tape head is cancelled by the falling response of the tape). In 1953 the correction of these two barriers to a flat frequency response became the American NAB standard. By the time the Europeans had caught up, tapes had evolved, slower tape speeds were common and the IEC (CCIR) standards reflected the improvements. Domestic reel-to-reel machines and new formats such as cassette muddied the waters, and new cassette tape formulations required EQ revisions, so it's all a bit of a mess! Cassette tape machines had front panel EQ for machines that can take the advanced tape products, and some of these machines could detect the correct EQ settings by indents in the back of the cassette housing.

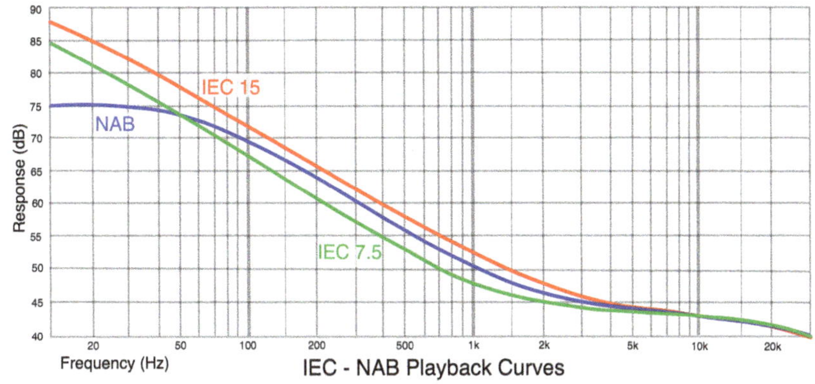

Figure 7.4 Tape playback equalisation

That leaves reel to reel. Again, there are shades of grey; however, classic American machines are normally NAB replay EQ, and European *professional* machines are IEC. Domestic European machines may be either type, based on country of manufacture and the manufacturer themselves. In the 1970s/1980s, the classic Revox B77 Tape Recorder could be ordered in either NAB or IEC, and whilst playback EQ was adjustable on the front of some Japanese machines, the German-built Revox required several internal circuit board changes to swap standards. Today the standards are worldwide, NAB = IEC2, yet they lag a little with tape improvements at high frequency.

So, you may ask, what does this matter to me? I am writing in the UK, and the machines recording the heyday of the Baby Boomers (and of magnetic tape) were, in my experience, 50:50 IEC/NAB. US tapes sent to our shores were NAB, and tapes from Europe predominantly IEC. Old master tapes should be played back on the correct EQ machine to be as intended. Is it the end of the world? Not really; people in the know have filter bodge boxes to modify the sound from one to roughly the other, and others may use EQ such as a Graphic Equaliser (crude – not advisable) to tweak the sound. The NAB curve carries a low-frequency bass boost which, *depending on material*, may be excessive when played back on an IEC machine; but if you know, you know! Changes in level in the treble boost from around 3.5 kHz are relatively subtle and can be outpaced by other factors such as what tape the recorder bias was set up for, as against the type you are playing.

Why use a tape recorder at all? Well, apart from the hypnotic turning of the spools, the magnetic tape recorder is tolerant of peak overloads, some people even using the magnetic compression of high frequencies as a production tool. This is a reason the VU meter worked in practice, as the actual peaks (way off the top of the VU scale) were considered a lower priority in the time of tape recording. This is of course in stark contrast to digital, which has zero tolerance. In conclusion, all tape-based recording involves many variables, which if not appreciated and thought through, can lead to errors in the playback of the source, changing the sound from that intended by the original studio production. If you are simply listening, set it as you will and enjoy! If remastering, ensure your monitoring is true, and from it, that the audio fidelity is preserved at all frequencies, and sounds clean and natural.

BIBLIOGRAPHY

BBC Recording Training Manual (1950) *Recording Training Manual.* London: British Broadcasting Corporation.

Brown, C. *Introduction to Hi-Fi* (Editor, Hi-Fi Sound Magazine).

Dolby, R. (1981) A 20dB Audio Noise Reduction System for Consumer Applications. *Journal of the Audio Engineering Society,* 31(3).

Hartley Jones, M. (1988) *A Practical Introduction to Electronic Circuits,* 2nd edition. Cambridge University Press.

Hellyer, H. (1974) *Stereo Sound.* Hamilton, Scotland: Gilmour & Dean Ltd.

Pannett, W. (1962) *High Fidelity Pocket Book,* 1st edition. London and Edinburgh: Morrison and Gibb.

8 Introduction to Acoustics

It seems crazy to me to spend a lot of money on equipment and give little thought to the properties of the room where either a recording or a mixdown is created. Huge problems can manifest themselves in a badly thought-through location. A sensitivity to the environment from the sound professional (that's you!) is crucial whether you are involved in a one-off recording (see Chapter 13, "Planning for a Location Recording") or planning the design of a top banana slick sound studio. The concepts follow simple science and for the most part solutions are common sense. Let's start at the beginning...

The study of sound waves is called "acoustics"; however, straight away we must divide this study into two discrete sciences.

THE TWO BRANCHES OF ACOUSTICS

The first is *isolation of the noises we want from those we don't*, sometimes called "sound proofing". Whilst measures to keep out unwanted sound might alter the acoustics in the room, it is entirely a different proposition, and the two requirements must never be confused into one. We might not want to allow sound out of our local space, "Bludgeon's Night Club", as it would be annoying to neighbours. Conversely, we don't want the local neighbourhood noise leaking into our studio, "Subtle Sounds", and picked up on our microphones, ruining the divine sound we are trying to record. Both needs are about isolation of the sound stage from the outside world and usefully follow identical science.

The second science is the *behaviour of sound within an environment*, or "room acoustics". If our recording is made in the wrong "acoustic", it will be ruined before it hits the microphone, and that damage is mostly unrecoverable when we get it back to base. Equally if the room where we listen and adjust our masterpiece alters the sound dramatically, yet we trust it; when the precious recording is taken elsewhere it will sound nothing like we intended. The rules are similar, with the same physics; however, there are substantial differences between the studio performance area and the "Control Room" cubicle where the sound is checked, mixed and recorded.

ISOLATION

We have split isolation from room acoustics, and now we must split again. Sound is obviously carried through the air, but we learnt in Chapter 1, "Introduction to Sound" it is also carried through structure, particularly at low frequencies. We must enter a third criterion that affects both air-borne and structure-borne

DOI: 10.4324/9781003717065-8

sound. It's resonance, which if not considered destroys the science behind the first two.

Sound that travels through the air, as we know, has two factors, its "amplitude" (how loud it is) and its "frequency" (how low or high pitched). The difference in amplitude between the isolation of the studio and the outside is measured in decibels and the higher the number, the better and more expensive the solution. But hang on, if I close the bedroom windows and talk or sing near a directional microphone, maybe the difference will be enough? Yes, in many situations that is enough isolation. However, if next door's lawnmower changes the equation and renders our precious session lost; we might need to upgrade and upgrade we can! Sound isolation through the air is most often a problem of the middle and high frequencies. The sound waves are small enough to squeeze through tiny gaps between window surrounds and floorboards, however, that can be remedied by door and window seals (identical to draught excluders) and silicone sealant between boards, and that (crudely) is "job done".

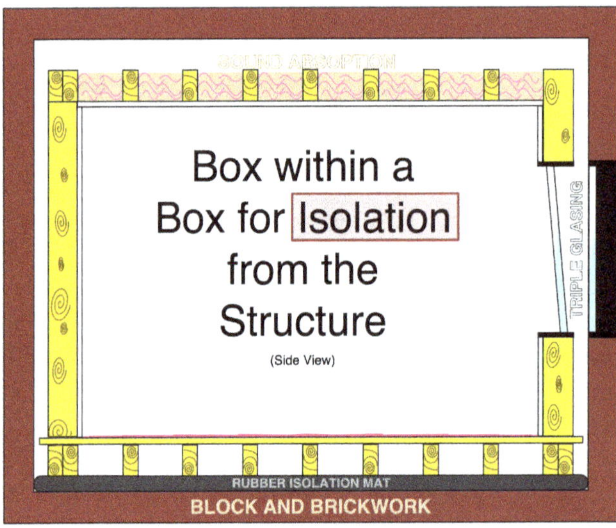

Figure 8.1 A side view of a "box within box" studio

Sadly, resonance will ruin our plans on occasion, and that is dealt with below. The upmarket solution for isolation involves building a new room inside our studio space, a "box within a box" which is isolated from the surrounding structure, giving another barrier to the sound, and which, if a proprietary standard, builds in smart ways to seal the gaps (Figure 8.1).

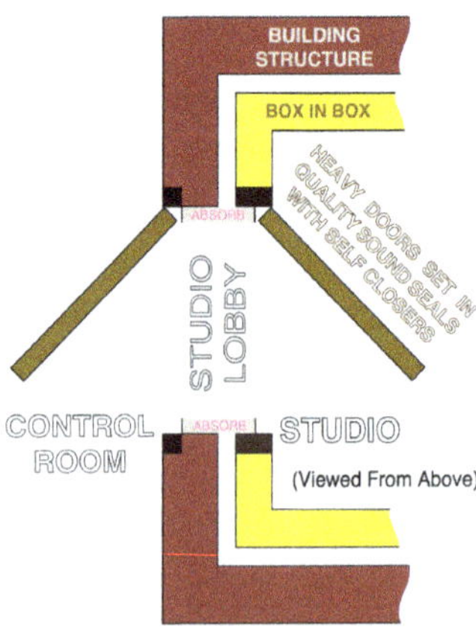

Figure 8.2 A door to enter a "box within box" studio

The problem of needing a door to get into the studio can be addressed by having a "studio lobby". We build in two substantial doors, with just enough space for a person to stand between them. One door is in the outer space, and the second is in the isolated box built inside it, so the second barrier to sound is maintained. The first door can be allowed to close completely before the second is opened, always maintaining a decent level of isolation. We can count on the high mass of each door to stop low frequencies vibrating through, however, effort needs to be put in to stopping the medium and high frequency noises that squeeze through the small chinks in our door armour. Good quality door seals are used, often magnetic and of a similar type to those on a fridge door. The cavity between the doors is lined with sound absorbent material, and often the doors are covered with carpet tiles, all to soak up as much unwanted sound as possible near the doorway (Figure 8.2).

STRUCTURE-BORNE NOISE

Structure-borne noise is exactly what it says on the tin, but there are several flavours. A state-of-the-art studio in a London building was completely useless, as despite many precautions, the sound of stiletto heels on an external wrought iron staircase cut through the steel structure like a knife through butter. A carpet on the steps would have solved the problem, but it was outside and the landlord did not want to change the architectural appeal of the structure, so bye-bye studio… More often, structure-borne noise is very low frequency, literally the whole building being shaken, perhaps by a passing juggernaut. This is where mass will come into play as a deterrent. Building thick brick walls with care to make sure the "frog" or indent in the top of the brick is filled with mortar, and a quality render on the surfaces, are both very good for isolation of all audio frequencies.

Simply banging in a nail on the top floor of a high-rise building can travel through the steel girder construction to the bottom floor on the other side of the building. Again, a workable sound studio solution is "box within a box"; however, a nearby jackhammer or "Kango" will still destroy a studio session. In the BBC we set up a "red light" system, so when a DJ came up for a link, the Tech Op could signal the workmen to stop. I remember they got so fed up they carried on anyway, and if nothing else, it gave the DJ something to talk about on air! Finally, the "structure" can be the floorboards, the walls, the table or desk you use or even the windows. Resonance remains the wild card in all sound treatments…

RESONANCE

Everything in nature has a natural resonance. Sound resonance is that buzzing door panel in the car which kicks off at a particular engine speed; it's the hall window rattling when an artic goes by or the rumble from the otherwise quiet air conditioner that is literally shaking the wall panels. We need two items to come together: a reasonably pure source of sound of a certain frequency (a strong fundamental) and an "object" of a mass, size and thickness that through nature is excited to move much more than usual at a particular frequency. This can be directly tied back to the "bandpass" filter in electronics; in fact often in acoustics, electronic symbols of inductor, capacitor and resistor are used to explain the acoustic effect of an air flow as it passes from one medium to another – a change in air impedance!

Double glazing is partly sold on its ability to block sound as it has two sheets of glass. Yet if we make two glass panes the same size and thickness they will resonate at the same frequency and that single frequency will not just pass through, it will be amplified by the "excited" glass. Resonance problems are equally a problem with air-borne and structure-borne sound and need to be tackled in our isolation design. If the homemade glazing had two sheets of glass which were either a different size or thickness the problem is gone, and that goes for the plasterboard sheets that form our box and for the timber or steel structure that sits behind them. If there is a cavity wall, one of block and one of brick is better than if both walls are built the same (but all air gaps in both walls need to be sealed).

The other solution is to set structures (of every size) on flexible mountings such as rubber or springs where the resonance is damped. This only works to isolate if the flexible material is partly compressed. If it sits as though the weight is negligible, or is squashed flat, the troublesome vibrations will pass through. To state it a different way, the total mass of the isolated unit should compress or expand (pad or spring) the flexible mount to roughly half of its travel. The mass and mount need adjustment to make this condition valid. Bear this in mind if you are trying to isolate a microphone stand from a desk or a turntable from its cabinet, and this sort of isolation rule works across the whole audio frequency spectrum, so is a golden rule to know. It's OK to double the isolation layers if the system resonance is clearly different for each half, i.e. a turntable may have its own springy feet matched to its weight which gives a level of isolation. This can be much improved by a higher mass platform with a stiffer neoprene pad working at a far lower system resonance (Figure 8.3).

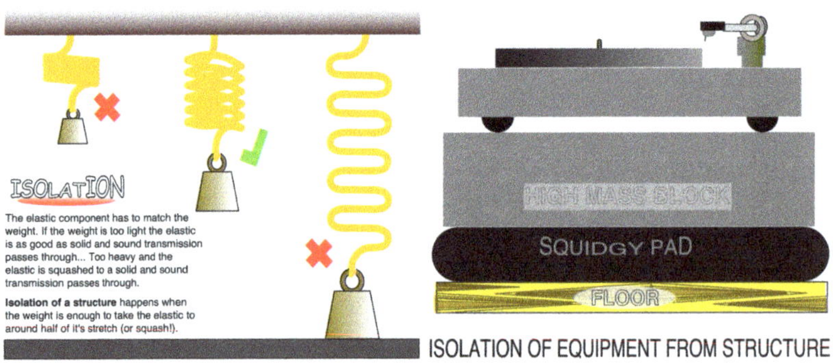

Figure 8.3 Springs and pads tackle structure-borne noise

A note for audiophiles. Metal spikes and cones are solid objects, as are the metal cups they usually sit in. In this combination there is no damping in the coupling, and so little to no isolation for your system. At best, resonances in this construction create a comb filter, which can slightly subdue some frequencies, but at the cost of unpredictably enhancing the passthrough of others (particularly bass) to above using no system at all.

MORE ON SPRINGS AND PADS

Let's take a little more time on springs and isolation. A car uses a system to isolate an uneven road from the passengers, normally featuring springs working in harmony with a damping device, the shock absorber. The engine will be isolated from the chassis with rubber "shock mounts". Motors in industry and the home are likely to sit on flexible mountings; the washing machine, the shower pump, the air conditioning fan all have "mounts" to be less intrusive in terms of noise. My point is that isolation of noisy components is a widespread mature industry, although its use in the sound industry is subtle and often out of sight. The most obvious daily use is in microphone mountings, where springs or flexible cords are used to isolate the microphone from the desk, table or floor. This structural isolation must be separated from the acoustic isolation which is achieved by:

1) The microphone pickup pattern.
2) Sealing the room from external noises.
3) Intelligently damping the room with acoustic foam or fibre to suppress the reflections of sounds produced within the room.

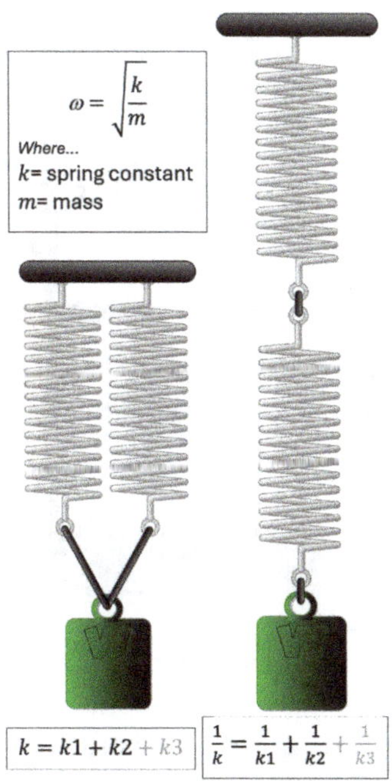

Figure 8.4 Using multiple springs to isolate a structure

The microphone shock mount suppresses hand and arm movements on the desk, table leg kicks, keyboard strokes and so on. Note that some microphones feature capsule isolation rubber mountings internally to the housing which helps to achieve improvement in cases where no external shock mount is used at all (a vocal mic on a stage stand is an obvious one). You will also find mass and springs at the centre of the acoustic membrane absorber.

The likelihood is that more than one spring will be necessary to isolate an object, say two on a shock mount, or perhaps one at each corner for a turntable or speaker. Perhaps we want to float an entire floor in a "box within a box" design! Fortunately (and proving the world of physics is a small one) springs behave like the electrical component, the capacitor when joined together (Figure 8.4). Unwanted energy is dissipated in a spring, like the removal of AC ripple by a capacitor in a power supply. Clearly, we can use the same, or differing springs together, in series or parallel to create the desired response. Hooke's law defines the constant, k, as a measure of the stiffness of a coil spring which varies by material, diameter, length and wire thickness. In the formula above, given as a teaser to further study, $\omega = 2\pi f$ so the resonant frequency of a system, (f), can be calculated. That is beyond the needs and the maths of this book, but the electrical parallels are intriguing! We will return to the "spring and mass" principle in "Appendix 3 – A Guide to Home-Built Absorbers", where it is used in the low frequency membrane absorber.

MANAGING VENTILATION

If you have read this far you clearly are serious about optimising your space and will probably be contemplating excluding outside noises with window and door seals. In addition, consider gaps around skirting boards on wooden floors as sound can pass around the rough plank ends, unseen as the skirting board covers the cuts. (This is another reason to have a fitted carpet!) An airtight box is excellent for sound isolation if breathing is not an issue. We have our superbly sealed box and the last thing we need is to take a can opener to it and let in noisy fresh air! Seriously, if we want to place people inside the space we need to supply them with ventilation, and this is an expensive and complex specialisation in professional sound spaces, often costing as much as the technical kit in the cubicle! I won't be surprised if this is not the way you want to spend your money on what for many is a hobby which must be funded from money earned elsewhere. Just bear in mind that is why professional sound spaces cost a lot to hire! I can help, but I can't stop the daily problems of construction workers renovating next door, and the council digging up your street every other month. It is one of the blights of the industry, you are not alone! Sound can enter our studio through two paths, the acoustic path, carried through the air, and the structure-borne noise carried through the ground and up through the fabric of the building. This book has shown techniques for structural isolation techniques, which can be (and are) scaled up to whole rooms, however deleting noise in the acoustic path is an easier proposition.

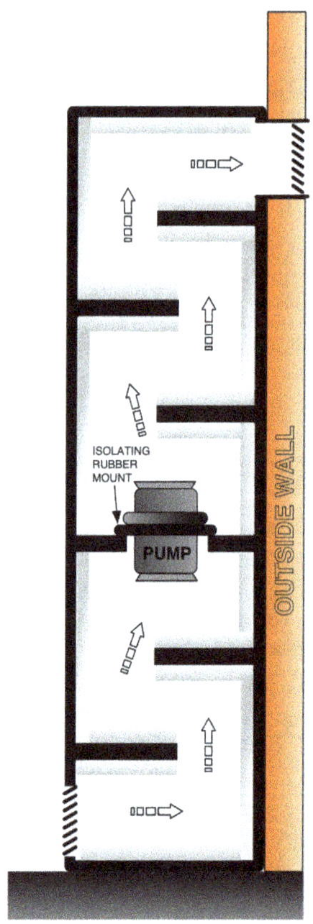

Figure 8.5 A ventilation duct with sound isolation

Above we learnt that sound can be absorbed by porous materials, and that if the impacting sound hits the absorbent at a decent angle the lower frequencies are better attenuated. We can use this grazing effect by creating a long tunnel for the outside air to pass through, lined with absorbent fibre that picks off the sound through friction, weakening it as the distance from the outside wall increases. Note we require two ducts, one to bring the fresh air in, and the other to let the stale air out. It may be that these are dual straight ducts that run in the room corners from the outside wall corner to the inside. If the length is unsatisfactory to curb the noise, the duct can be folded to increase its length as in Figure 8.5. I have seen quite elaborate ones, doubling back to make a long duct within a small area, but I hope "simple" will work well enough for you.

Some pointers, a pump is necessary to move the air, particularly on smaller diameter ducts. I would have a 4-inch (10 cm) minimum cross section for the air space throughout a duct in a domestic space, scaling up size by room size and the maximum capacity of people you envisage in your space. I would place the "in-line" pump on the outflow, so it is pulling the stale air out, and in a domestic setting the inflow duct is unlikely to need help. The next point is important; fans make their own noise which we need to address in the design. Slow fans with large blades rumble, and that sound is more likely to be carried structurally. Small fans with fast blades are likely to whine, which is more likely to be heard acoustically. Your choice is likely to sit between the two extremes, however, paying special attention to the method of fan mount isolation will be time well spent! Pliable rubber mounts are my favourite, being careful if bolts hold it together, that they don't bridge or compromise the elasticity of the mount, and that the holding frame is solid so it can't vibrate in sympathy with the fan motor. An inspection hatch for the fan must be airtight with seals, and again solid so sound is isolated. Adequate movement of air in smaller rooms is achieved by the purpose-designed "in-line" domestic ventilation fans (bathroom extraction, etc.).

The sound deadening lining of the ducts can be done with most open pore fibre mats, quilts or foam, but if the material can shed, it must be contained behind an acoustically open weave mesh to stop fibres entering the room, and then your lungs!

Heat rises, so the collection of the stale air at floor level and input of fresh air at the ceiling level circulates the air and minimises the loss of heat. If you live in the middle of LA and add filtration, the duct size and fan capacity will need to increase to manage the more restricted airflow. If you are in a hot climate modern split-pack professional installations of air conditioning can be used, which, if sourced from a good brand, have quieter fans. Please don't buy one that doesn't publish a noise figure. The indoor unit noise is in the specifications from the manufacturer, measured in A Weighted decibels (dBA). Quality units are capable of a sound pressure of 40 dBA or less on lower settings, and this should be your target. You should size up rather than size down; be generous here to have a good system ticking over at a slow fan speed, rather than a small one going at full speed to gain the target temperature. Beware in your room positioning of air-con outflow draughts hitting your microphones. I have used "skirting" units over wall or ceiling mounted units with good results. Ensure the connecting pipework is well sealed at the entry points by the contractor to avoid sound bearing air gaps and that the internal unit is soundly mounted on a solid wall so it can't resonate against its mounting.

I think that's about it; home-produced ducts are simple woodwork, common sense and materials that are cheap enough that you can experiment with different designs in your space to optimise the result. I can't foresee every room and application, so these are basic guidelines to get you on your way, and I must leave it to you to create something you feel is appropriate and safe. More bodies will obviously use the fresh air more quickly, so bear that in mind; a headache or drowsiness are signs of stale air. Take a break and ensure you are safe in your hobby. I know you will be proud of your unique solution!

THE CORRECT ROOM ACOUSTIC

I can point to many interpretations as to what is "correct" in terms of the acoustic within a room. There are two basic needs: that the recorded audio sounds natural to the end customer (i.e. the home listener in his/her lounge) and that the imperfections in the room don't overly damage the recording. I make no apology for repeating a diagram from the first chapter to remind you that sound (at least above 300 Hz) travels in straight lines and reflects off hard objects like walls at the angle of incidence (Figure 8.6). Parallel facing walls are a problem, and low frequencies (below around 300 Hz) will resonate to grow against other frequencies and create multiple reflections, which can be annoyingly focussed by the shape of the walls with respect to the sound source (the vertical sound path in the diagram). These are known as standing waves or "eigentones", and there will be more on this important subject later.

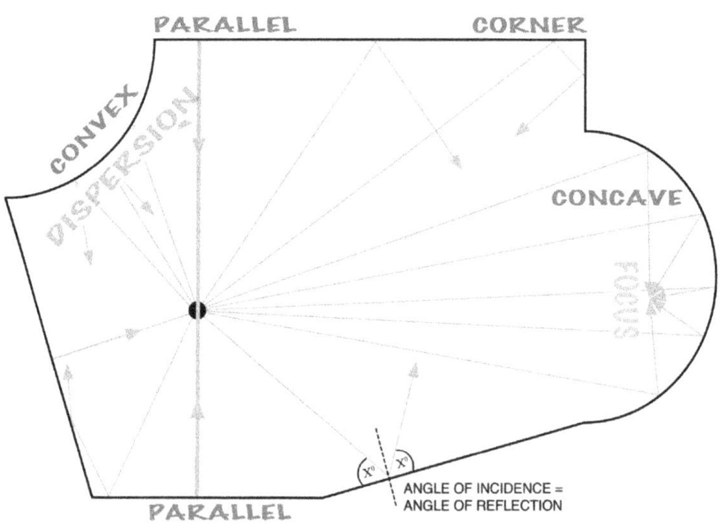

Figure 8.6 Sound within an enclosed space

The BBC, where I trained, likes to record speech so it sounds close to the situation of people sitting in a comfy room at home. There will be hard surfaces that reflect the sound like brick walls, windows, doors, and solid ceilings. There may well be carpets on the floor, curtains at the windows and soft furnishings that absorb sound and so take down the amount of reverberation and tame an echo that you might get in other rooms like the bathroom or hallway. So, what is the difference between "reverberation" and an "echo"?

My favourite definition is that whilst reverberation-*n* adds to the sound it remains part of it. Conversely, echoes-choes-choes-choes sit as distinctive sounds from the original. The industry standard is to measure the time for a sound to die away to 1/1000 of its initial volume (a loss of 60 dB), and we have dedicated equipment to measure it. (An experienced sound engineer can simply clap his hands and can get a good idea!)

The measure is called the "Reverberation Time" or "RT" (RT_{60}, or even T60) and for a "talks studio" dedicated to recording speech, it would be quite low or "dry" at RT = 0.3 seconds, slightly below that of an average sitting room of RT_{60} of 0.4 to 0.5. If you were trying to play a musical instrument in such an environment it would be hard to hear yourself, and so music and "Light Entertainment" studios and venues are nearer an RT_{60} of 1 second. Dedicated live music venues strive for between an RT of 1.5–2.0 seconds, which is a lot more "wet" to listen to. Opera is a little drier at RT = 1.2 seconds, and a cathedral can be as much as 6 seconds! You get the picture! In closing on this subject, I have designed several DJ or "Self-Op Presentation" studios and whilst they are nominally a talks studio, I find they are much better for working in and listening to, at RT = 0.5–0.6 seconds.

So, we have chosen an RT for our space, but what is implicit in that is that the *reverberation time remains broadly constant across a wide range of frequencies*. Further it must work whether the space is empty or full, as people are also sound absorbers. Next time you are in a high-end theatre have a look at the bottom of the fold-up seat. In professional installations, padding on the bottom of the empty seat squab absorbs the same amount of sound as an "average" person sitting in it. The acoustics will sound the same in theory, whether the theatre is empty or full. In music studios there are often tall stand-alone screens with a hard side and a padded side, which the recording engineer can move around to get the sound she/he wants.

ABSORB OR DIFFUSE?

Unwanted reverberant sounds can be managed in two ways. The traditional way is to use open cell acoustic foam or mineral wool (like loft insulation) to soak up the sound by friction, turning it to heat. These are called "velocity action" absorbers. The deeper the box, the lower the frequency it will control. Varying facing material on the boxes will lead to differing results; a canvas cloth facing will ensure virtually all the sound will pass through to the absorbent material behind. Perforated hardboard (normally 25% holes to 75% board) will reflect some sound and absorb other sound, but the "velocity action" absorber is not good at low frequencies. Here, the well-tried and tested solution is a "Panel Absorber", utilising a membrane cover like roofing felt, thin ply, hardboard or more recently mass loaded vinyl (MLV) placed over an airtight wooden box like a drum skin. The trapped air acts as a spring against the mass of the membrane.

The membrane is excited to resonate at low frequencies (like the car door panel above) to wear down the energy and so attenuate it. A similar action can be created by a "Helmholtz" resonator which looks a bit like a bird box. The hole or port in the front is sized for a particular frequency, and the sealed box is lined with porous absorbents to tame and disperse that frequency. Helmholtz resonators are hard to design and build satisfactorily. In my experience, panel absorbers are easier to build, a little less predictable in use; however, the frequency of operation can easily be modified by changing or doping the panel material. We call these "pressure action" absorbers. What if you want to break up and scatter the sound without taking down the RT? My initial answer is not to overdo the porous absorption in the first place, however, many will!

Diffusion is where we use hard wooden profiles or plaster wedges and blocks of differing sizes, designed to disperse and scatter sound waves at different calculated frequencies, so the reflections cease to be a problem. The methodology of dispersion has been used in large buildings since the birth of cinema, yet it was not headlined in acoustics books for regular room sizes such as studios until Schroeder's work on diffusers in the 1980s. So, this is a comparatively young science in smaller spaces and is gaining popularity, but it must be said there are a worrying number of products which just don't work in practical high street studio sizes or have been adapted well past the initial scientific principles to be useless for the reason they have been purchased; but perhaps do look pretty on the wall!

A health warning that a badly designed diffuser can throw out specular reflections that are worse than nothing at all, and in a simple "2D quadratic diffuser" much loved at the time of writing, each strip is a resonant plank, and unless built like a battleship, it will act like a resonant absorber and have unexpected results. There are several large and/or intricate 3D structures which, though well-designed, tackle a restricted mid-band and so serve to improve the loudspeaker imaging and not subdue the wider room problems. A diffuser may well delight in a studio setting, providing an uncoloured acoustic with string and woodwind instruments; however, it cannot be directly adjacent, as it will have unpredictable results.

In a typical domestic setting, a certain star in the bargain armoury of the home acoustician is to obtain several deep bookcases and place them across the back wall, filled with an artistic spread of different sized old books. This, in my experience, often makes an excellent lower-end diffusor (don't knock it till you try!). In a home environment a thick felt "jellybean" rug, hung on a wooden batten a couple of inches from the wall, can be effective at absorbing and controlling the RT at higher frequencies. It's also decorative! Pleated quality curtains, particularly hung at a handspan's distance from a wall, are great across the all-important middle frequencies, and a nice carpet on the floor for the high frequencies – already a well-thought-out space, and no ugly and expensive acoustic boxes.

You will note I have not mentioned egg boxes or the foam sheets which resemble them. It's amazing how many people mistakenly buy these to "soundproof" an area. They don't! Whilst they do take down the reverb time through both absorption and diffusion, in my experience they can disproportionately affect the middle frequencies, leaving a pinging of over-reflective highs and a stark imbalance in the more major problem of low frequency "standing waves" as a tragic elephant trumpeting in the room…

STANDING WAVES (EIGENTONES)

Let's get away to the beach! Sitting watching the waves, have you ever wondered why every so often there are giant waves and occasional calm patches? The flow of the water either occasionally adds waves together creating some giant ones or cancels them out, leaving flat water. Where there are hard parallel surfaces opposite each other (walls, windows, doors, etc.), and dependent on the dimensions of the room, certain low frequencies (below 300 Hz) will be much enhanced by resonance, bouncing back and forth between the walls, floor and ceiling and adding multiple reflections to the original sound (Figure 8.7). The points of maximum particle velocity are at the crossing points and are called nodes (five nodes show the frequency in the diagram is the 5th harmonic). Sound pressure builds between the nodes to peaks called antinodes (blue arrows). These standing points of resonance mean that *at the wall there is the maximum sound pressure*, which in turn is a big mass to dissipate! This addition and cancellation create a "comb" filter, where some low frequencies are hugely enhanced, whilst others adjacent to it can audibly cease to exist. This is why a monitoring room must not have parallel hard surfaces opposite each other, and in general, for all frequencies, the sound in the room must be tamed by either (1) absorption or (2) diffusion.

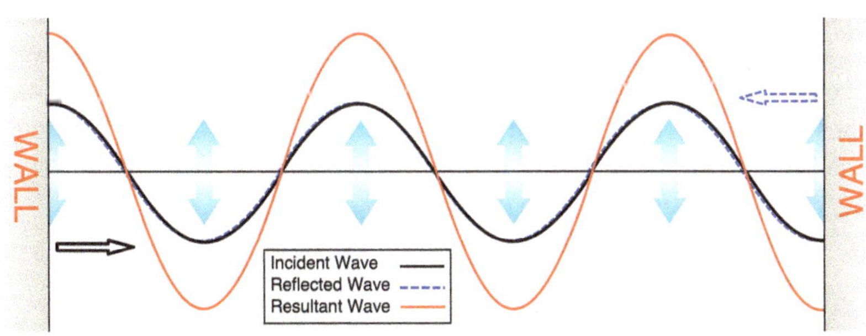

Figure 8.7 Low frequency standing waves

At higher frequencies (above 300 Hz), the same thing happens but is not so profound, and the effect is often a reverberant honk or ping heard off the microphone known as flutter echo. Sadly, both male and female voices have crucial content below 300 Hz, and the colouration can be particularly oppressive at the low end, so even these simple recordings need to be made with respect to the modes of the room. The situation can be much improved by well thought-out acoustic treatment.

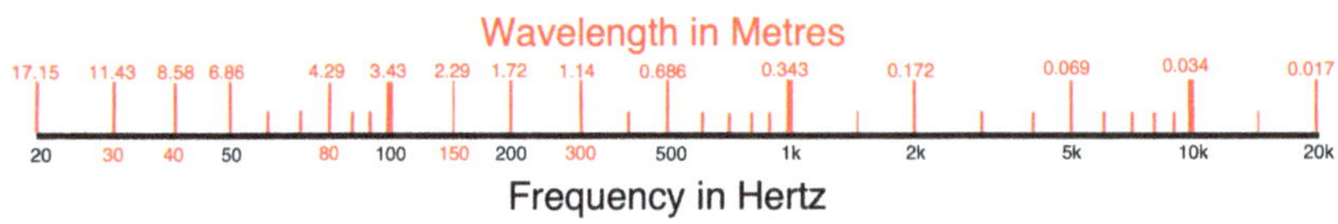

Figure 8.8 Conversion from frequency to wavelength

In Figure 8.8 there is a plot of wavelength against frequency (at room temperature). We know the wavelength of human hearing stretches from a tiny 17 mm at the top end to a whopping 17 m at the difficult low end. A porous absorber requires $1/4\lambda$ to start to tame a sound frequency; a diffuser requires a structure $1/2\lambda$ deep. Compared to an average room size, you will be quick to see very low frequencies are difficult to tackle, simply from the sheer depth required at a given wavelength to get porous absorbers to work!

If we were in a public hall, the enhanced frequencies from eigentones (standing waves) from speakers to microphone can cause the system to overload at the resonating frequencies and cause a "howl round" (also known as "feedback") at the troublesome pitch even at comparatively low volumes. In small studios, the monitoring speakers can sound completely wrong in the space, seemingly having too much bass, yet still missing crucial low frequencies. Sometimes the mid frequencies can be overpowering; I would describe them as honky or even brutish. Not a room to do a mix in!

With room dimensions in mind, you would never choose to build a studio from scratch with parallel walls, but often we must take the space we have and make the best of it, and careful attention to ensure the width, height and length don't share common base multiples. A room 10 metres long by 5 metres wide by 2.5 metres tall would be excruciating to record or listen in, whatever you did with it! Again, common sense. The maths has been well researched, and "ideal" room dimensions are readily available in online calculators to match your needs. A rule of thumb with absorption is to treat one surface and leave the wall opposite it bare or more recently diffused. That may mean several absorbers on each opposing wall, just not normally positioned facing each other. In passing I must tell you that standalone porous material "bass trap" absorbers placed at the wall or corner, despite what some suppliers will tell you, cannot cure standing wave problems.

PATH LENGTH

Helmut Haas published a paper in 1951 that looked carefully at levels and reflection lengths in sound. He placed a primary loudspeaker at a central position and then placed a second one at an increasing distance to see the effect the addition of a secondary path had on the listeners' perception. He found at a roughly 50 mS delay a separate sound could be just heard on speech, the difference between reverberation and an echo. His pioneering work created "The First Wave Front" used as the basis of public hall designs for decades. It's important to note that in general a Music Control Room has more than one sound source (at least two stereo speakers); multiple sources interact with each other and so have additional timing problems we will come on to later. Haas used speakers to mimic source and reflection, but his work assumed a stage as a *single point source* in the auditorium (Figure 8.9).

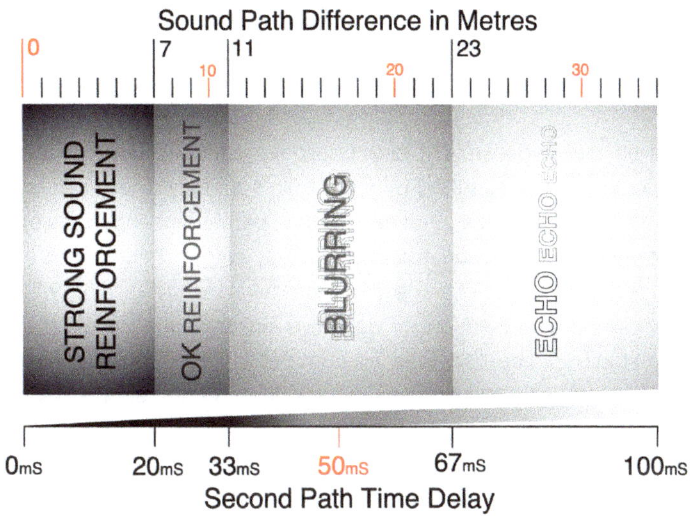

Figure 8.9 A secondary path delay confuses the primary sound path

It was found that up to 11 metres (36 feet) the second path sustained the source and often made it richer! Above 11 metres path length difference the brain could not focus, and the sound became blurred and difficult to decipher. At a path length of 23 m (75 ft) and more, the distinct echo on both speech and music became disturbing and unpleasant. Architects of public buildings use the shape of the structure (and added design details) as reflectors to carry the natural direct sound to parts of the building that otherwise would fall silent (Figure 8.10). Where the original shape of the building does not lend itself to simple modification, deflective and reflective structures need to be introduced to bring order to the chaos. The large domes added in the 1960s to the roof of the Royal Albert Hall in London are a spectacular example of the change that is sometimes required.

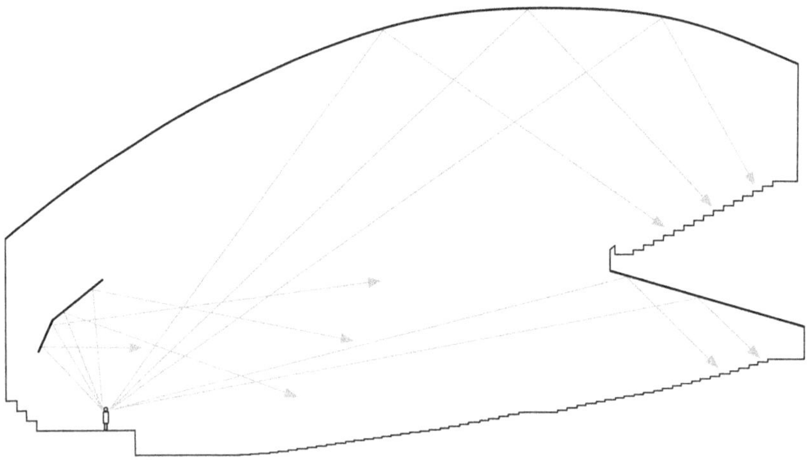

Figure 8.10
Creating a natural sound spread in a large building

Haas gave us the golden rule that with a point source multiple audible sound paths should not be separated by more than a maximum of 11 m path difference. It's a simple rule, but a great one for both permanent and occasional sound reinforcement, and I've used it with consistent success in churches and public halls. I'm pleased to note from experience that in small spaces, whilst the human ear can be confused, larger delays in indirect sound, say off the back wall of the studio, can be distinguished from the source and the very early side wall reflections, and be subjectively rejected from a good sound mixer's balance.

BEING PRACTICAL

An easy way out of poor acoustics is to use well-tried good quality headphones (I always take a pair to a new studio location) to use alone or in comparison to the existing speakers. However, if you don't fancy that option you need to break up the walls with absorption and diffusers, and in very bad cases the commercial solution of tuneable bass traps (Helmholtz resonance principle). Reflections pile up in corners, particularly where the floor and ceiling meet, and these are best kept out of the audio layout whether it be recording or listening. Keep speakers out of corners and away from walls, and ideally site your listening position at around the centre of the room if you can, or at least a metre or more from the nearest wall.

Remember absorption cannot sort low frequency standing waves in average sized home rooms. This is because the maximum particle velocity occurs at $1/4\lambda$ from the wall (it is zero at the wall!). To take a practical example, a wavelength of 35 Hz is approximately 10 m. To tackle that frequency, it would take a wall mounted absorber 2.5 metres thick! That won't leave much space for your equipment! Membrane absorbers and dispersion through angled panels (see "Appendix 3 – A Guide to Home-Built Absorbers") are the answer here.

There are mathematical formulae to calculate how much absorption is required such as "Eyring modified Sabine's Theorem"; given the volume of the room and the knowledge of the wall construction and the fitted furniture (textbook absorption of each item can be looked up in tables), the amount of absorption set against frequency can be calculated, and so panels specified to bring the RT_{60} down to the wanted level. However, in my experience the formula does not work reliably at RTs of 0.5 seconds and below. My honest advice is to use the common sense that runs through this piece. Thin absorbers like carpets absorb at high frequencies, and the thicker you go through curtains to sofas, the more the absorption tackles the medium and eventually low frequencies. Add treatments to all walls but look to leave the directly opposite surface free of absorption. That does not mean diffusion cannot be used, but bear in mind diffusers only work as they should if they are on axis to the sound sources (i.e. most often the rear wall). A glancing blow gives a very different result! Treating opposite glass is particularly important, and whilst out of fashion, a decent use of thick pleated curtains on a runner is a great way of having discreet control of the environment.

Always treat behind the microphone, particularly near walls, and listen carefully after each change until you are satisfied. Treat a computer screen like it is reflective glass and try to keep it out of the way. Stay away from overuse of oversimplistic commercial foam products which have negligible bass attenuation yet build in absorption as frequency rises. These are unhelpful in the quest to produce a clean sounding, resonance-free balanced listening or recording area, if not backed by honest science and alternative strategies. Finally, don't forget the ceiling! The further the walls are (over) damped for a dead acoustic (a common mistake!), the more the shiny ceiling sticks out as a reverberant surface! I have said elsewhere that treating the opposite surface yields good results, in this case of course the floor. If the ceiling is going to be bare, the floor must be carpeted with a decent underlay to avoid the room "honking".

The field of acoustics is often misunderstood and is confusing to new entrants; I hope this chapter has got you started! I'm sure you now realise that the acoustics are as important as the technical equipment with which we traditionally seem more at ease. For this reason, for the intrepid, there is further reading in two appendices on acoustics at the rear of this book. The aim of "Appendix 2 – Typical Acoustic Environments and Their Design" is to give a practical guide to understanding problems across many everyday rooms, hopefully supporting where the smart use of sound control is necessary, and a decent outcome is essential. In "Appendix 3 – A Guide to Home-Built Absorbers", the aim is to provide our audio enthusiast readers with the basic knowledge to construct both velocity-based and resonance-based absorbers using readily available, affordable materials and to use them wisely in each situation.

BIBLIOGRAPHY

Alton, F. (1984) *Acoustic Techniques for Home & Studio Everest*, 2nd edition, 4th Printing. United States of America: Tab Books Inc.

Architectural and Civil Engineering Department BBC Engineering (1980) *Guide to Acoustic Practice*. BBC Engineering Training Department.

Burd, A.P., Gilford, C. and Spring, N. (1966) *Data for the Acoustic Design of Studios*. BBC Engineering Monograph No. 64. London: BBC Engineering Division.

Davis, D. and Davis, C. (1980) The LEDE™ Concept for the Control of Acoustic and Psychoacoustic Parameters in Recording Control Rooms. *Journal of the Audio Engineering Society*, 28(9).

Mackenzie, G.W. (1964) *Acoustics*. Bungay, Suffolk: Focal Press Limited, Richard Clay and Company Ltd.

Moore, J.E. (1978) *Design for Good Acoustics and Noise Control*. Macmillan Press Ltd.

Parkin, P.H., Humphreys, H.R. and Cowell, J.R. (1979) *Acoustics, Noise and Buildings*, 4th edition. London and Boston: Faber and Faber.

Sengpiel, A. and Sengpiel, E. (2025) *Standing Waves*. https://sengpielaudio.com/StandingWaves.htm.

Varney, N. (2022) *Clearing Up Misunderstandings Regarding Loudspeaker Feet*. https://www.enjoythemusic.com/magazine/equipment/0422/Ten_Misconceptions_About_Loudspeaker_Spikes_Review.htm.

9 Digital Audio

WHAT IS DIGITAL AUDIO?

In earlier chapters we learnt that to move sound around we must change it into an electrical signal. That signal needs to sit in the sweet spot between being too high when it distorts, or too low where the background noise becomes noticeable behind it. Each time we copy or reproduce a normal AC audio signal (also known as an "analogue" signal) the next stage of electronics adds its share of noise and distortion, and so with the signal passing through multiple devices, the sound signal deteriorates to the point where the degradation makes the signal inviable in a professional situation.

What if we could send or store our sound signal as numerical samples, each representing the original signal level at a point in time? Now a simple number or code would replace a voltage level and if we took enough samples in a second, we could capture up to the highest required audio frequency. Our signal is no longer a voltage vulnerable to outside corruption; it can be copied and stored without end, as the copy is always identical to the original.

The binary code (noughts and ones) used in computers lends itself to the scale of capture required, and so the concept of "Digital" audio capture has evolved. A smart engineer named Nyquist worked out that the number of samples captured each second had to be more than twice the highest frequency (number of wavelengths a second) required by the system. An early BBC system worked up to 15 kHz and so had a sampling frequency of just over double at 32 kHz. Such a digital system is called Pulse Code Modulation, or "PCM".

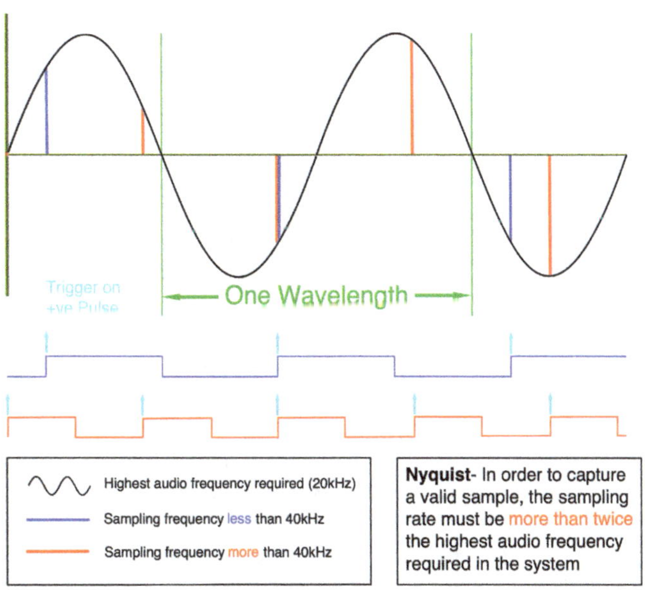

Figure 9.1 Nyquist valid and invalid samples in an ADC

DOI: 10.4324/9781003717065-9

ANALOGUE TO DIGITAL CONVERSION

In the Nyquist diagram of Figure 9.1 an "analogue to digital converter" (ADC) is capturing samples of a 20 kHz audio signal. If sampled above twice the highest frequency, enough of the red samples are taken to have one in every half wavelength. A lower sampling frequency (shown in blue) is just under two times and fails by missing the third half cycle completely, and so the ADC will assume it did not exist with dire results in the digital output.

I indicated above I worked on the first BBC PCM system installed in the late 1960s (not that I was!) to carry all the BBC national stations from Broadcasting House to the transmitter sites across the UK. It took up around five full bays of equipment for 13 audio channels. It still makes me laugh that many eminent audiophiles muttered about the purity of the FM (analogue) Radio 3, claiming "perfection compared to 'digital' sources", when every national BBC signal heard on the radio had been "digitised" for many years.

In the early days the conversion from analogue to digital and the recovery of the audio back from digital to analogue required hundreds of discrete components, many of which had to maintain a high level of ongoing precision and so systems were too expensive to buy and keep running for the consumer market. Storage of such a signal was problematic as digital storage devices had a tiny capacity compared with modern devices. The earliest household computers had typically a 10 megabyte hard disc which would be full up after the capture of around a minute of just one song! This explains the slow start of a technology which now allows thousands of tracks on a slim mobile phone.

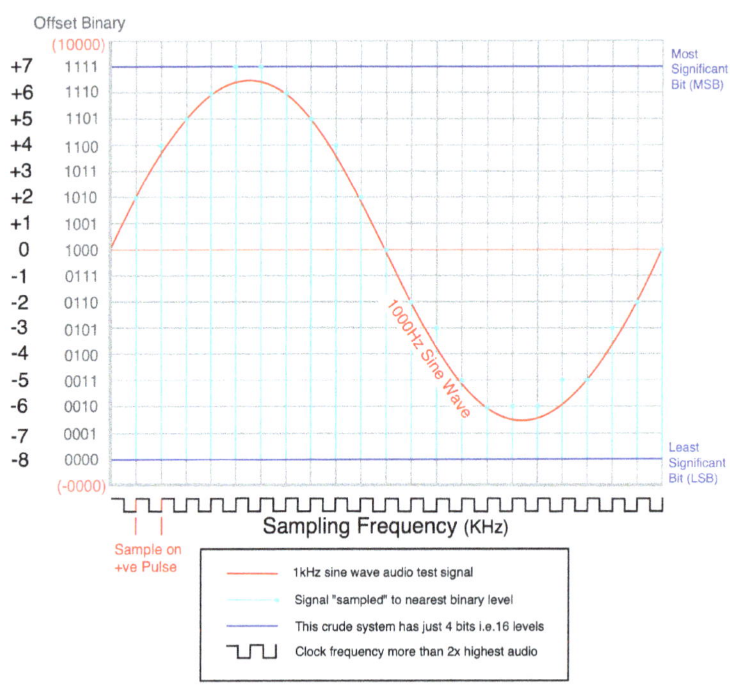

Figure 9.2 A crude "4-bit" ADC captures a sine wave

Digital audio at its simplest requires 1) the gathering of levels of sound samples and 2) the capturing of enough samples to recreate the sound to the highest sound frequency required. The first is signified by the number of measurement levels, the more the better! In Figure 9.2, our crude "4-bit" system (4 binary digits = 16 levels of sampling seen on the X axis) contains a pure sine wave. If you look at the light blue dots close to the intersect with this red audio signal, they are collected "samples" you can easily see the difference between the dots and the original. The difference between the sample taken and the true analogue level is the quantisation error.

This system would sound terrible, but at least it would sound. Whilst you can get away with "8-bit" (256 levels) for communications, high quality audio (like CDs) uses a 16-bit system (65,536 levels) which gives a decent quality. Studio mastering is mostly to a higher number of levels "24-bit" which gives a discrete 16,777,216 levels of sound level analysis. The overriding advantage of adding the extra bits is a larger system dynamic range which when correctly utilised pushes the digital noise floor to a usefully lower level.

All analogue systems add a constant low level hissy noise which can become annoying if low levels of wanted sound come too near it. Digital systems also have an unwanted residual which is called "quantising noise" and its approximate level is easily calculated by multiplying the number of bits by 6 dB. That makes 16-bit CD have a noise floor of 16 × 6 dB or measured at 96 dB... Not bad! That's like the best analogue electronics, however because quantising noise is annoyingly buzzy compared with analogue hiss, we say it sounds twice as poor (6 dB worse) and so I would describe the system as 90 dB s/n (signal to noise ratio) in comparison to its analogue equivalent. However digital designers do have other tricks, and by using "dither" and "noise shaping", the impact of quantising noise on the listener can be reduced. Just to note, taking Figure 9.2, if a loud signal is input above the 4-bit 1111, the next number is 00001. The system does not understand the fifth digit. Is it (0000)1 or 0(0001)? Original systems would squeak or thump, many now just mute to silence; however, regardless, the sound you wanted is lost if you exceed the most significant bit (MSB)!

The moral here is to take care to keep a decent headroom between your chosen maximum recording level and the MSB (the maximum level) of a digital system.

Significant under-recording (e.g. not using the dynamic range available) not only takes you nearer the quantising noise floor, but less of the available levels are being utilised in the restriction of "dynamic range", so the quantising levels of your sample will be coarser, and thus the quality lower. What if we could look at the whole problem a different way? At the end of the chapter, I introduce 32-bit Float recording, which, when manufacturing costs allow, is the future!

HEADROOM

Headroom? From experience I would say at least 8 dB for a known tested situation like a studio and at least 12 dB for a location recording where the unexpected could happen. It's worth pointing out the obvious: each dB in level lower on the recording brings the wanted sound closer to the quantising noise floor, so please don't go overboard, or choose 24-bit when you can for mastering, to get those decibels of dynamic range back!

As explained earlier, the sampling frequency sets the maximum sound frequency that can be captured. The CD system uses an awkward "sampling frequency" of 44.1 kHz, which of course does meet the criteria of being over twice the highest human ear frequency of approximately 20 kHz. Whilst many "prosumers" and home users continue to adopt this original standard, broadcasters tend to use a much more rounded 48 kHz, and high-end recording studios may well use 96 kHz. Great when dogs and bats are your customers!

Seriously, there is an important point to make here: the industry gets material from many sources, so it pays to standardise. Yet, in an imperfect world often a piece needs "resampling" to a different standard to match other digital material. If a 32 kHz signal needed to be resampled to 48 kHz, there is a common factor of 16: $2 \times 16 = 32$ and $3 \times 16 = 48$. Therefore, to go from one to the other is as simple as multiplying by three and dividing by two. From 96 kHz to 48 kHz is a simple division of two...

So, what's my point? A common need is to go from 44.1 kHz domestic to 48 kHz professional, and so what is the common factor there? Uh-oh... We have a problem. The lowest common frequency is 100, so $441 \times 480 = 211{,}680$ is a whopping multiple! Making this technical leap is complex using high frequency clock rates; it is done, but in my experience, it can result in aliasing or other artefacts that can affect the audio quality, particularly in the high-frequency range.

A professional sound engineer finds out the file format, sampling frequency and bit depth *required by the customer before making the recording* and sets the recorder to meet it.

DIGITAL TO ANALOGUE CONVERSION

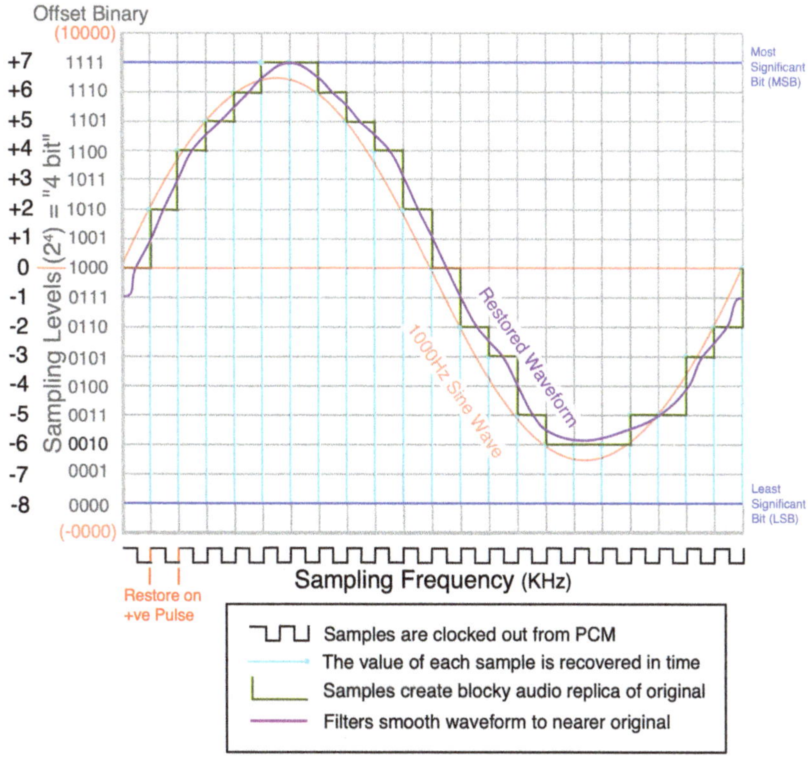

Figure 9.3 A crude "4-bit" DAC attempts to reconstruct a sine wave

Once a signal is in a digital state it is very robust; techniques exist such as checksums to spot and correct potential errors, and the digital waveform can deteriorate a long way and still be recovered with specialist electronics. This is why CDs are so reliable, unless, of course, dirt like fingerprints obscures the surface! To get our audio back to an AC signal requires a digital to analogue converter (DAC) in the receiving equipment, such as a CD player.

The samples are clocked out at the same frequency as the encoder (Figure 9.3) and the step values of each ADC sample are returned to a voltage level. This resembles a staircase, and so a low pass filter is used to smooth the waveform to the restored wave you see above on our crude fictional 4-bit system. The more the bits and the higher the sampling frequency, the smoother and more accurate the waveform, and the closer it comes to the original audio signal. A quality 16-bit system (65,536 voltage levels) is enough to create a seamless smooth replica of the original audio signal, even to critical ears.

I've touched on sampling rates above 44.1 kHz and 48 kHz. The professional sound industry often uses twice the lower sampling rates, namely 88.2 kHz and 96 kHz, which can yield a flat frequency response up to 45 kHz. Clearly the storage of the files doubles in size, and it is difficult to imagine why frequencies above 20 kHz are necessary; however, in these days of plentiful cheap digital storage, I guess we could say, "Why not!" Well... One answer is that in multichannel recorders, the number of simultaneous channels that can be recorded is halved, from 64 to 32 channels, but I would say that is an exceptional problem.

BRICK WALL FILTERING

A problem in early designs was in the design of the high-pass filters. In Chapter 3, "Introduction to Electronics", we learnt that traditional filtering has a turnover loss of just 3 dB at the chosen frequency. It then has a slope of just 6 dB/octave. There are 2nd and 3rd order filters, of course, but even the latter would only be 21 dB down at 40 kHz (already 3 dB down at 20 kHz damaging our upper frequency response).

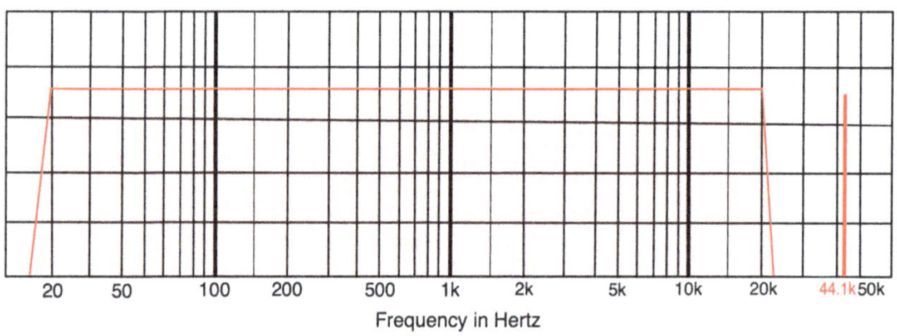

Figure 9.4 "Brick Wall" filtering at 20 kHz on CD frequency spectrum

Early designers required an audio response that is flat to our precious 20 kHz and completely gone by 22 kHz to meet the demanding Nyquist criteria. Special filters were constructed, called "Brick Wall" (Figure 9.4) and of course the CD players functioned, so all was well! Not quite! Some believed they could hear artefacts such as distortion or ringing effects because of this brutal audio "amputation" of the signal, and the debate over brick wall filtering continued as another supposed nail in the coffin of digital throughout the 1980s. I mention this as more recent techniques, such as oversampling converters and FIR (finite impulse response) filtering, allow more gentle slopes retaining excellent anti-aliasing properties, which can be shown to negate any difficulties with modern digital designs. There is more reading on this subject if you look up Audio FIR Filtering on the internet.

DIGITAL COMPRESSION

The digital system we have looked at to date (PCM) has been "linear"; we use our storage to record by the second, and the silence between the soundtracks uses as much data storage space as a full orchestra! Remember how much hard disc space audio tracks took up on early restricted storage? I'm sure you can understand a historic wish to compact the data to allow greater lengths of recording back in the days when storage was costly and the capacities severely limited when compared to today. Ideas were discussed such as not using data when silence occurs, assumptions on what can be discarded (quiet sounds obscured under loud), and research into what amount of distortion "average" people will tolerate (the science of psychoacoustics). These purely mathematical changes are either applied at the point of capture or can be later applied to a recorded linear PCM signal (such as a CD) to squash the amount of data used to a tenth or more, so ten songs are stored where you previously only had space to store one! These coder-decoder pairs or "codecs" which transform the signal to take less space take place in software now, and so are just a function of a computer programme within a hardware device such as a computer, tablet or phone. This form of compression is known as "lossy" as once the signal is squashed and had the corners knocked off it, the damage cannot be undone. The leading creator of "lossy" sound is the Motion Picture Experts Group (MPEG), responsible for producing MPEG-1 (later 2), Audio Layers 2 and 3. These are respectfully known by the suffix, .MP2 and .MP3.

Can you hear the difference against linear PCM? We do have choices in managing the amount of compression, measured as "bit rates" to create a slide of values between high quality (high storage space) and low storage space (low quality). MP2 is used widely in DAB radio, and originally, it did not stack up against FM broadcasts at the bit rates used in Britain. Some Europeans used higher rates, particularly for classical music. DAB+ improves on earlier codecs, but for me the jury is still out – FM rules! The ubiquitous MP3 has fared far better over the decades and still has its feet firmly under the table. The most common middle road solution is 128 kbit/s, giving a listenable stereo sound with an 11:1 data compression, fine for, say, dance music. 192 kbit/s gives a good sound in all but the most demanding situations at 7:1 compression, and the highest 320 kbit/s should not be distinguishable to average consumers from CD, yet still allows a compression of data of 4.4:1. I'm aware some professionals may disagree with that proposition. What to listen out for in a naff "lossy" codec is personal, but I'd say primarily a loss of excitement in the music, which sonically can be a loss of depth and height information giving a flat stereo image, a muffling veil is drawn over the sound, and it can be muddy and splashy in places. I personally store at 256 kbit/s, which gives a decent data compression ratio of 5.5:1 from a CD source or similar, and with older ears I don't hear additional artefacts when listening to background music at this setting.

The original linear PCM almost always has the suffix .WAV and occasionally .AIFF. To recap, the popular "lossy" codec is the original and universal MPEG 1 Layer 3 which has the suffix .MP3; there is a Windows version called .WMA which it's said will never play on Apple Mac products, and an Apple Mac version called .AAC which (it's said) will never play on Windows computers. One to watch out for is Ogg Vorbis which outperforms the others with ease, in my opinion, but (as I write) sadly is not so widely supported. Whilst you may be able to play it, it is likely when you go around to your friends, they won't! MP3 at a decent bit rate can be successfully exchanged with anyone.

There is a middle road. The data can be squashed down from PCM, yet the toolkit supplied in the codec is designed to recreate the original pristine sound upon decoding. This sort of codec is called "Lossless" as the data is recreated back to the PCM coded original. Codecs that can achieve this are "Windows Media Lossless", "Apple Lossless" and one I like that is unlicensed, and so open to all, is called "FLAC" (Free Lossless Audio Codec). With the current cheap price of audio storage, unless I wanted an elite home music server, I'd stay away from "lossless" and work and send finished files in PCM (.WAV) and use "lossy" 256 kbit/s (.MP3) for sending demos by e-mail. The reason for MP3 demos is that the e-mail attachment maximum file size at the time of writing means you can only send half an average song in linear PCM, but several tracks in MP3. Dropbox or similar file sharing overcomes the hurdle of sending linear tracks to a would-be publisher, but you run the risk they won't bother to download.

People get very hung up on the quality of digital systems, arguing subtleties into the early hours. I personally think this is snobbish and unhelpful, but it is healthier than arriving at a point where quality doesn't matter at all, where the aural expression of the height of fine cuisine is limited to beans and chips. So much emotion and expression are lost that way; life is flatter, less, and if you don't care, time to move on to another book...

If you wish to train your ears to hear digital codec imperfections, try the acapella by Suzanne Vega from the album Tom's Diner where often a poorer DAC can be heard to hunt rather than capture, or try a decent CD (piano only version) of Trois Gymnopedies composed by Erik Satie. I've not heard a system yet that can cope with this subtle piano work without a slight twanging distortion of the higher chords and a sort of overlaid firework "fizz" on notes in places. Hear this and you will in time train your ears to recognise the same problems on more subtle digital material. Good hunting!

LEVELS IN A DIGITAL WORLD

In Chapter 6, "Sound Levels and Connections", we learnt that professional equipment is lined up on a suitable meter (some sort of PPM, VU meter or bar meter that draws from one, the other or both worlds). The European broadcast line-up level in the analogue world is 0 dBu which equates to 0.775 V and the US is 1.228 V (+4 dBu). The maximum peak recording level in both is a much higher voltage, typically allowing a good margin above the maximum meter reading to save the engineer's blushes if there is a sudden loud blast like a hearty laugh. The BBC standard (used here as it has many countries that support its parameters) allows an extra 8 dB, (maximum peak/ nominal level of PPM 6) regardless of the equipment used, which creates an approximate AC output voltage of 2 V.

A reminder from the earlier chapter that the true maximum level should be much higher through most equipment as it is set by the power supply rails. The larger the power supply voltage to the audio equipment, the bigger the margin for a decent "headroom". If the audio voltage swing exceeds the power supply rail voltage, the audio signal is clipped off and the sound is permanently distorted.

In the case of a properly designed digital system the supply rail voltage plays no part in the dynamic range. The number of bits in a system provides the dynamic range. To take an example, a CD uses an uncompressed 16 bits which, at (approximately) 6 dB a bit, gives (6 × 16) 96 dB between the least significant bit (LSB) and the MSB. Given that around the LSB is where we will also find the abrasive quantising noise and the MSB is where the system gives up and fails, we chose to stay within these binary number bounds. In a CD, this suits as a noise floor of –90 dB is outstanding and 0 dB at 90 dB (whatever we chose that to be) is suitable for reproduction of loud music. The BBC got away with just 10 bits for sound distribution in the seventies (60 dB dynamic range) but made it appear better by using compandors, (much like the Dolby A™ system explained in Chapter 7, "Analogue Sound Storage"). The electronic component technology to do better simply was not there! Another way to make a system appear better than it is would be to treat it in the background before sending the signal on. An early one was BBC NICAM.

NICAM stands for Near Instantaneous Coding and Multiplex. The system again transmits only 10 bits; however, it achieved (then state-of-the-art) 14-bit encoding and decoding by an ingenious twist. The "near instantaneous" bit is that the codec (ADC) looks at the incoming signal and selects from a scale of "loud" to "quiet" after analysing the signal. In quiet passages the 10 least significant bits are transmitted (including LSB); in loud passages the four least significant bits are ignored, and the loudest 10 bits are transmitted including the MSB, with the not unreasonable expectation that the loud passage will mask the large jump up in the quantising noise. A simple digital marker is transmitted with the digital audio to tell the decoder (DAC) at what level to set the decoded audio. Ingenious!

By the mid 1980s cutting-edge electronics almost allowed for "16-bit" linear systems (some early CD players were shipped with 14-bit DAC chips). But of course, technology has moved on, and reliable 16-bit DACs are easy to manufacture, particularly when using oversampling converters (Delta Sigma technology). The speed of current digital electronics makes even "24-bit" systems normal today. To equate back, that allows a theoretical audio dynamic range in the digital domain of around 144 dB, which is wildly past our needs as human beings! I need to temper that statement as the reality is the replay device system thermal noise, even on low impedance designs, will mask the least significant bits like a warm blanket of fog, making our 24-bit converter more like 20-bit on a good day! An everyday practical limit is around 18 bit, a 12 dB better noise floor than CD; audibly, it sounds better as thermal noise (white noise) is kinder on the ears than the prickly quantising noise. This has proved good enough and, like Dolby A in the old days of tape, suits the requirements of the multichannel mastering process of many channels, each playing its part in the mix, each needing manipulation, which in turn uses additional levels of our digital range before they come together as a single defined output, whether it be stereo or multichannel.

REFERENCE LEVELS FOR DIGITAL AUDIO

In the early days of "digital", when we struggled to create sufficient dynamic range in the digital domain to match analogue audio electronics, choosing a professional reference level was not difficult. We would match the clipping level typically found in our audio systems to the digital MSB. A modest headroom above our metered maximum signal allowed for most slip-ups to be contained, although each decibel more in headroom compromised the noise figure by 1 dB. With a nod to existing analogue metering systems, digital systems were mostly set up to have similar properties, so line-up tone at 0 dBm became a level of dBm0 (dBm zero) in the digital domain and in the decoder dBm0 at the output was adjusted to read 0 dBm (test tone level) on the analogue meters. As bit rates have increased, the opportunity to have either very low "silent" amounts of quantising noise in the background, or alternatively high digital levels which would transcend the ability of even the most ambitious analogue equipment to reproduce has arisen. How do you standardise an abstract concept? It's like measuring an aircraft's progress in acres...

Common sense has prevailed, and the firm link between analogue audio reference levels and the peak levels managed in the digital domain is severed. The digital scale now relates to the maximum signal level possible. The early equipment may have referenced that level to a kinder nominal level chosen below MSB to provide some headroom, but in the modern world, expect the metered maximum level and MSB to be one and the same. This measure is dBFS (dB Full Scale), and its reference to 0 dBu (where there is one) varies with different organisations and equipment. You can easily construe that any figure below the MSB (0 dBFS) is bound to be negative, stepping away from the brink of the system. Sadly, it is even more complex because sine waves (like line-up tone) measure lower in the digital domain than complex audio waves which can approach a saturated square wave. There are also differences in codec "law" (the difference between RMS and peak if you will, that make differences of 3 dB or so). The content presented here is to start a conversation with the studio(s) you work with to find which standards are used locally, what "line up" means and how the finished recordings are presented to external agents, so they may get the best out of them. There is more on this topic in Chapter 6, "Sound Levels and Connections".

IS DIGITAL AUDIO PROBLEM FREE?

It does pretty well! All signals are degraded over long lines; the cable capacitance and, to a lesser extent, inductive elements react with the cable resistance to round off the original sharp square signal. Further, radio frequency interference (RFI) can be induced as an overlaying signal which mixes with and modifies the waveform shape. (Note balanced lines work well to suppress problems here as they do for audio lines!) However, unlike audio, the digital signal can be passed through a comparator and then to a pulse generator to recreate a neat sharp waveform almost indistinguishable from the original (Figure 9.5). However, as in all things this is not completely infallible.

Figure 9.5 Recovery of digital degradation

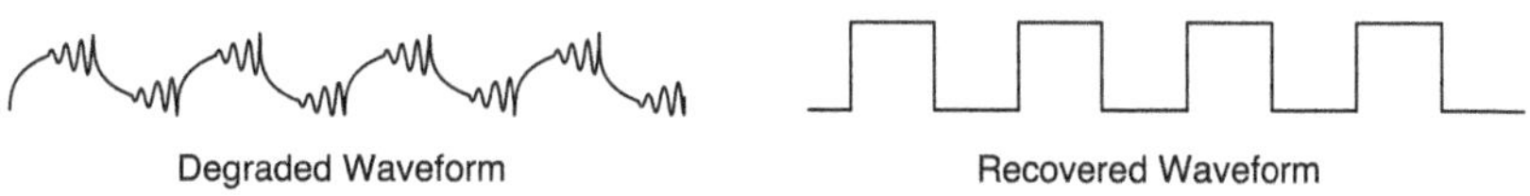

Digital sound systems make use of the defining fact that they live in a world of maths, rather than the physical world and its limitations. Since the earliest systems, simple checksums monitored the passing data and spotted a bad apple from a good one. If an error was found, the system would repeat the last good data until things came right and hoped you didn't notice. If "good" data failed to materialise, in time the system would simply mute.

As far back as the introduction of CD, *parity* was a simple and effective way of error concealment, adding a single bit to the digital stream which simply noted even and odd numbering. This strategy quickly revealed samples which didn't ring true, so they could be suppressed. A next step, called *scatter codes* corrected the error (remember this is maths!) rather than repeating earlier samples to cover the void. In this, the original digital code is mixed with a known secondary code, the output of the two is scattered across the medium. If a sample is found to be corrupted, the system can search for valid data elsewhere on the medium and insert a corrected version. CD itself uses Cross Interleaved Reed-Solomon Code (CIRC). The science of digital manipulation is constantly evolving and is simply beyond the everyday needs of the sound engineer and this book. However, if your fancy is suitably tickled, the internet will fill your boots on this subject! I suggest a decent supply of aspirin (other painkillers are available) before you start.

Digital is generally robust, but like early four-wheel car drivers, whilst the limit of failure is much higher, when it comes it is profound and absolute! The need for headroom against audio peaks is not advisory, it is obligatory! Also digitally encoded audio cannot be heard; failure can happen and be completely invisible to the user. The sound practitioner must make "confidence recordings" to be sure it will be alright on the night!

CD, MINIDISC AND DAT

The origins of our digital standards are rooted in our recording hardware. I could write a book on this history alone; however, I will leave that to others and jump straight to those that achieved commercial successes. The earliest digital recorders used a "PCM Adaptor" (a bespoke ADC/DAC interface) that sat alongside a traditional "video recorder" (most often "U-Matic" but including the domestic Betamax and VHS) as the recording medium, and the distinctly odd choice of a 44.1 kHz sampling standard was chosen because it was the best technical match to writing the digital audio data into the existing TV line and frame periods required by the analogue video recorder.

Such hybrids quickly became popular in recording studios because of their cheapness and quality against the best analogue tape recorders of the day, and so when it was decided to make digital quality available in the home with the 1984 Sony/Philips CD (Compact Disc), the 44.1 kHz sampling rate was used to retain the digital mastering process from the initial recording through into the home. I was invited to the UK launch of the CD in London by then Radio 1 Breakfast DJ, Noel Edmonds. Sadly, neither the 1st generation Sony nor Philips machines (they jointly developed the system) could be made to function on the day.

The early digital recorders (PCM adaptor plus video cassette machine) outshone the traditional analogue reel-to-reel machines with a much better signal to noise ratio, meaning less hiss, no muddying of the midrange through incorrect bias adjustment and a guaranteed clean extended top range as there was no concern over tape head problems such as azimuth adjustment or ingrained dirt. If this is too much detail, it's easier to say tape machines generally played back their own material well, but quality often suffered when played back elsewhere on another machine (see Chapter 7, "Analogue Sound Storage"). Digital in general terms either worked "perfectly" or not at all, and so offered great gains in comparison with the outgoing analogue tape systems, particularly when the studio recording needed mass copying to move from the studio to distribution warehouses.

It was little time before manufacturers started placing the "PCM adaptor" in a box with the tape recorder, making an integrated digital sound tape recorder (hard disc or flash storage capacities were still way too small at this point). Sony were leading the way in small format video cameras with helical scan (rotating) heads at this time and so used this technology to create the R-DAT (Rotary Digital Audio Tape) system. When the rival "static" head system (S-DAT) failed, they simplified the name to "DAT". The system could record using sampling rates of 32 kHz, 44.1 kHz and 48 kHz, and eye-wateringly expensive professional machines carried all three. The cheaper "domestic" machines* tended to miss out 44.1 kHz capability to ensure the public didn't have an easy route to rip off CDs at source quality, yet otherwise they sounded as good as the expensive ones.

Professionals including studios and broadcasters were happy to buy the cheaper DAT machines and simply did without the 44.1 kHz rate. This led to industry norms of 32 kHz and 48 kHz, with the bonus that it was cheap to convert between them, and when technology allowed, DAT (along with its competitors) could manage the sampling rate of 96 kHz. These sampling rates have survived decades in use and are still industry standards. The vintage standards of CD and DAT are uncompressed PCM formats, whilst minidisc was the first domestic "lossy" recorder which meant disc size could be dramatically reduced for the same recording time.

JOINING DIGITAL EQUIPMENT TOGETHER

If we want to connect domestic analogue audio equipment, we may well use phono connectors or jacks, and with professional equipment balanced jacks and XLR connectors. What if we want to connect professional digital equipment? Like CD, the standard interconnects are always "dual channel PCM" (normally used as stereo). In a domestic setting the standard is the Sony/Philips Digital Interface (S/PDIF) and can use either an electrical connection using a coaxial connector (an RCA/phono plug) or an optical one called TOSLINK invented by Toshiba, using a red laser to pulse the audio data. Both can work up to cable lengths of 10 m; the optical cable is likely to die if kinked, and the wired one is prone to radio frequency interference (RFI) if too cheap or passed near computer cables. A poor connection in either can kill an otherwise viable signal. Dirty electrical connectors can rectify and add noise to the signal, and a poor or soiled fibre endpoint can create reflections back down the fibre which beat with the existing signal and corrupt it past recovery.

The domestic S/PDIF standard can also carry compressed 5.1 or 7.1 surround sound for cinema systems and carries the SCMS copy protection system developed for DAT.

The professional standard is the precursor to the domestic one and is very similar to S/PDIF, both utilising bi-phase mark code. It is called AES3 (or AES/EBU) and is also available in two flavours. One suits permanent installations where a cable run can be up to a kilometre! This uses professional standard 75 Ω coaxial cable terminated by BNC coaxial plugs. These cables are identical to ones carrying video signals in broadcast venues. The other, called "AES3 balanced", uses XLR sound cables in an identical way to analogue, with cable runs up to 100 metres. Outside of these standards sit manufacturer-specific

* DAT never became popular with the home enthusiast as they had great sounding "special formulation" cheap "Compact Cassettes" by this time and the emergence of the new small lossy "minidisc" digital players.

solutions for the carriage of multiple channels, like the links between a digital recording studio mixer and a multitrack digital recorder. A digital link (particularly multitrack) is a much higher frequency than audio, and so resistance and capacitive effects of cables limit the length of transmission.

MADI INTERFACE

Pioneers saw digital audio as natively two channels, most often used as a stereo pair. Three points on that: (1) patching multiple source paths to differing destinations on a switcher would not suit AES3, (2) getting multiple stage microphones to the sound desk would be as cumbersome with AES3 as would a traditional thick audio "snake" and (3) the rise of multichannel audio listening experiences, such as Dolby Atmos, Digital Theatre Systems (DTS) X, Sony 360RA or MPG-H, are not supported in AES3.

A new product was required and has continued as an "open standard" development since 1991 managed by the American-based Audio Engineering Society (AES) backed by manufacturers AMS Neve, Sony, Mitsubishi and Britain's own Solid State Logic. It is called AES10 and is also known as "MADI". This "Multi-Channel Audio Digital Interface" has features in common with AES3. It is scalable in levels up to 24 bits, to sampling rates of 48 kHz (96 kHz with reduced channels) and up to 64 channels at the time of writing. Like AES3 it can be carried over coaxial cable or fibre-optic links. The system is inherently robust, "sample accurate" and can run over very long distances. It is not surprising to see this powerful tool grow rapidly as the years pass. Further design improvements have yielded unit boot-up and sync times of just seconds, it has low latency, the technology is fully scalable and faults are easily traceable with simple digital tools. The communications tool Musical Instrument Digital Interface (MIDI) is easily carried over MADI allowing automation and organisation both in venues and between venues.

INTERCONNECTS

Like microphones, high frequency connections obtain the best performance when impedance matched. This includes the cable, which exhibits a "characteristic impedance" of 75 Ω (Figure 9.6). The higher the frequency, the more degradation is likely through resistance/capacitance issues between the core and the screen of the cable. This combination forms a classic low pass filter (see Chapter 3, "Introduction to Electronics") limiting the bandwidth of the link and so the number of possible concurrent digital streams. Whilst inductance has much less of a role, it is still a mild concern particularly where cables are coiled. So why have I shown inductors at the cable ends? Like standing waves at a wall (Chapter 8, "Introduction to Acoustics"), if the patching is imperfect at the cable ends (length, connector quality, impedance matching) the electromotive force rises, and part of the signal is bounced back from the ends.

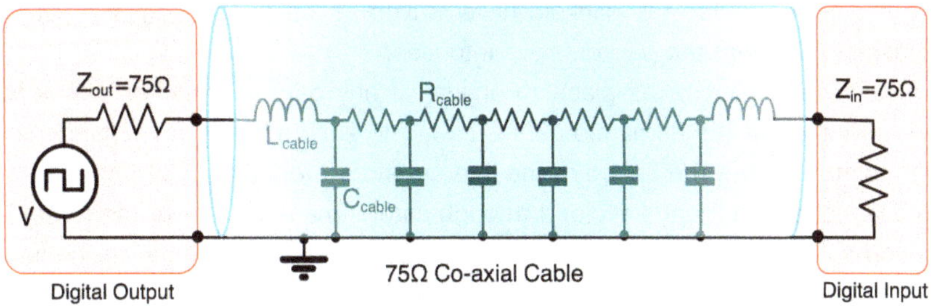

Figure 9.6 Digital audio will suffer on poor or lengthy coax leads

These reflections interfere with the valid digital signal, impairing the link. As we stretch the bounds of what is possible in the digital domain, the purity of the interconnects becomes more critical, and we must factor quality into our budget. An optical connection is not immune to this. If poorly terminated, or if the connection is frayed or kinked, reflections from the damage to the light flow interfere with the digital signal and compromise it.

To come full circle, we initially went down the route of digital to stop deterioration of audio signals by external factors such as RFI, cable losses (loss of upper frequencies in particular) and distortion and noise from line amplifiers. MADI achieves a smart cost-effective solution to all these problems for professional audio, and in doing so, uses a cost-effective single connection in place of a bunched bundle of audio cables. Distances of 40 km have been successfully transmitted over single mode optical fibre.

NETWORKING DIGITAL AUDIO

We have established the close link between the ascension of personal computing and the birth of digital audio, both using the same binary maths system and many of the same computer chips. To date we have looked at moving audio between devices using digital interconnects, but I note these are simple transport systems. What if we want to utilise digital audio within a computer and then communicate with other computers using the worldwide network of the "internet"? Computers use a universal set of rules to communicate with each other; they are called Internet Protocols (IPs). These rules ensure computer data can be sent, received and understood correctly between differing systems. Each protocol serves a specific role, whether it be sending, transferring or simply security. To be in the game, each computer device needs a unique address, the "IP address", which is both a source and destination for the "data" we want to send, which may pass through many devices like routers, switches and websites with their own IP addressing in a journey from one computer to another.

The data, in our case, a digital audio file in a valid format, is divided into packets, which are sent from our source IP address to a destination IP address; through many devices on the internet, which exist to help speed its way to the correct home. Once the packets have arrived at the destination, the data is reassembled from the packets back into files.

Housekeeping takes place to ensure iif any data is missing, a protocol will require that the damaged data is resent to ensure the file is accurate and complete. Further the speed of the internet at different places will need to be managed, and collisions avoided through careful management of the route. In a home or business our internet-enabled equipment is connected to the local area network (LAN) sources by an RJ45 (a bit like an old landline telephone socket), and of course we are all familiar with the airborne networking version, WiFi. The connection from our business or home to the world is through a modem connected by a local telecoms provider to the internet, normally today by a fibre link, although in some areas the internet data may travel at slow speed on the old telephone copper pair close to home before switching to fibre at a local box or in the exchange.

Several internet protocols (IPs) exist to support the sending of professional sound over the internet at high quality, in real time and with low latency. Internet-borne audio has the advantages of digital transmission, and lends itself to multiple channels, sent over long distances to multiple destinations. There are several products specialising in providing live audio over Ethernet links, including Cobranet, AVB and Ethersound. Here are two short examples...

INTERNET DELIVERED AUDIO – CLEANFEED

Others exist, but I'd like to mention "Cleanfeed" as a commercial product developed by ex-BBC Engineers, which joins together a small number of contributors to create a high-quality live broadcast mix, which can be passed on to "air" or recorded locally on a tasked computer. It replaces the Integrated Services Digital Network (ISDN), which used two phone lines to create the bandwidth for a fixed, reasonable quality voice digital contribution. The contributions can be mono or stereo, optimised for speech or music. The system provides a return cue path for communications.

INTERNET DELIVERED AUDIO – DANTE

One of the best Ethernet enabled audio products is Dante (Digital Audio Network Over Ethernet). It was developed by a government-sponsored Australian company called Audinate and is a combination of software, hardware and network protocols. Dante has a great penetration in professional audio (600 companies and over 4,000 products); it offers native gigabit support, higher channel count, lower latency and automatic configuration. At the time of writing it is state-of-the-art for the movement of complex professional sound.

BLUETOOTH WIRELESS CONNECTION

Bluetooth was first designed in 1989 as a short distance wireless technology to exchange data between a fixed and mobile device, and it took a further 10 years to develop into a product range. A Swedish concept, the technology is named after Harald Bluetooth, who famously united the Danes behind him. The system is regulated by the Bluetooth "SIG" (Special Interest Group), which has around 40,000 members. Working (like WiFi) in the less regulated 2.4 gigahertz (GHz) band, there are two flavours: the original (Classic) and the later low power "battery" version "Bluetooth LE" (Low Energy). The connection distance inside is normally limited to around 10 m; however, a range of over 1 km is feasible outside with line-of-sight specialist providers. A system (master) device can communicate with up to seven other devices; one is always master and the other(s) slave in a Bluetooth connection (a piconet). However, it can be confusing as the master device can simultaneously be a slave in a different piconet.

The early iterations of Bluetooth had an audio latency that was clearly audible. This has steadily improved, and in Bluetooth 5 is good enough for people to play along in sync to a track; or better, to use Bluetooth headphones whilst playing electronic drums! (Better for the neighbours!)

The system has advanced since its conception, and its functionality has widened far further than its inventors could have predicted. To use the additional tools in its box, each must conform to a "Bluetooth Profile", and these profiles are designed (1) to work seamlessly with the other profiles, (2) to use suggested interface formats and (3) to manage options and parameters in the Bluetooth stack. It will come as no surprise that one of the earliest was the Headset Profile (HSP), which is still the most common in use on mobile phones and gaming consoles and defines both the simple audio encoder and controls to ring, call, answer, hang up and adjust volume at a headset. As the system grew it could incorporate high-quality audio streaming in real time and equipment remote control. The ones of most interest to the audio tech are the Advanced Audio Distribution Profile (A2DP), the Video Distribution Profile (VDP) and the Audio/Video Remote Control Profile (AVRCP).

Advanced Audio Distribution Profile (A2DP)

This profile sets out how Bluetooth can be used to carry audio from one device to another; its nickname is "Bluetooth audio streaming". It paints with a broad brush, covering music streamed by a mobile phone or laptop to a headset, car or hearing aid. Voice can be streamed from a microphone to a recorder on a PC. This profile works closely with AVRCP for remote control of the devices taking sound, such as Bluetooth speakers, car audio or, of course, headphones. The system may drop back to HSP or "Hands-Free" Profile (HFP) to take a phone call.

The audio is typically streamed as two channels (stereo obviously) and can go in both directions. It must have certain other profiles on board to function and at least must support the low-quality SBC codec. Others are optional, but increasingly standard are MP2, MP3, AAC, HE-AAC, ATRAC and aptX. Rights management can force a lower quality in some headphones, whilst some sellers disable A2DP completely to avoid connection problems. It is worth looking at the small print and some of the better reviews before making your next purchase! This profile is under upgrade at the time of publication for further functionality.

Video Distribution Profile (VDP)

Here for completeness, VDP simply allows the Bluetooth transport of a video stream. It could be a video server to a portable device or a camera straight to a TV. Again, the profile contains a baseline codec, H.263; however, optionally MPEG4 and higher versions of H.263 are supported.

Audio/Video Remote Control Profile (AVRCP)

This profile exists to bring together standard interface devices to control either one off, like a TV or HiFi, or control several devices at once working as a universal remote control. It is designed to work closely with both A2DP and VDP. As you can imagine, this profile needs constant updating to remain relevant to the latest products. The more obvious controls are the basic transport controls, but the profile allows for the destination to update its status. There is access to metadata such as track information and cover art, and some limited search functionality.

32-BIT FLOAT RECORDING

If you have followed the science to date this will surprise you! To recap, 16-bit audio has a sufficient 65,536 levels of audio, and 24-bit a whopping 16,777,216 levels of sound level analysis. These yield excellent signal to noise ratios of 96 dB and a theoretical 144 dB respectively. If used wisely, they are "good enough"; however, both under-recording and over-recording create unrecoverable fidelity loss. Any excursion over 0 dBFS is catastrophic!

Recently several recording solutions have entered the market which use "Floating Point" mathematics. Whilst the approach has varied, the results are similar and simply astounding! Modern Digital Audio Workstations (DAWs) have been quick to exploit the new 32-bit Float file format, and so making use of this product in post-production is a practical proposition.

One system uses dual ADCs, one which only records the low-level signals and the other, the high-level signals. Anything too quiet for the high-level ADC is captured by the low-level ADC, ensuring it is well above the quantising noise floor and is utilising a wide number of levels for high precision. Anything too loud (including the apocalyptic 0 dBFS) reaching the high-level

ADC is accurately captured by the low-level ADC. By rescaling the levels in the DAW, any levels at risk of degrading the signal are simply excluded, yielding a perfect sound.

The resultant dynamic range of 32-bit Float ADC is so extraordinary (over 1,500 dB) that there is no need to set levels at all! Whatever extreme of level you throw at the recorder it will easily cope. The dynamic range is essentially infinite! A comparison is shown in Figure 9.7. A remarkable aspect of this technology is that the file size of 32-bit Float is only 33% larger than a 24-bit file. I must finish by saying that it is only the recorder that can manage this range. Any hard-pressed equipment before it – microphone overload or sound mixer preamp clipping – will have their distortion faithfully recorded. The concepts of sound staging introduced in Chapter 6, "Sound Levels and Connections", hold valid in the more complex recording situations.

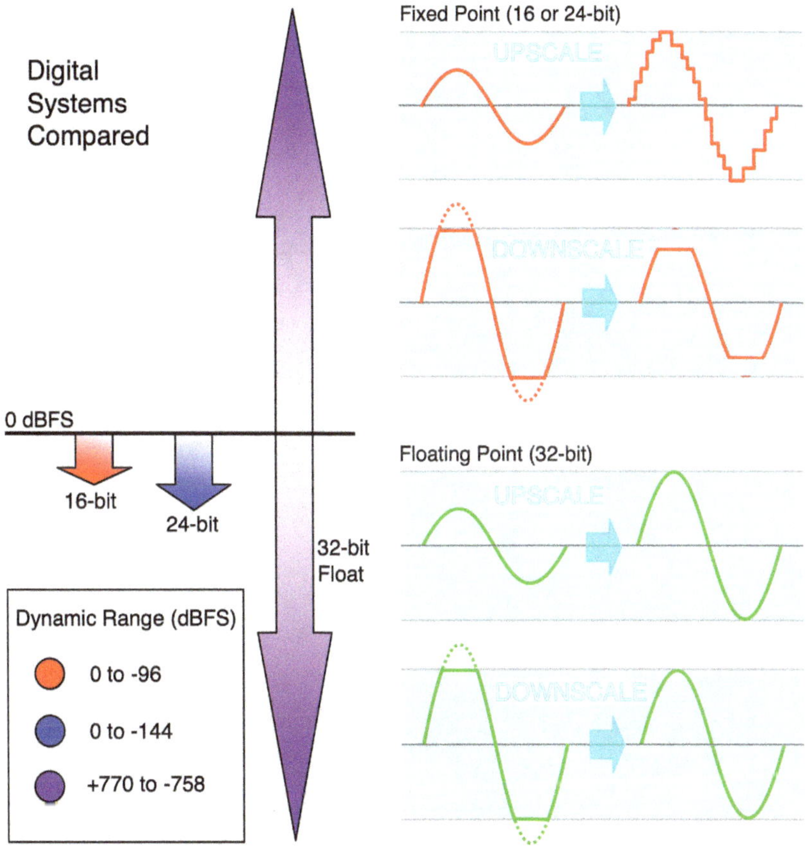

Figure 9.7 32-bit Float changes the rule book on digital recording

BIBLIOGRAPHY

Bluetooth Organisation (2025) *Bluetooth Network Encapsulation Protocol 1.0*. https://web.archive.org/web/20100225214851/http://www.bluetooth.com/Bluetooth/Technology/Works/Data_Transport_Architecture.htm.

Davis, Y.P. (1993) Digital Audio Compression. *Digital Technical Journal*, 5(2).

Hartley Jones, M. (1988) *A Practical Introduction to Electronic Circuits*, 2nd edition. Cambridge University Press.

Horowitz, P. and Hill, W. (1989) *The Art of Electronics*, 2nd edition. The Press Syndicate of the University of Cambridge.

Nyquist, H. (1928) *Certain Topics in Telegraph Transmission Theory*. http://www.ieee.org/publications_standards/publications/proceedings/nyquist.pdf.

Sound Devices (2024) *32-Bit Float Files Explained*. https://www.sounddevices.com/32-bit-float-files-explained.

10 Introduction to Sound Desks

MONO AND STEREO

To date we have assumed an audio signal is a single channel. It may surprise you to know multichannel sound experiments had already started in the late 1800s! In 1931 Alan Blumlein filed the famous British patent 394,325 and much of his work is still in use today. Multichannel sound provides width to a seamless soundscape and should also provide audible depth and height when a decent recorded sound is laid out in front of the listener. A system requires *at least* two discrete channels of sound from source to destination (normally two loudspeakers with a healthy distance between them) and the listener can then pinpoint sounds on the stage at any point between. This system was classified as "Stereophonic", shortened to stereo, and to distinguish the old single channel system, it was christened "Monophonic" or plain mono. We have lived in a stereo world for many decades, with practically all playback sources expected to provide at least two channels (left and right), and with cinema sound, multiple channels. However, a microphone is typically a single channel device, and so we need to factor this into the stereo world within the mixer.

MIXER INPUTS

We have talked of the rough levels we expect to appear at the input of a sound desk; it's time to start talking about what happens inside the desk and how we manage the levels with meters, "clipping" lights, speakers and headphones. We learnt earlier that a sound desk is likely to have limited input options. There are normally just two sound levels, microphone or line, and either a single channel input (mono) or a two-channel input (stereo). Most sound desks have mono channels which can be switched to be either microphone or line level, and stereo inputs which are normally line level only. The mono channels will often have different input connections for the low- and high-level sources; XLR is popular for the low-level microphones and stereo (balanced or TRS) jack for the line level sources, but there are no hard rules so expect to be surprised on occasion. There is even a single socket which can take both XLR and TRS connectors and so reduce the amount of panel space taken up.

The following description is for a typical analogue sound desk (as you can follow the logic), but it is not exhaustive; there are many differing solutions to the same problem. An obvious one is to take the audio signal into the digital domain at the point of entry to process the signal. This is a digital mixing desk, yet the logical points of process through the chain are not significantly different, even though the access is! Figure 10.1 is a schematic of a simple

DOI: 10.4324/9781003717065-10

mono input channel; it is a common design, but note they do vary. Let's take the microphone input first. Does the microphone need powering? Capacitor or condenser microphones do (not battery electret), dynamic and ribbon microphones do not. Phantom power can be switched on at each channel on better mixers and one switch common across all the microphone inputs on cheaper ones. The next stage is a "pad"; pads reduce the sound level (measured in dBs of loss) where "amplifiers" increase the sound level (measured in dBs of gain). The pad is switchable to allow for a large range of input levels from the incoming microphones, from say a rock singer to a flute!

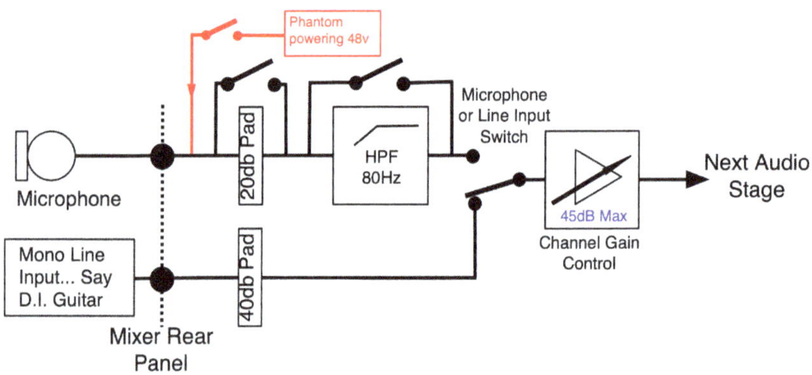

Figure 10.1 Input section of a sound mixer

A rock singer may well be ten times louder than a flute, so switching in the pad (by opening the bypass switch) will make the singer more equal in level for the next stage. Whether you do so or not is a matter of common sense, not a hard and fast rule. Note that many professional microphones have a "pad" switch on the body of the microphone which serves the identical purpose of preventing input overload (but in addition protects the internal amplifier if there is one). Don't use both; the one on the microphone is favourite.

For simplicity, at this point in the book, I have not shown a channel phase reverse button. If fitted, this button effectively swaps the two wires of the incoming signal from the microphone, in the words of our Introduction to Microphones, making this input "suck" when the others all "blow". It's highly useful if you encounter a rogue microphone, but has dire consequences for your mix if operated in error. Note a legitimate use below in creating a "Mid-Side" microphone array. The same often exists on stereo line inputs, as though it is rare, recordings can arrive in the studio with a phase reverse that needs correcting. Again, a button to stay clear of, unless you are sure you have this fault. See more on this in Chapter 6, "Sound Levels and Connections".

The next block in the chain is the high pass filter (HPF) which is normally set around 80 Hz. This exists to remove thumps and rumbles, particularly of microphone handling at low frequencies, and as speech has no content at this frequency, whether it is out or in should not be noticeable on the sound you *do* want but gets rid of spurious rubbish low down where you *don't*. Whilst typing this I'm listening to a light piano solo from the last century. The low frequency rumble from passing lorries has destroyed the work! Maybe the monitors/room combination in the location could not go low enough for the recording engineer to notice? Take headphones! The bypass position of the switch exists for instruments with significant content below 80 Hz like tympanies, double bass and grand piano to ensure their full sound is captured. If in doubt simply switch the filter in and out during rehearsal, if you can't hear any change in sound quality, you can leave the filter in.

The next switch chooses between the mono microphone input and (mono) line input, a line input being a mono keyboard or perhaps an amplified direct inject guitar. Now let's follow the much easier line level input. Manufacturers tell us "line level" varies between *approximately* 1.2 V nominally (professional equipment) and 300 mV (domestic equipment), a range of four times or nominally around 12 dB (see Chapter 6, "Sound Levels and Connections"). If we take the sound level down to around the same level as we are getting from the microphone input using a pad, we can adjust it further up the way at an appropriate point. This signal then approaches the microphone/line selector switch before meeting with the input pre-amplifier.

A lot of attention goes into the pre-amplifier stage as it is the most sensitive part of the chain. It needs to be smoothly adjustable (gain scale measured in dBs) to cope with the huge variety of input levels; it needs to have a lot of headroom to cope with unexpected peaks of sound, and at high gains it must be designed not to drag up the inherent noise of the mixer circuitry. You are unlikely to design it; yet you should appreciate it, as that's where the money is spent and what separates an average mixer from a great one.

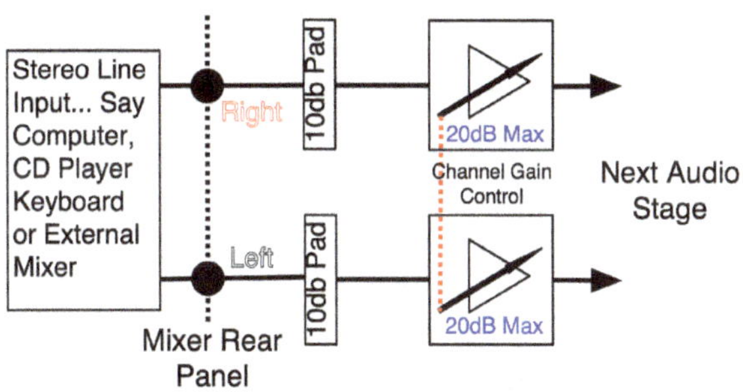

Figure 10.2 Stereo line input to mixer

Stereo inputs to the mixer look similar to a mono line input, and the two channels share one knob to adjust both channels at the same time. This is typically shown by a dotted line (in Figure 10.2 red) linking the controls.

EQUALISATION

The next stage in the mixer, whether it be mono or stereo, is equalisation (the stereo controls are linked like the gain control). There are controls to adjust the bottom frequencies known as "bass" or "low", to adjust medium frequencies known as "mid" or "presence" and the upper frequencies known as "top", "high" or "treble". The designations change with time and fashion. Middle controls can be broken out to choose the middle frequency of the boost from several presets on some desks, and often the "Q" or sharpness of the peak (or trough) is usefully adjustable. A normal range for a control is between 15 dB and 20 dB of "boost" or "cut", simply more or less than the unchanged "flat" position. A typical range of settings is shown in Figure 10.3.

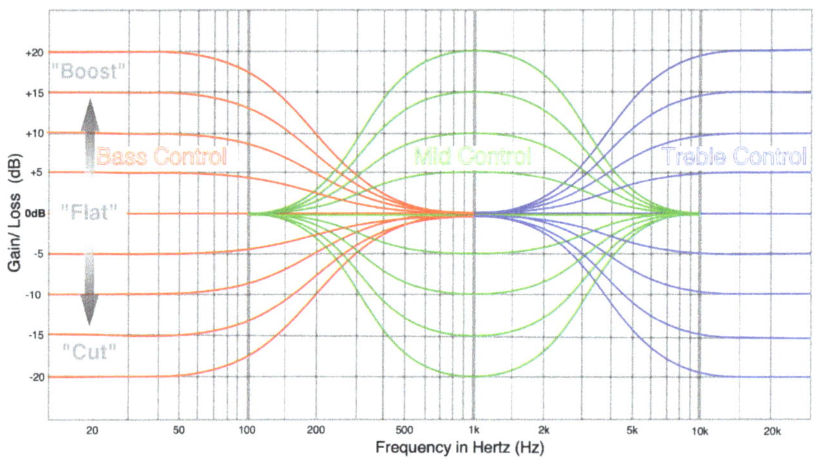

Figure 10.3 Typical mixer equalisation curves

When sound desks were first invented engineers did not give enough thought to the misuse of equalisation. The idea was that the control was to correct a problem or subtly enhance the sound, so the circuits were often at a nominal "line level" of around 1 V. People became creative and chose, on occasion, to lift an already loud sound for dramatic effect. 20 dB of lift on a 1 V signal is 10 V, which was enough to drive many systems above the headroom and into clipping. Now signals nominally enter equalisation lower, and the required makeup gain of up to 20 dB is applied later in the chain. I put this comment in so you can think about all parts of the sound stage, keeping the signal well out of the noise, but away from accidental clipping.

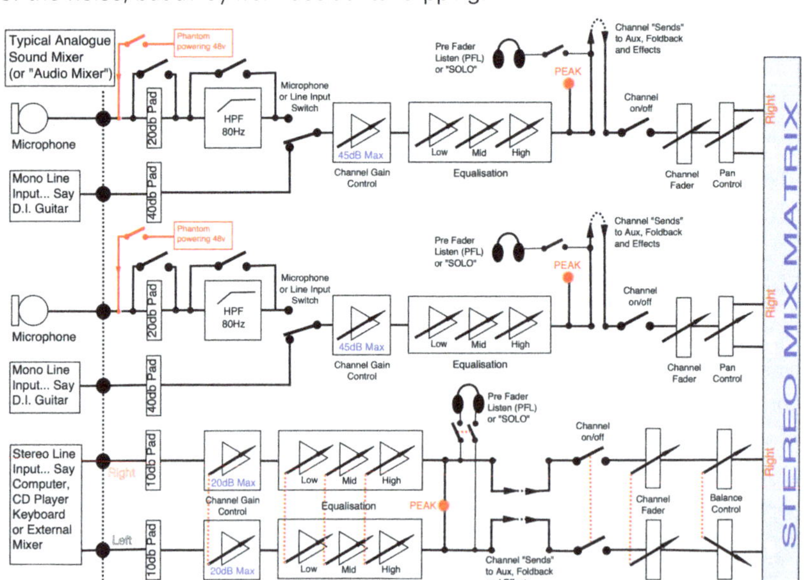

Figure 10.4 A basic mixer schematic, input to mix matrix

This should become your instinct, both within your equipment and between equipment. It is not by accident that many mixers have a simple "clip" light at the output of the equalisation which flashes if any of the foregoing stages are pushing the signal into distortion. Back off either the channel gain or equalisation as necessary to ensure the output will be clean. Better still, use your ears and have comfort in your choices! In Figure 10.4 we have a full diagram of the early stages of a simple practical sound mixer.

A PRACTICAL SOUND MIXER

The last breakout point of a channel is also known as the "Channel Output" and is a great place to listen to the quality of the individual channel sound on headphones or speakers prior to mixing it in with others. The button may be called "solo" as you are listening to just that one channel or "pfl" (pre-fader listen or, in the BBC, Pre-fade listen) and using it doesn't affect the sound going into the mix. The next bits do.

The channel fader takes the incoming sound of the channel on to the main, or perhaps group mix. Sometimes these faders have "gain in hand" of around 10 dB or so and shouldn't be taken all the way up in normal conditions; others expect to be fully open for the correct level. The channel on/off isolates the channel away from the mix, so this is one little button to look for if you think a channel is "broken!" The pan/balance controls are basically the same control (for a mono or stereo channel), moving the sound image across the stage.

On a mono channel the output of the "pan" is where the sound is split into two so you can place the channel into the stereo sound field just where you want it (say interviewer to the left of centre, interviewee to the right). On a stereo channel the control is called "balance", again you can alter the sound position; however, you are less likely to use it, unless for dramatic effect. If this is a DJ mixer, for instance, people expect the stereo music sources to be centre stage, but if it is a stereo keyboard in a band you may wish to offset it against the other instruments and singers to achieve a pleasing sound stage.

MID-SIDE STEREO RECORDING

We learnt in Chapter 4, "Microphones", that a carefully placed pair of crossed cardioid microphones is a firm favourite to make a stereophonic recording, regardless of space or complexity. An exception is the world of film where the scene will often be carrying dialogue from actors in the centre of the stage. A stable "M" or monophonic recording is key to ensuring compatibility between scenes and shots. This is achieved by a mid-side or coincident pair of microphones, which often will be housed in a single casing. Both the crossed pair and coincident technique were first described a century ago by Alan Blumlein, and it is a testament to his forward thinking that these are contemporary today. Whilst the crossed pair are simply placed in adjacent microphone channels, one panned left, the other to the right, M-S recording of discrete microphones is more complex at the desk, so it is worth taking the time to talk through.

In Figure 10.5 two microphone capsules, the "mid" and the "side", are stacked vertically and brought closely together to minimise comb filtering effects caused by a difference in arrival time. The "mid" capsule is normally a cardioid facing the sound source (although Alan Blumlein described the usage of an omnidirectional transducer in his original patent as cardioids where yet to exist!). The "M" is placed on a microphone channel of the desk and simply panned to the centre of the sound stage. The second microphone is a bidirectional microphone (a figure of eight polar pattern) facing sideways to the cardioid. The bidirectional output is paralleled across two microphone inputs to create the side or "S". One is panned extreme left and the other to extreme right. As with the description of the creation of the cardioid microphone earlier, we then *phase reverse one channel* to create the "difference" signal.

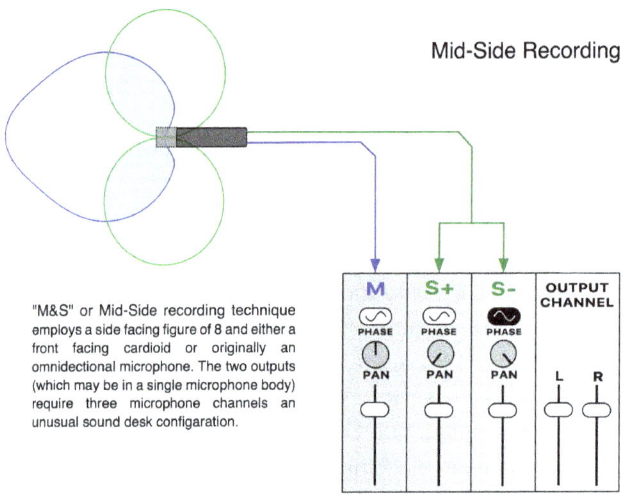

Figure 10.5 A guide to mid-side recording

In Figure 10.5 we see the coincident technique uses the left and right channels as a simple matrix: Left = Mid + Side, Right = Mid – Side. This configuration produces a completely mono-compatible signal and, if the "Mid" and "Side" signals are recorded separately in a sub mix (rather than just added together at the time of recording), the stereo width (and with that, the perceived distance to the sound source) can be changed to suit a scene after the recording has taken place. You can *zoom in and out with the sound* as well as the picture. Clever!

THX CINEMA AND HOME THEATRE SETUP

In this chapter, I think it's worth having a paragraph or two on home cinema as the references we have discussed in this chapter and others can get confused with the comparatively modern THX or other multichannel sound references. We have considered the voltage reference of 0 dBu and 0 dBV; we have come across digital 0 dBTP and 0 LUFS and have found audio sound pressure levels (SPL), weighted and unweighted, designated as dBA and dBC (see the end of Chapter 1, "Introduction to Sound"). It's not surprising that these standards become wrongly meshed with reference levels for modern cinema reproduction, but let's try to find a way forward with a fresh page!

THX reference levels are used in the movie production process, including post-production, and in reproduction in screening rooms and theatres. THX considers the non-linear response of the ear, especially the bass and the sound balance between ambience, speech and dramatic effects. *A reference of normal listening level set against the maximum volume is critical, and it is set 20 dB less than the peak sound (–20 dBFS).* This reference of THX "0" is a SPL measured off a cinema's systems loudspeakers with a Sound Level Meter with C weighting. The source of sound is a pink noise generator, and its level is measured by a microphone situated at the prime listener's seat (and ideally several other close positions) to create a room audio level (SPL) of 85 dB (see Chapter 5, "Loudspeakers"). This is a moderate listening level in a cinema, and the system is designed to reproduce film dialogue as a comfortable listen.

When running an automated setup, the system is adjusted to this level and the amplifier or receiver volume display is set to "0". After a successful setup, the volume can be raised or reduced and will show +dB or –dB set against this moderate reference level as appropriate. To be clear crescendos such as musical highs or perhaps explosions are allowed to peak up to a maximum of 20 dB higher than the reference, and the system must be able to achieve this level without fatigue.

The Cinema THX system equates entirely to the Home THX system; however, a *lower 75 dB testing level is used* (–30 dBFS) at the seated position in the home, as distractions are deemed more controllable in a domestic setting. The THX stamp on the front of a receiver attests that the unit can deliver audio peaks of 95 dB (75 + 20) with precision and clarity. You may choose a higher reference level in the home for personal entertainment, but do this by listening long term, so you have experienced several films with varied material, and of course, ensure your amp and speakers can take the punishment and still deliver faithfully!

As part of a setup of the THX system the subwoofer (or subwoofers) is (are) set internally by the receiver to be an SPL of 10 dB higher than the general system speakers. An SPL of 115 dB in a cinema normally requires multiple subwoofers; however, one quality home unit in an average lounge should achieve the less challenging 105 dB. The THX preferred crossover frequency is 80 Hz.

MULTICHANNEL SOUND AND FORMATS

THX™ was developed by Tomlinson Holman at George Lucas' Lucasfilm for the soundtrack of the third "Star Wars" film, the 1983 *Return of the Jedi*. The "TH" of THX are his initials, and "X" stands for "experiment". The base technology of THX technology arrived in the cinema with the "THX Deep Note" at the start of the *Star Wars VI* film. This vivid short soundscape was the creation of Dr. James "Andy" Moorer. The earth-shaking audio bends out from the cinema screen on multiple speakers surrounding the listener, enveloping them in the action of the film. The experience was astounding to young viewers like me, and we wanted to take it home!

The first system to bring this experience into a domestic setting was the analogue Dolby Pro Logic Surround Sound carried on some stereo home VHS TV-recording cassettes. A further two audio channels were matrix encoded into the stereo tape track, creating the usual left and right channels, and in addition a centre channel to go under the TV optimising actors' dialogue and a mono bandwidth-restricted rear feed which was often fed to two rear speakers to give the surround effect. The great news is any normal VHS player could output this matrix, so an upgrade was unnecessary. However, a Dolby Pro Logic receiver was required to decode the matrix hidden in the analogue stereo, and this normally contained power amplifier channels to directly feed the speakers. Frankly VHS soundtracks were poor anyway, and the matrixing did not improve the sound quality. It wasn't great, but many diehards filled their sitting rooms with bulky speakers and invited their neighbours to squeeze into the remaining space.

Dolby released a *digital* version of their surround sound with the 1992 cinema release of *Batman Returns*. Called Dolby Spectral Recording-Digital (SR-D), it was a huge improvement on the analogue sound, although it used a lossy codec. The digital code was sent 26 frames in advance of the picture and read by a "Penthouse Soundhead" mounted above the projector. Unbelievably, the code sat vertically between the sprocket holes of the film, with the normal stereo sound sitting in its normal position of two optical tracks. Dolby Digital AC-3 (Advanced Codec 3) surround sound set the trend for many years to come with its six-channel format. This was described as "5.1", as whilst five channels were used to provide the full bandwidth surround (left, right, centre, rear left, rear right), the sixth channel contained low frequency content below 80 Hz. This is still the base standard today, and in the home, Dolby Digital remains the favoured system, although in differing flavours of Dolby Surround 7.1 in 2010 and the immersive 3D "360°" Dolby Atmos in 2012. Competition first came in the form of Digital Theatre Systems (DTS); its first cinema outing came with Stephen Spielberg's *Jurassic Park* in 1993, and Sony's Dynamic Digital Sound came later. These companies are still very much in competition with Dolby, with Atmos variants of Sony 360RA (Reality Audio) and DTS X. These are both excellent alternatives; however, this book is about building a solid general knowledge base, and I can't cover all technology alternatives and remain honest to its core purpose.

The requirement for a low frequency speaker created a new market which suited the emerging class D amplifiers which were cheaper to make and where the subtlety of precision was unnecessary. So, a powerful active subwoofer became a standard below 80 Hz, and the "receiver" carried the smaller power amplifiers for the surround channels above 80 Hz. A famous quality HiFi brand manufacturer created a subwoofer the size of two adjacent box files, and satellite speakers the size of oranges to be mounted high on the walls, but sadly the frequency suck-out between the bass and top frequencies was disconcerting to hear. The many clones that copied this compact concept were truly horrible! As a result, multichannel sound systems have tended to be conspicuous by their size in smaller homes, and so not popular. People with an empty garage or barn may well do a cinema conversion, and they can be amazing; but multichannel sound is yet to gain the simple acceptability of positioning a second speaker to provide an upgrade to stereo. A popular compromise "cinema" lounge solution is a 2.1 system, where high-quality stereo speakers provide the front soundscape, and a suitable phase-adjusted subwoofer in an unobtrusive location provides 80 Hz and below at a THX inspired 10 dB boost over the stereo sound. Personally, I am yet to hear a soundbar that convinces me, either in terms of audio quality or of sound stage.

Clearly 7.1 adds two more speakers over 5.1, splitting the positions of the rear pair (often parked in the back two corners of the room) into side (often high wall position) and true rear. Dolby Atmos™ introduces the idea of a sound field above the listeners, a true 3D sound experience. There is no mandated layout to achieve this; the "Atmos" decoder works out the best use of your system in the initial setup. That said, professionals favour a 7.1.4 layout, which many claim is the minimum (an astounding 22.1.8 system is in use and apparently sounds great!) You may ask what is this new number after the "1" of the subwoofer? These are speakers either mounted on the ceiling or sitting on top of other speakers, pointing at the ceiling. So, 7.1.4 sees additional small upward-pointing speakers perched on top (or integrated into) the front and rear pairs. The listener hears the rebound of the sound from the upward-facing speakers in the listening position. With immersive audio the height of the sound is key, and somehow that must be captured. It's relatively early days, but experiments with both AB and XY pairs have had some success integrating with other sound sources; however, the smart money is on 3D microphone arrays such as PCM-3D, OCT-3D and 2L Cube. These three arrays feature nine microphones, so require nine groups set up on the mixer to present to the Dolby Atmos encoder. One popular recording format is MPEG H Joint Object Coding. If you wish to know more, I can recommend papers by Dr Hyunkook Lee of Huddersfield University (references). Some say nine microphones are not enough, and I am aware of a 71 microphone "spheroid" currently in use for Atmos recordings. Who said modern recording had got easier!

MONITORING, METERS, SPEAKERS AND HEADPHONES

Before we go further, I'm pleased to tell you that you already own the best music testing instruments in the world; they are your ears! Everything else, including dour colleagues and the finest technical facilities available, pales beside them. Treat them well and they will look after you! Now to what else is available. I guess we are all used to seeing cheap metering as strips of light these days, many changing from green for OK to orange for beware and red if you have a ruined recording. Many manufacturers are kind and give a margin for error, some "headroom" in the desk to save your precious recording. These meters are highly variable in quality and in the ability to give confidence to the sound operator about what is going on. We've covered this in detail in Chapter 6, "Sound Levels and Connections", and Chapter 5, "Loudspeakers", where we also cover headphones. I'd just like to reiterate the point that sounding "nice" is not the same as sounding "correct", and a competent sound engineer must have "go to" solutions, like trusted headphones, to ensure the monitors and room are not playing tricks on the listener.

The words "Studio Monitor" are used far too often and wrongly, as is the term "Professional" for tacky microphones and sound equipment which will mislead you on your mix and sound terrible when placed on a quality professional system. They mean nothing! It is fundamental to the success of a sound engineer (please read Chapter 8, "Introduction to Acoustics") to understand that even with the best studio monitors, reflections in the room can muddy the sound and destroy your work. The world's finest monitor speakers, and I am fortunate to have used many over the years, can sound terrible in the wrong acoustic; and trust me, you won't believe which world-famous brands own some of the worst studio designs!

Studio technical faults occur all the time from the analogue curses of dirty patching cords, "noisy" pots and switches. Digital is subject to reflections in the fibre, to jitter and decomposing chipsets. Ageing capacitors and damaged semiconductors create oscillations outside of the audio spectrum and unusual distortions such as shot noise. All these faults have signatures that experienced ears can detect and work around. The only constant in the sound world is your own ears; educate them by critical listening, looking to gain experience in many situations and eventually quietly trusting that you are right over some strong views of others. That said, we never stop learning if we truly care about our craft!

BIBLIOGRAPHY

Avid (2025) *Dolby Digital. The Audiovisual Identity Database, The Motion Graphics Museum.* https://www.avid.wiki/Dolby_Digital.

Houghton, M. (2020) Introduction to Mid-Sides Recording. *Sound on Sound.* https://www.soundonsound.com/techniques/introduction-mid-sides-recording.

Lee, H. (2021) Multichannel 3D Microphone Array: A Review. *Journal of the Audio Engineering Society*, 69. https://aes2.org/publications/elibrary-page/?id=21013.

Mellor, N. (2013) *THX Reference Level Explained Acoustic Frontiers*. https://acousticfrontiers.com/blogs/articles/thx-reference-level-explained.

THX (2025) *About*. https://www.thx.com/about/.

THX Home (2025) *The Entertainment Experience You Want Wherever You Want It*. https://www.thx.com.

11 Introduction to Sound Synthesis

Sound Knowledge is a primer on audio technology; invariably, it will stray into music technologies in explanations of sound theory. This chapter describes the evolution of sound synthesis in the context of the audio engineer. It is not purporting to be a guide to the selection of equipment currently used in music production. There are great books that make this their sole remit and do it well! Moving on...

In Chapter 1, "Introduction to Sound", we touched on seven sources of sound, the electronic oscillator being the last. Whilst the six others require a physical medium like a pipe, string or drum to create sound by natural resonance, this last one differs because the property of the medium is a comparatively recent discovery: electricity. In Chapter 3, "Introduction to Electronics", we learnt that a capacitor and inductor in parallel will resonate, the capacitor passing a charge back and forth between its plates, sending energy through the inductor, creating a single AC frequency of resonance dependent on the two component values. This resonance can easily be made to overdrive an amplifier with positive feedback, falling into a permanent state of oscillation and creating a circuit called an oscillator. If the frequency of oscillation is within the audible range, we can hear it using a speaker on the circuit output.

Developments of this simple idea include the Hartley and Colpitts oscillators, and more common are RC oscillator circuits such as the Phase Shift and Wien Bridge, which use a positive feedback loop to push an amplifier into oscillation (think of a sound system going into howl round; it's the same idea). From these simple beginnings we can create useful audio waveforms including the sine wave, square wave, triangle wave and sawtooth wave. Each shape produces a different sound (Figure 11.1); the sine wave is a pure fundamental, whereas the others represent a sine wave with additional harmonics which give the sound its character. The waveforms with vertical sides produce an edgier sound, the ultimate being the square wave, which has odd harmonics added to the fundamental to create the blocky shape of the envelope.

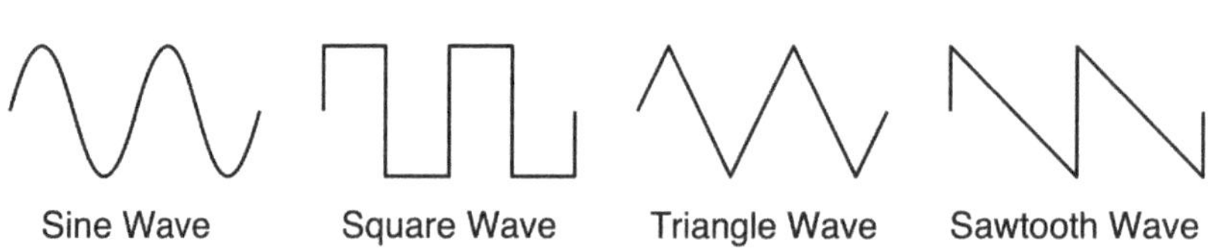

Figure 11.1 Waveforms easily derived from a basic oscillator

DOI: 10.4324/9781003717065-11

WAVEFORM MODIFIERS

The Voltage Controlled Oscillator (VCO) is a simple development of the fixed frequency oscillator, where an applied voltage at the input varies the frequency of sound at the output. If we use a keyboard to provide a rising scale of voltages to the oscillator (through a ladder of resistor potential dividers), the selected sound will have an increasing scale of frequencies (notes) we can play. Further if we create multiple outputs from our oscillators, we can mix them together in differing amounts to create an endless variety of sounds and can play our modified audio envelope over a wide range of notes. This is the basis of subtractive synthesis, but there are others:

1. Subtractive synthesis – we take the basic waveform envelope and alter it with white noise, pink noise or boost and cut audio filters to use subtractive synthesis to alter the timbre. This technology was used in the Minimoog and Korg MS-20. It also found a strong toehold in early personal computers and home consoles including the Commodore 64 and Atari ST.
2. Additive synthesis – we take a fundamental sine wave from an oscillator and add in further tuned sine wave oscillators creating overtones (integer multiples, just like the harmonic breakdown of the clarinet in , Chapter 1, "Introduction to Sound", but in reverse!). This is called additive synthesis. Varying the levels of the harmonics contained within the additional oscillators creates a different timbre and can mimic traditional instruments. Additive synthesis can utilise aperiodic sine waves (not direct multiples of the fundamental) to produce inharmonic sounds which rarely occur in musical instruments; however, the bell is an interesting exception!
3. Frequency modulation synthesis – an additional modulating frequency is added to the fundamental waveform; the harmonic content of the oscillator is altered by the amplitude of the modulating signal. As the level of modulation is increased, the waveform grows more complex. If it has a harmonic relationship to the original waveform, the result is harmonic. If not, the resultant waveform is inharmonic. This type of synthesis lends itself to digitalisation, and was made popular by Yamaha in the 1980s, where it also found favour in the growing arcade games market.

Most synthesisers utilise an Envelope Generator as a modifier to a given sound envelope at the point it is played out. The most common is the attack, decay, sustain and release (ADSR), which composer Herbert Deutch suggested to Robert Moog, and he designed in 1965 (Figure 11.2). This circuit controls four attributes of a pressed key; other designs may do less or more.

- Attack – the time taken for the signal to rise from zero to the peak level.
- Decay – the time taken for the peak level to reduce to the sustain level.
- Sustain – the level maintained until the key is released.
- Release – the time taken for the signal to decay to zero.

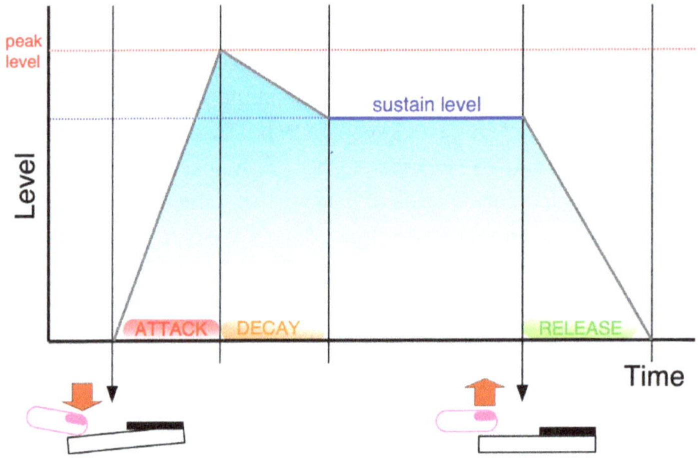

Figure 11.2 The Envelope Generator creates a more natural keystroke

There are many flavours of these analogue modes, later synthesised into the digital domain which brings better stability to the sound. Understanding the basic concepts will help you successfully navigate the evolution of the synthesiser in this chapter. Whilst these form the basic principle of different sound synthesisers, we can choose to add further VCOs to make more complex sounds, save certain waveforms, so we can recall them as a "voice" when we require it, and create a sequence of notes to replay as and when required.

The Telharmonium, producing sound by additive synthesis, was patented way back in 1897, and whilst other electronic sound producers were built in the intervening years, the first practical Sound Synthesiser was the RCA MkII, installed at Columbia University 60 years later in 1957. It used thermionic valves (transistors were yet to make their mark) and took up a large room with bays of electronics and physical plugs and cords in patch panels to route the sound sources around. It was a huge attraction to composers because a writer could create compositions using a paper tape recorder/player (analogous to MIDI today). Tunes were replayed using the paper tape instructions to the synthesiser and then could be recorded to disc on a nearby lathe. To be clear, this was not an early computer, yet I'm sure you can see how computing would suit this sort of repetitive precision delivery. Tempos and rhythms that were impossible on acoustic instruments could be written, making this instrument a complex but intriguing source of music. In addition to 12 fixed oscillators (rather than VCOs), the RCA MkII also had a white noise source, where again the envelope could be modified by the other sources. Electronic music had become a new art form!

ANALOGUE SEMICONDUCTOR SYNTHESISERS

The transistor revolutionised electronics, in both power requirements and requiring a fraction of the footprint of valve equipment. Robert Moog was quick to use this technology in the Moog Synthesiser created in 1964. He pioneered new VCO circuits, noise generators, envelope creation and filtering. The Rolling Stones, Beatles, Doors and Grateful Dead all beat a path to his door! The first practical portable instrument (keyboard included) was the Minimoog in 1970, which found favour with progressive rock bands like Emerson Lake and Palmer, Pink Floyd and Yes. Let's not forget Gary Numan and Kraftwerk! Last, but certainly not least, 1970s jazz and funk suited the instrument completely, and in many senses, the artists made the instrument their own. These included Stevie Wonder, Chaka Khan, Herbie Hancock and James Brown.

Following the impact of the Moog and Minimoog, other manufacturers entered the market in the early 1970s. EMS in the UK and ARP in the US. The EMS made a huge impact with users Pink Floyd and Brian Eno. Brian was responsible for the highly successful signature sound of Roxy Music and later produced and played on David Bowie albums including the eponymous "Heroes" in 1977.

The 1980s introduced commercial access to integrated circuits (ICs; electronic microchips); these were discrete circuits miniaturised to be used as building blocks in larger circuits (see Chapter 3, "Introduction to Electronics"). The repetitive nature of the synthesiser circuit systems suited the use of microchips, and so new devices could be constructed with far higher circuit complexity, yet the same outer physical size.

Early synthesisers could only play one note at a time; they were "monophonic". With the 1980s came new entrants to the market. Tom Oberheim introduced several polyphony-capable units (starting with two-note in the early 1970s), increasing in complexity to the eye-wateringly expensive OB-X in 1979, crammed with the new microchips on a few large circuit boards that ran the length of the keyboard. It was available with 4, 6 or 8 voice cards; these numbers indicate the number of keys that could be held down at one time to make a chord. It found friends with Queen (*The Game*), Eurythmics (*Sweet Dreams Are Made of This*) and Japan.

The OB-X and the possibility of eight consecutive notes, albeit at a price, were likely an answer to the 1978 Prophet-5 polyphonic synthesiser, created by Dave Smith and John Bowen from Sequential. This allowed for five simultaneous keyboard notes. It was also the first unit with recall of musician-created sounds stored in "patch memory" using microprocessors. It was an overnight success! The list is long, but a few highlights are: Eurythmics, Pet Shop Boys, Madonna, Vangelis, Giorgio Moroder, Rick Wakeman and Michael Jackson.

In 1981 Oberheim, together with Roland's Ikutaro Kakehashi and Dave Smith of Sequential, invented the MIDI format, which was quickly adopted by all major manufacturers. The market for synthesisers grew massively in the 1980s, and with the introduction of MIDI in 1982, the electronic instruments in the studio (or perhaps bedroom) could finally be connected. A mention must be made of the Fairlight CMI, released in 1979, where samples could be replayed and their pitch altered. It was brilliant but very expensive, which drove new entrants like E-mu Emulator and Akai S-series to compete on both price and facilities.

DIGITAL SYNTHESIS

There was about to be a new kid on the block! In 1983, in the middle of the analogue synthesiser price wars, Yamaha released the first commercially successful digital synthesiser, the DX 7, using designs by John Chowning of Stanford University, the father of Frequency Modulation Synthesis (which was used in this design). The sound was clearly brighter, some said glassier, than the analogue sound, and this suited the synthesis of complex transients like piano, brass and bells. It was just the sound the bands of the time were looking for! It's staggering to note the Yamaha DX7 appeared on no less than 40% of the Billboard No. 1 singles in 1986! The unit went on to sell over 200,000 units, yet it's interesting to note that the DX7 is now seen as of its time; the sound appears dated, maybe even stilted, to modern performers.

It was not just the sound of digital synthesis; the DX7 had a raft of 32 installed preset sounds (patches) onboard, which could be instantly used! This was in addition to a generous 32 sound generating algorithms to personally design sounds to name and use yourself. The DX7 changed the way musicians viewed the synthesiser; the sleepy restaurant now did fast food to go! The installation of preset sounds featuring traditional instruments became synonymous with digital synthesisers, making the technology far more friendly for working musicians; the gigging piano player was now a one-man band. Strings and brass on tracks no longer required additional musicians; the keyboard player did it all!

The Synclavier (originally additive synthesis) later emulated the DX7 with an optional FM synthesis module to create similar sounds and, in addition (for a price), came with cutting-edge 16-bit sampling and digital recording. A standalone genius of the time, it was equally at home as a digital recorder, where it radically changed techniques in TV and film sound and effects, moving them "tapeless".

In 1987, the digital Roland D-50 (subtractive synthesis) added digitally created chorus, delay and reverb effects. The 1988 Korg M1 also used subtractive synthesis, had all the Roland effects and, in addition, tremolo, distortion and EQ. It further featured sampled transients and the introduction of loops, both in 16-bit quality. The sound patches were sensational, featuring the normal range of house instruments and exotic ones like sitar, kalimba and wind chimes. This was the synthesiser for the 1990s and went on to sell over 250,000 units, the biggest hardware synthesiser seller of all time!

The analogue synthesiser had understandably lost sales in the onslaught of the digital age. You might have thought that was it; however, sales rebounded in the last years of the millennium. Suddenly the old patchable modular units were fashionable again, often selling for more than the original ticket price, as composers went "old school" looking for new creative sounds. The industry responded in the 2010s with new offerings from Moog and Korg, and there were even some new entrants in the discrete analogue synthesiser marketplace. It was a good money spinner, as the basic circuitry from the old designs could now be reintroduced using cheap surface mount technology to slash the price to market.

COMPUTER EMULATION

In 1997 Propellerhead Software released "ReBirth" and Seer Systems released "Reality". These were the first synthesiser software emulations that could run under MIDI. In 1999, Cubase launched a software revision which allowed users to run software instruments, which included the emulations of synthesisers. With these two developments the monolithic music box in the corner was no more, your PC or Mac could now be your software synthesiser or "softsynth". The screen designs emulated the original hardware look, some softsynths even imported the original sound patches, so the audio experience was identical. Virtually all the synthesisers named in this chapter have high quality softsynth versions right from the first Moog to the fabulous Korg M1.

The differences between hardware and software are few but noteworthy. On the plus side, the softsynth can communicate with other computer-based elements, particularly a DAW and/or sequencer to make an end-to-end production process. The softsynth version can include much of the capability discovered since the original legends stalked the earth. Memory is no longer a barrier to holding large numbers of high-quality samples, now potentially indistinguishable from the live instrument. Sequencing software like Ableton Live, Logic Pro and GarageBand allows the creation of music directly from a softsynth, and Nyquist and Csound allow better creation of software instruments.

On the downside, the obvious point of missing the subtlety of physical contact with tactile controls changes the vibe. Further, with the use of large patches, latency can be a problem; in extreme cases, the project can crash and burn. The industry has answered this with a "Draft Mode" for the initial setting down of the project score in a lower quality. This is a light draw on computing power, so is unlikely to suffer a meltdown. When the draft is satisfactory it can be rendered in "Production Mode" (offline) to generate the high-quality final version.

MUSICAL INSTRUMENT DIGITAL INTERFACE (MIDI)

It is likely that many aspiring musicians reading this book will have practical use of MIDI and will be far more advanced with it than my humble introduction will allow. For other readers, here is a short run-through of what it is and does, to whet your appetite! The "outboard" devices I have listed to date could be used for speech, vocals or with musical instruments. They can be used together to create a chain of units to modify a sound, but importantly they do not, in general, communicate with each other. However, separately, there is a means of communication between differing electronic musical instruments and allied computer systems, such as DAWs. MIDI exists to provide a standard by which different computers and musical instruments can communicate musical events, and to allow one device to control several others. MIDI files contain no music; they are simply a set of instructions to create a certain sound. Imagine a keyboard containing a MIDI controller chip. A key is pressed, and the controller sends an "event message" which corresponds to the pitch of the note. This message is relayed and received by other components connected by the MIDI data cable. However, MIDI is much cleverer than a "note-on note-off" pitch message; it can convey how hard the key is hit (the velocity), aftertouch (how hard the key is pressed), the volume, pan, pitch-bend and so much more. Each time a change is made, a data message is sent.

The standard was invented over 40 years ago, yet is still relevant today; indeed, it stands out as a rare burning bush of co-operation, much like USB in the computer industry, bringing together music-makers and their diverse choices in equipment make and model under one common shared umbrella. Before MIDI, certain manufacturers had a level of co-operation between models in their own ranges, but this was crude in comparison, utilising analogue control voltages and triggers. The advent of MIDI corresponds to the introduction of personal computing, and this is no coincidence. Digital equipment did away with the physical knobs and sliders of the analogue age, and as soon as we had virtual controls, we could send the positional information between devices in a digital stream. In use, a single MIDI controller (most often a keyboard) can "play" multiple electronic instruments (each often contained in a 1U rack mount case), meaning a keyboard is not needed for each. MIDI instructions can be used to drive quality samples of instruments such as guitar, violin, flute or maybe a full orchestra, meaning a talented beginner can build full musical works from a simple keyboard.

Given the age of the standard, it is amazing it is still relevant today, however, it does have several issues. Whilst it is very compact as a data stream (bearing in mind how slow and crude PCs were back then), it does have beat timing problems through jitter and latency issues, creating pauses between enabling an action and what happens down the way. As more units are daisy-chained together and more commands are sent, the time delay can build into discernible milliseconds. Sadly, these weaknesses are not addressed by using a modern "fast" USB connected MIDI instead of the good old 5 pin DIN connectors, as the system cannot leverage the greater bandwidth.

I'd love to say that a new universal protocol is in the wings at the time of writing; however, smart new protocols, using vastly increased computing power, tend to be manufacturer-specific, moving away from the simplicity and cooperation of MIDI to create complex, efficient, powerful protocols which only work on closed systems such as dedicated music studios. Still, MIDI is everywhere, and still a valid means of building and sharing the feel of your latest creation, as perhaps you might send an MP3 as an outline sample of your work when your masterpiece back home is a 24-bit WAV.

BIBLIOGRAPHY

Avid Technology (2023) *Ultimate Guide to Using MIDI in Music Production*. https://www.avid.com/resource-center/what-is-midi-musical-instrument-digital-interface

Crute, C. (2019) *The History of Sample-Based Synthesis*. https://musictech.com/guides/essential-guide/history-sample-based-synthesis/

Kuseki, A. (2020) *Basics of Synthesis and Sound Design - A Beginner's Guide*. https://medium.com/@kusekiakorame/basics-of-synthesis-and-sound-design-a-beginners-guide-9c3d0314c6d5

Reid, G. (2000) Synth Secrets, Part 9: An Introduction to VCAs. *Sound on Sound*.

12 Outboard Devices

In recent chapters we have grown a greater understanding of a sound desk and the sources that can be connected to it. On occasion a simple mix using the "inboard" equalisation of the desk is not enough to get a sound that is wanted. As an example, back in the day, the set-up of a professional radio DJ would certainly require careful control of level and use of the available "EQ" to perhaps provide a bit of punch or bite in the midrange, and maybe a bit of top to provide crispness and finally careful use of the bass control to either wind out "bass tip up" if the DJ was working too closely to the microphone and in danger of distorting, or perhaps wind in a little on a weaker voice to give it a little more soul. But that was not where it ended. A touch of reverberation was often added to give excitement and structure to the voice, and a strong form of compression, called "limiting", to help the voice punch through to compete with the limited audio dynamic range of commercial music, as otherwise music peaking at the same meter level would drown out the DJ. It is normal to lower the music level anyway during a voiceover, so she/he is not lost under the track. Without limiting, we would have to reduce the overall station level, making it sound quieter than its rivals. Both reverb and compression effects were not traditionally part of the audio desk and so needed to be inserted into the sound chain at some point.

AUXILIARY SENDS AND RETURNS

In Chapter 10, "Introduction to Sound Desks", Figure 10.4 introduced the idea of taking a feed of the channel output to use for other means, the "Aux". For simplicity, a feed was shown taken from after the fader (called post-fader) so any sound return carries no sound unless the fader is open. Equally it could have been taken from before the fader (called pre-fader), which is favoured for creating monitor mixes to provide stage vocalists with their own mix. These audio feeds are used to make up one mix or more which terminate on a desk output called an "Auxiliary Send".

To save space on the crowded rear panel sometimes a stereo 1/4-inch jack (TRS) is used, with the send to the outboard device on the tip and the live return from the device on the ring, so special breakout cords are required.

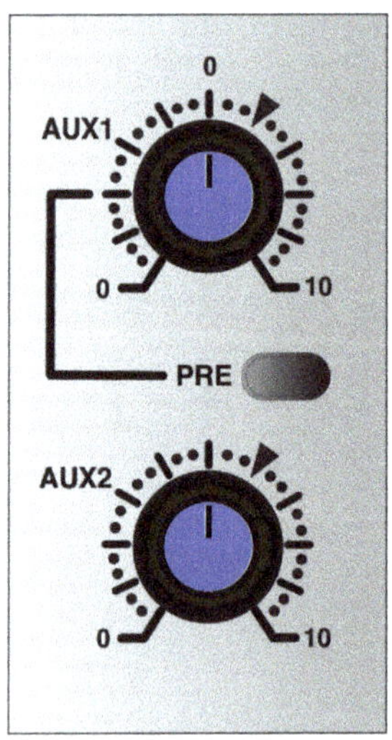

Figure 12.1 A typical breakout point to add external technology

DOI: 10.4324/9781003717065-12

What will catch many by surprise is that the levels on these TRS breakouts often have a nominal consumer level of –10 dBV, even though the desk inputs and outputs are at the professional level of +4 dBu.

Where the breakout is separated into balanced output and input jacks on high-end analogue sound mixers, you can also expect these will be at +4 dBu. So, sadly there are shades of grey in sound operations, and the engineer needs to be on the alert, using the "–10/+4" switching on the outboard devices with skill to avoid either clipping in the outboard unit or excessive noise. In Figure 12.1 a section of a sound desk channel strip is shown, which normally sits between the pan control near the fader and the equalisation.

This is as it should be as the path of the sound signal is assumed to start at the top of the panel and work down to the fader. The mix of "Aux1" is switchable between pre-fader and post-fader, whilst the mix "Aux2" only carries sound when the fader is up (post-fader). The arrow marks a nominal value which, if the source is at "line level" (see Chapter 6, "Sound Levels and Connections") it should be a good match for the desk.

If the knob is turned fully clockwise, a useful but dangerous 10 dB gain is introduced to help match to professional products, and of course if an outboard device is overdriven the input can be attenuated. The output of the external unit is returned to the mixer to be added to either a group output or the main out with another control, the knob colour-coded the same and marked similarly as the Aux1 or Aux2 "Return". Let's look further at what some of these outboard devices are, and what they do.

OUTBOARD DEVICES IN DETAIL

I'm purposely taking an old-fashioned look at effects to distinguish the traditional desk outboard from the many digital hybrids that now can give access to a myriad of effects within the mixer. I am also aware that all the effects we are to step through have digital software clones that look to emulate the often expensive, but certainly complex rack mount boxes of a golden age. I do this to give a baseline, a purity of the original concepts, which I believe is getting lost in the maelstrom of hybrid digital products. Like microphones, many of the types of equipment have a unique sound that transcends their cost and frankly sometimes their sound quality. They get used because they have a signature sound that has been in use for decades, and will be, in hardware or digital form, available well into the future.

Analogue Effects

Yes, but are they? I'm looking first at the original units that were available to the professional studio, often at an eye-watering price which modern digital software frequently recreates as free add-ons. And that's the point; only professionals could create sound effects that we all can now download for free, but what goes with that is the experimentation and practice of years as a sound engineer, which in turn is difficult to achieve without support in the home environment. Highly experienced 1980s photographer David Bailey, armed with his trusty Olympus OM2, created masterful magazine covers every week, yet in the home we had inflexible box brownies (look it up!). My analogy is, mobile phones can create wonderful pictures, yet does the average owner have a clue how to take a decent photo?

EQUALISATION

We opened the subject of varying volume at different frequencies in Chapter 3, "Introduction to Electronics". A simple capacitor can be made to remove top or bottom frequencies in a passive filter circuit. In 1952 PJ Baxandall published a design in the popular magazine Wireless World for an active circuit (a powered amplifier) which, by utilising negative feedback, could boost frequencies as well as cut them. The bass and treble controls were born and quickly made their way into the home entertainment system of the day, the "Radiogram". A later circuit application called a "gyrator" allowed expensive inductors (tuned coils of wire) to be replaced with cheap capacitors to produce affordable "active" bandpass/stop circuits at certain frequencies, and so the "Mid" or "Presence" control was slotted between bass and treble on mixing desks.

This cheaper technology also opened the door to having multiple filters stacked next to each other, managing frequencies across the range of human hearing, and the "graphic equaliser" became affordable for domestic systems, as well as complex installations in studios and PA companies. "Graphic" in terms of using vertical sliding controls for boost and cut, set out on a panel to mimic the frequencies they controlled, creating a graphical representation of a frequency response that the movement of the controls would create.

Cheaper domestic systems may have had just six sliders, later ones ten channels for each stereo side, and professional ones up to 32 channels, controls spaced just ⅓ of an octave apart. The graphic equaliser had a fixed "Q" on every filter, based on the number of channels needed to cover the frequency range of the unit. The more channels, the tighter the Q. The graphic equaliser was popular in the home and in the professional studio for over 20 years. It was used to correct anomalies in a recording in post-production, to shape voices and instruments for a mix or to notch out single frequencies that were causing a howl-round (feedback) in a PA system, so it could go louder. But even the best ones created artefacts; something just didn't sound right, and it was eventually supplanted by a newer technology, the parametric equaliser.

If we go back to the humble presence or mid control, it was a filter that could simply boost or cut the sound at a fixed frequency, within a fixed width of frequencies either side, the Q of the circuit. Advances in circuit design allowed for a filter that could vary its frequency of operation with the turn of a knob, and (in some) the sharpness of the tuned peak (Q) and so the range of frequencies that are affected. Whilst one filter could solve a problem, such as notching out a whistle or a hum, it became usual to daisy-chain several units in the same box, so the device could be more versatile in tackling several problems with a single unit.

Parametric equalisers come in many flavours; some allow an adjustment of the turnover frequency for bass and treble controls (as well as mid), and some are all singing and dancing rack mount units. However, the most common example is to add a single knob to each sound mixer input channel to select the frequency at which the mid (or presence) control operates.

This is not to be knocked; the ability to decide on which middle frequency dominates opens up real versatility, in particular much better shaping of vocals. Expect further mid-band parametric controls in the strip, so more than one change can be made at the same time. Don't let the extra complexity fool you into thinking the science has grown with the number of knobs; stay calm and assess what each function does. On cheaper desks there is no doubt an outboard plug-in will enhance your recordings. Just keep the gain you apply at a particular frequency within sensible limits, or the channel will distort at that frequency and your session will be badly impaired.

COMPRESSORS

Another "old school" technology, the humble "compressor" remains a studio favourite. Early ones used electronic valves and often were controlled by a humble light bulb which changed its resistance with its brightness (driven by the sound volume). The compressor is used to manipulate the loudness of the incoming sound to achieve an objective. Most often, the aim is simply to make the sound appear louder for the same peak volume by restricting the dynamic range. Simply put, that means making the quieter bits louder whilst not allowing the loud bits to increase to a volume where they could distort. Unlike many other outboard units, the input level from the previous equipment is important in that it takes a certain volume to activate the compressing action; this level is called the "threshold". If the level is too low the threshold will also need to be set low for the unit to do anything, which risks dragging up the noise floor with the makeup gain (see "Output/ Makeup gain" below).

Feedforward/Feedback Designs

A compressor works by taking a sample of the sound signal and creating from it a DC control voltage (This is called the "sidechain"). The control voltage acts on the main sound path by altering the voltage applied to a Voltage Controlled Amplifier (VCA) to vary the gain depending on parameters set up by the user. Note the feed of the main sound chain can be taken before the VCA or after it. These are called respectively a feedforward design and a feedback design.

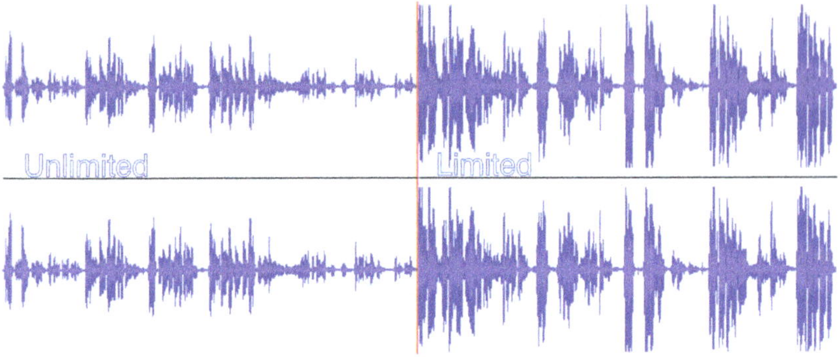

Figure 12.2 The dramatic effect of speech limiting

Figure 12.2 shows, on the left, a speech extract which peaks at times near maximum level, so in theory cannot go higher. The right of the picture shows the same speech recording; now it has soft limiting to make it louder. Note that some peaks seem truncated; however, on listening there is no discernible distortion.

Like microphones, compressors come at all price points, yet the way the industry perceives them is not always consistent with the value or quality of the design. Some "just sound great" is an often-voiced comment, particularly on designs using thermionic valves in their construction. The controls found on a compressor are typically:

Threshold (input gain): typically −40 dB to +10 dB. This sets the level at which the VCA starts to act on the passing sound signal.

Compression ratio (or "slope"): this is the ratio of decibels in, against decibels out, often from a gentle 1.5:1 up to a whopping 20:1 and beyond in some designs. The extreme compression of 20 dB (or more) input for a rise of just a single decibel (or less) at the output is known as a "Limiter", a tool much used in broadcasting to (1) protect the transmission chain from unexpected overloads, (2) increase the loudness of the DJ's voice to match that of commercial records and (3) get the radio station output as loud as possible to compete against other stations. In Figure 12.3 the blue arrows show that below the threshold, the input and output levels track, hence there is no compression of the signal.

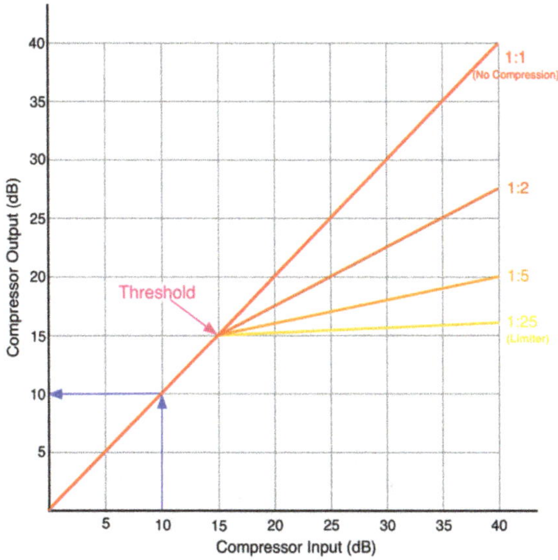

Figure 12.3 The effect of a compressor on differing levels

Attack time: the time taken by the control circuit to act on the sound path, from single milliseconds to perhaps half a second. A "Delay Line" limiter often used in broadcasting passes the main sound through a time delay, so the VCA in the sidechain (feedforward design and so not delayed) can act on peaks before they get to the VCA… Smart!

Release time: the time taken for the control circuit to allow the sound signal to return to normal. The range is typically from around 10 mS to 3 seconds or more.

Output/ Makeup gain: the foregoing compressor controls are likely to have changed the input level, so the last control in the chain allows the output of the compressor to be trimmed to match the input level, so there is no level difference if it is bypassed.

Metering often indicates the level of compression labelled "Gain Reduction" and a bypass switch to take the unit in and out of the chain. Stereo compressors require linking of the two sidechains to keep the image stable, and a switch is often provided to link the channels for stereo or allow them to be used as two separate mono compressor channels. Filters are often incorporated in the sidechain; a high pass (this bass cut is often labelled "Pre-Emphasis") is common, otherwise a strong bass beat can swamp the compressed sound in the self-explanatory "mid-band gain ducking", this is not nice to hear! Other filters can be applied; a treble boost in the sidechain can be used to bring in check a distortion of the mid/high frequencies, particularly of vocals, which is called sibilance. A stand-alone compressor with this tool is called a "De-esser". Finally, we talked of a DJ artistically knocking back a music fader during a voiceover so she/he could be heard. Some use "automatic voiceover" to duck the music, where the DC feed of the microphone limiter sidechain is also sent to the sidechain of the music limiter.

On some units, expected controls are missing or are just buttons, so flexibility is lost to the user, or on others the knobs are not calibrated, perhaps just labelled 1 to 10. I guess this because a compressor can be a crude tool in the wrong hands and cause more damage than it solves. In Figure 12.4 I have drawn a typical compressor design so you can get a feel for the layout. This is less likely to be the physical unit at the time of publication. It may well be reproduced on screen in a DAW simulation or just labelled sliders to drag with a mouse. On this panel the "attack time" has become a simple two-way switch to reduce potential operator error. Note that the choice of compression ratios is an arbitrary part of the design process, and so expect to see differing compression ratios marked on different manufacturers' designs.

Figure 12.4 A mock-up of a typical compressor front panel

My advice is to play in studio downtime (or on your computer DAW software) and try out different scenarios. On vocals I personally like a fast attack time of 5 mS or so and a moderate release time of 150 mS. Compression ratios vary with voice, but I'm guessing you will find greater ratios (therefore greater loudness) help match a voice to modern instruments better than lesser ones. I would suggest 5:1 or more in your experimentation. Do listen out for "mid-band gain ducking" as it is a fatiguing and annoying sound where a dominant bass can force the mid frequencies to pump up and down in level with the beat, and consider slower attack and release times where the sound you are compressing has a high background noise, as otherwise you will hear the noise pump up with pauses and down again when the sound resumes; again annoying to listen to!

There are other specialist uses of compressors worth mentioning. An "Expander" or "Noise Gate" or even "Noise Reduction Unit" is a sort of reverse compressor. Whilst sound is present it is passed through (virtually) unaffected. When the wanted sound stops, the background noise is reduced to a much lower level (below a chosen expander threshold), so it doesn't add to the mix. In other designs mid-band gain ducking is overcome (particularly at broadcast outputs) by breaking the audio signal into individual frequency bands (just like a graphic equaliser) and then compressing each band separately, before adding the signals back together at the output.

Perhaps the King of compression was Ray Dolby of Dolby™ Laboratories. He was key in the development of the first video recorder in 1956. He came to London in 1965 and tackled the problem of annoying levels of background noise on early multitrack recorders. In the early 1960s, the Beatles had created amazing soundscapes using a very restrictive four tracks, but the demand for further channels (and so narrower tracks on the tape) brought the serious problem of excessive tape hiss. The Dolby Noise Reduction System revolutionised work in recording studios, which in turn directly contributed to the complex, dynamic and exquisite sound of music albums of the 1970s and 1980s. The Dolby system compressed the high frequencies on each channel when recording the sound and then decompressed them to the exact same level on playback. The noisy high frequency tape hiss added by the tape machine was suppressed. The first system was the discrete multiband circuit Dolby A™, giving 10 dB (15 dB @ 15 kHz) of noise reduction. Dolby then created a simpler microchip version, which brought the price way down, and so Dolby processing became affordable in compact cassette machines. First Dolby B (10 dB NR) and then Dolby C (20 dB NR) from around 2 kHz and above. Ray's pioneering work continued into film sound. Dolby processing made early multichannel home cinema systems viable, and his legacy is still creating brilliant new products in the digital age.

ECHO AND REVERBERATION

In the chapter on Acoustics, we learnt that with multiple sound reflections coming back to the listener, things started to get confusing at a path difference of around 33 mS (11 m) and at twice that, distinctive echoes could be heard. The effect between a time delay of 33 mS and 66 mS is called "reverberation". The addition of echo or reverberation to music (especially vocals) became a cult with the advent of "Rock n' Roll" in the middle of the 1950s. (try "Be-Bop-A-Lula" by Gene Vincent for an extreme example). Many decades later, the use of time delay effects is still in vogue, as a special effect or simply to thicken and brighten a voice.

Reverberation was first made by putting a loudspeaker in a bare-walled room and picking it up with a microphone in the opposite corner. Obviously, these "rooms" were not in prime real estate; they would be in dank basements, which would be fine unless the London Underground thundered past a few metres below, as it did at BBC Broadcasting House! Frugal studios used the toilet with a temporary sound rig in it, or even recorded the singers in the loo to get that... certain sound. Many "girl group" 1960s hits were made in this manner and hopefully left the singers flushed with success. Using rooms occasionally just for echo was a waste of space, so a new approach involved installing a metal plate (often the size of a door!). A sound signal at one end excited a transducer to vibrate the plate, and the modified vibrations were picked up at the other end with another transducer. Several plates could be placed side by side in a single room, so much better on space. Overuse tended to create a "pinging" effect, which can be heard on several 1960s and 1970s tracks. The cheap version of this for local bands in the 1960s replaced the plate with a tensioned spring (or springs) around a foot long. It could be more successful than the science might suggest; however, it could be twangy or splashy in extreme, as is evident on many early 1960s guitar instrumentals.

Other systems created echo by using a tape machine, sending the (go) sound to the recording head and picking it up on a playback head (or more than one) some 100 mS later, to mix in with the original sound (return). These days we simply write a sound into a digital memory and read it back out again at a chosen time a little later. Although technically on the same technology path, they are mostly treated as separate effects by professionals. Reverberation "reverb" adds to or thickens the original sound whereas "echo" produces distinct repetitions that sit behind the original sound.

These effects are often labelled by the size of a structure where such a change could be naturally heard: small room, large room, hall, arena, etc. You get the picture! A lack of added time delay is known as "dry", and the increasing addition of reverb/echo in the mix as making the sound more "wet". These effects can be added to the front, as well as the back of the sound as "pre-delay"; the stereo effect can be modified, and even tonal differences to the frequency response. Whilst the concept is simple, sound engineers have created many terms to explain subtle changes, so in this universal text it is difficult to evaluate and support the reader, as could be done with compressors. I think the bottom line is "less is more": use time delay effects with caution; however, they remain great when they are subtle.

DISTORTION (FUZZ)

Strange to take a perfect sound and destroy it for art, but that's what we do, and we love it! Much of the music we admire would not have the richness and bravado gained by purposely distorting the sound. To be clear we love the "warmth" of the sound wave clipping of *even* harmonics, but the ear hates the clipping of *odd* harmonics, so simply over-driving semiconductor technology into clipping is not the solution, and again vintage valves have a place. Digital designs rule now; they don't just emulate the effect but ape the musician's favourite manufacturer's designs from over the decades.

FEEDBACK SUPPRESSION

In "Graphic Equalisers" above we learnt that feedback (or howl-round) was often subdued by "notching" the channel where the frequency of the howl round was present. The advent of operational amplifier ICs in the seventies allowed much cheaper production of formerly complex and expensive devices by replacing the many coils with "inverted" capacitors. They were still technically analogue devices. One of these was the "frequency shifter", which did what it said on the tin. The circuitry simply shifted the frequency of the input by 2 Hz to 6 Hz up or down, so a random vocal frequency of 784 Hz might be shifted to 790 Hz. The effect counts on the relatively small change in frequency not really being noted as a change of key, but it is enough for a sound reflection to be modified so when it is returned to the room it will not add to the original to cause a build-up to a howl round.

VOCODER

This analogue device was created by Homer Dudley to synthesise human speech in 1938 at Bell Labs, the hothouse of technological invention at the time. It was developed for secure encrypted communication but has found its way into music production. A sound is deconstructed into frequency bands and analysed to create plots of energy; it can then be reconstructed to be understandable. In music production, it is altered to have a highly mechanical feel. Listen to Autobahn by Kraftwerk.

PHASER/FLANGER/WAH-WAH

Phasing is another early effect, this time from the 1960s. Combinations of capacitors, inductors and resistors are used to create several "all-pass" filters, which whilst they don't alter the frequency of the sound (too much), they do create a time delay. When the delayed sound is mixed back in with the main sound it creates peaks and troughs in the frequency response, which are heard as an imposed rushing electronic wave. A great example of phasing is "Itchycoo Park" by the Small Faces and more recently, Daft Punk love using phasing on their albums. The effect was particularly popular in 1960s psychedelic music. A variation is the Flanger, where the notches are spaced in a linear fashion to create a more ordered and obvious effect and "Chorus", which thickens a single sound to have more body and impact. Examples are easily found on the internet for comparison. Wah-wah or Wah pedal is also an old analogue effect, which unlike the others, I have only heard on electric guitar. The effect sweeps a bandpass filter up and down the frequency range and can be likened to a strutting bird call or the guitar "crying". All these old analogue effects are now easily created in the digital domain.

DIGITAL EFFECTS

Eventide Harmonizer

Several effects only became possible with the advent of digital electronics, and in particular digital signal processing (DSP). Suddenly time delays were cheap and easy, and extended times were available. Phasing and flanging were cost effective and reverb simple and versatile. In my mind the earliest I knew of and used was the Eventide Harmoniser H910 in 1981. The unit offered pitch shifting, delay and feedback regeneration in a US$1,600 box. In time the versatility of the effects has increased, the quality of the effects has improved, and new entrants have made the originators up their game. Many products have become available as digital software for use in DAWs. It's a great time to be creative!

Aural Exciter

Originally made popular by Aphex, a hugely complex analogue device, the advent of DSP made the technology available to mere mortals. The original business model involved hiring the units out to a recording studio and charging by the minute of recorded time; they were not available to buy! Now it's available as a digital plug-in! Used by the very best in the business, the Aural Exciter was initially technically complex – a combination of dynamic equalisation (based on analysis of the input signal), synthesis of higher harmonics, phase manipulation and the addition of subtle even harmonic distortion, all to create a bigger, brighter, louder, seemingly cleaner vocal. And it did! Also known as "psychoacoustic processors", improvements over the years, particularly in the low frequencies, with the "Big Bottom" have made the "Exciter" a must-have in the leading studios for that sparkling vocal.

Pitch, Tempo and Speed

Most analogue tape machines of the 1960s and 1970s came with more than one speed. This was a trade-off between the quality and the duration of the recording. Faster speeds had a better high frequency reproduction and lower noise but of course used more tape. The tape speed of domestic recorders was often 1 7/8 ips (inches per second) and 3 3/4 ips. In general, British professional recorders used 7 1/2 ips for speech and mono recordings and 15 ips for music recordings, although 30 ips was also common, particulaly in the US. Great fun could be had by making a recording and then changing the speed for a chipmunk effect. Some studio recorders could be ordered with vari-speed which allowed for adjustments either way. Changing the speed clearly changed the pitch of the sound, so the tempo changed too.

The advent of digital recordings and DSP has meant that tempo (speed) can be separated from pitch (key change), and so by a change in both, a track can be made to sound the same but run for less or more time. This is used in real time in broadcast playback systems, with the duration of the hour being adjusted to cope with the presenter's links, automatically bringing the programming to a close in time for the news on the hour. Clearly such a tool is useful for club DJs to beat match between tracks and is used to keep people on the dance floor.

Auto-Tune

The ability to change pitch without changing tempo led to a new device in 1997, the Auto-Tune introduced by Antares Audio Technologies. Like other units in this overview of "outboard devices" it is available as a stand-alone rack mount unit; however, it is more often encountered as a plug-in for a DAW. Whilst some might assume it is a variation of a vocoder, it uses different base principles. The unit is now found extensively in sound studios and has become an essential tool in music production. What it does, simply, is use a key sequence of notes, laid alongside a vocal (or possibly an instrument), and the note sequence is used to pull the vocal on pitch.

In use it can be subtle to the point it is completely transparent to the end listener, although at extremes the singer will seem to jump from note to note rather than a natural slide, a little like the vocoder sound. This effect was famously laid on thick on the 1998 song, "Do You Believe" by Cher, although at the time the producers sought to protect their new sound technique by claiming they had used a vocoder. Others tried, but of course could not match that unique style, the "Cher Effect". I'm sure you can imagine there has been a lot of music industry debate about a box that can make a poor singer sound perfect, and I'll leave that to your private research...

Normaliser (Loudness Normalisation)

This book has covered at some length the quest to have a finished sound cut through louder than the opposition. We have talked of the Commercial Advertisement Loudness Mitigation (CALM) Act in America, which forces broadcasters to use metering which perceives loudness accurately and so ensures inserts like channel adverts are the same loudness as the programmes. In many other countries there is still a family rush to the TV remote when the breaks happen! Online content providers are keen that there are no surprises either, and so sources of music and video apply "Loudness Normalisation" to ensure the content all sounds the same to our ears. This is another area for hot debate, as certain songs that have been heavily tweaked in the production process are seen to lose out to the more mundane in this process. The way the loudness has been traditionally assessed is by "peak normalisation", simply ensuring all sources peak to the same level on a decent meter such as the BBC developed PPM (Peak Programme Meter).

It can be quickly seen that items such as the humble compressor, and more recently the Aural Exciter, create greater loudness for the same peak reading. A different measure is required. "Loudness Normalisation" can go two ways; it can analyse the average "RMS" power in the sound wave, or more accurately still, devices can use data collected from human perception experiments as defined by the European Broadcasting Union (EBU) in paper EBU R128, to synthesise a sample level of "Perceived Loudness", which is then matched to the programme content. Phew! Note, dual mono content needs managing with care to be compliant. The Normaliser adjusts the sound to match the content provider's standard, and after conditioning all songs should have the same loudness as you listen. Unfortunately peaks which could distort are still generated, so the output of a Normaliser will have a protection limiter just like the compressors above, so signal to noise and dynamic range of certain material is changed. It's what that "certain material" is that is the subject of debate and concern.

BIBLIOGRAPHY

Antares (2023) *The Science Behind Auto-Tune: How It Works*. https://www.antarestech.com/community/the-science-behind-auto-tune?srsltid=AfmBOoproQsrky9TvaPGO59F1I79LxPIAP1m9LDkDW4d7nd6JZfTSWvu.

Audacity (2025) *Loudness Normalization*. https://manual.audacityteam.org/man/loudness_normalization.html.

White, P. (1993) Martin Caesar, Behind the Aphex Aural Exciter. *Sound on Sound Magazine*, August.

13 Planning for a Location Recording

This chapter offers practical support, built and refined over many years. It should be read in tandem with Chapter 16 on "Real World Problems With Audio Systems" to gain greater depth on the everyday operational and physical problems that can confront a responsible sound practitioner at an event. It's imperative to remember that good relationships with crew and, in particular, the artist(s) are paramount! Get people comfortable with "you", on your side, and you have won half the battle!

A WORD ON HEALTH AND SAFETY

Chapter 3, "Introduction to Electronics" has given you terms and explanations for the audio signals found in sound equipment. The levels of voltage used in sound desks, whether they are from power supplies attached to mains electricity or batteries, are not dangerous; however, the mains itself is! The residential US (and Canadian) mains supply is 110 V AC, which is enough to be life-threatening; however, it is unlikely to be fatal. Americans visiting Europe will find double the voltage, a standard of between 220 and 240 V AC single phase, which must be treated with great respect. In unfortunate circumstances where a person's lower body has good contact with the ground, touching a live contact will cause electricity to flow from the hand through the heart and can be fatal!

In most European homes there is an earth to one side of the generated power (created at the local substation) unbalancing the feed. The earthed side is called "neutral" (blue wire) and is not generally dangerous on its own, floating just a couple of volts above a locally sourced earth (green/yellow stripe) (literally a metal pole stuck in the ground, or a good connection to the incoming house water main, which is again buried in the earth). As stated above, the live, however, is potentially lethal, and whilst it used to be colour-coded red for danger, is now unfortunately coded (for colour blindness reasons) a benign brown. Be aware!

On location, please *refuse* to connect to any dodgy looking mains supplies (cracked casings, wires protruding), and if you use a mains extension be careful it is in great visual condition, it is not overloaded, it is not daisy chained by other extensions and it's tucked away so it's not a trip hazard. Tape down wires where there are people walking past. Be particularly alert if there is condensation or damp nearby, as working in the wet multiplies the danger to you and your colleagues, whatever the voltage! The worst are on beaches or near swimming pools! It has been unusual to find high voltage DC; however, new

DOI: 10.4324/9781003717065-13

electric cars are already charging at well over 100 V DC and are likely to go to higher rails in time. *Anything over 100 V DC should be treated with great caution!* Whilst AC tends to throw a person away from the danger, DC can paralyse the victim and so keep them connected.

Europe also makes great use of the Residual Circuit Device (RCD) trip, which breaks away the feed of live and neutral should a small current flow between *either live or neutral to earth – a fault condition.* This factor can cause confusion as a competent on-site engineer will isolate a circuit on the individual breaker in the fuse board to cut the local live feed and make working safe. (And it should be a competent engineer!) However, in making a change, say pulling temporary wires through into a new area, earth and neutral may briefly touch. Whilst this is not dangerous, the system sees it as a fault; the RCD trips out *all* the circuits that pass through that section of the fuse board and likely plunges the property into darkness – an embarrassing and unexpected outcome unless understood.

Portable equipment you carry around *which needs a connection to the mains,* like mixers and amplifiers, *must be visually inspected for damage before each time you plug it in.* Please don't use it if the case is split, or the mains lead looks compromised in any way. In addition, portable mains powered equipment taken out to public places should be regularly tested for electrical safety. PAT Testing (portable appliance testing) exists in many countries; however, (at the time of writing) it is not normally law; it is simply good practice. That said, your venue may insist on compliance! How often you test is a matter of the frequency of use (or is that abuse?) and of the hostility of the environment. My view is annual checks for audio kit used by in-house professionals in regular controlled conditions, more if it is used by unknown personnel (hired out) in unknown places. After testing, a label with an expiry date is affixed. If you are hiring in, it's worth checking the unit is in date when you pick it up, or if you find an expired label on a unit whilst on a gig, look for an engineer or electrician to get you up and running.

There are specialist safety courses on electricity, and if your work regularly takes you on location (other than using a portable battery kit) you should have been on one. If a shock is believed to be in progress, the first instinct *must* be to turn off the source of mains power and so remove the danger to all. If this is seemingly unavailable *do not touch the victim*; they must be separated from the electricity by a non-conductive long stick such as a wooden broom handle.

You must be particularly alert if you are on a film or TV set, as three-phase mains power may be required because of the high current required for lights, cameras and set machinery. Do not run mains extensions from one part of the set to another, as you might have adjacent connections which are of differing phases of the supply (same with putting mains over the garden wall to next door!). If this happens you may find yourself across as much as 440 V_{RMS}! Of course, the complexity of a set also means that you must follow marked walkways, and always be on the lookout for trip hazards such as cables and low-slung lights etc.

A small point about chargers: whilst at this time many devices (mobile phones etc.) can be charged by generic USB 5.0 V outputs, many still do require a specialist supply. It is unwise to use any other than the manufacturers' specified charger, as the least of your problems will be terminal loss of your equipment; a flare-up of fire is a possibility. Charge items in well-ventilated spaces (not under blankets or clothing) and well away from water and direct sunlight.

GOING OUT ON LOCATION

The technical content that took up much of the earlier part of this book is designed to make a creative activity reliable and rewarding by giving you the confidence to get involved. When you step out of the safety of your studio, all the same rules apply, and then some; this chapter prepares you for the adventure. However big or small, however many people are employed to make it happen, the risks out on the road multiply through environment, equipment needs and simple preparation.

In my history, the BBC defined roles for work where an "Outside Broadcast" takes place, whether it is a recording or a live transmission. A planned site survey prior to the event is normal to sort out permissions to park, safe storage space for equipment and recording, make equipment lists necessary for the gig, to source appropriate power supplies, gain access to line systems and so on. On the day there may well be specialist vehicles, sound engineers, lighting and camera engineers for video, presenters, producers, researchers, production assistants (PAs), catering, health and safety and so on. Or it might be one person, responsible for everything and that could be you!

The scale and needs of every "event" are different, yet despite careful planning, too many go wrong. I was present one year for the disaster of the mega live peak programme TV "Brit Awards" when no one took responsibility in the independent production office to book the lines to get the outside broadcast back to the live broadcaster, the BBC that year – No event... The simplest error is, in the excitement, to forget to press the "record button", and so the most carefully asked and answered questions are lost for eternity. Where someone does have the courage to go back and knock on the door to ask for the time all over again, the confidence, spontaneity and often the simple professional bond are broken.

For these reasons, it is worth me writing this chapter, and it is worth it for you to read it! However life treats you, it is likely that you will start off on a small scale, either on your own or with a small and devout team to support you. I cover these solo trips in detail below. It doesn't matter what role you may play on the day; *someone* has to take responsibility for pulling together the right people, in the right place, at the right time. That is the role of a producer, plain and simple, but they may well subcontract the equipment planning and the people to a senior technician such as a Technical Director or Sound Designer. The same as an impossible magical trick, it's planning for all eventualities that makes it look perfect on the day!

I'm aware that some readers will still believe they can just turn up on the off chance – see who's there and blind them with charm... Good luck!

When I ran a community radio station, we had a reggae presenter who week on week bagged the most incredible "A lister" guests, to the point where we thought he was getting people to impersonate them. Seriously! That does show his recording technique left something to be desired; however, he was able to show the guests were genuine. It was his hard work, charm and approach that won them over.

Another presenter started on the business show, but I could see his great love was rock music. I gave him the offer of 120 minutes of rock programming a week; he made initial contacts and secured local bands. The music community is a small one, and through insightful interviews, he secured by word of mouth increasingly bigger bands and outstanding interviews in Britain and America, eventually to the point where he has been nominated several times for national radio awards. Another radio presenter at the same small station who loves Americana used the contacts he gradually grew with concert organisers to meet stars backstage and again has grown a formidable reputation which, in turn, opens new doors. American stars look him up when they come to the UK, even though the station is a small local one.

Like so much in this book, passion and commitment will take you a long way. However, there is no substitute for research and planning prior to an event so you can use your time effectively and come away with the intended result.

RESEARCH

If you don't know the person, how are you fixed to ask the smart questions? If you don't know the event, how can you frame those questions in the correct context? If the venue is in Coventry, what do you know about the city? Once researched, these layers of context create a smart and deep conversation. It may appear easy, but then magic tricks always do!

With smart research turned into intelligent questioning, you are ready to think about the technical side. In a very long history of doing this stuff, it's never one error that takes you down, but an unexpected combination of two!

Build your kit in the office or home the day before, and methodically test it. Then pack it as one so you know everything is in the bag. Use fresh batteries and pack a spare set; I don't use rechargeable types as they become fickle over time. If this sounds extravagant, think about the money invested in time and transport, as well as the guest's time. Rechargeable batteries can seem "full" on the recorder scale but last a few seconds under load, so are a bad risk. After a session throw the partly used batteries in a separate box and use them in non-critical roles like TV remotes and toys.

Pack a wind jammer, even if you think you will be indoors. Pack an additional small recorder if you have one as an emergency spare. Pack a reliable pair of headphones or buds. Pack your research notes, and a small notebook and pen so you can write down fresh details. *Ensure a memory card is fitted,* has tons of space and works on a test recording you can play back, before you leave home. Think 1 gigabyte/hour as a rough guide. Take a spare because life is like that!

My rock music friend always takes two recorders; there's his "lucky" one that is always pulled out of the bag, but he has a spare in case one day it's not so lucky. The headphones ensure a test recording can be heard back to check quality. I had one digital recorder which sounded perfect until the audio peaked loud enough for the internal limiter to operate. This functionally had mysteriously failed; the distortion was terrible! My advice is to set up and test the recorder in the way you are going to use it on that day, and in general to test it to its limits, so you understand the broader picture of what they are.

External tie clip microphones are an ideal way to avoid handling noise, especially if you are shooting video, but be sure the wires don't get pulled, and that outer clothing can't rub against the mic whilst you are recording. If you are using the internal microphones of a handheld recorder beware of handling noise (keep your sweaty fingers still!), ensure the microphones are pointing in the right directions (yes really, people hold them 180 degrees out!) and don't rock the microphone between you as you will be caught out by your guest's replies! If they are louder, bring the microphone closer to you; if they are quieter, get closer to them. The only rule against this is when there is a loud background, such as traffic, when stuffing the microphone under their nose makes sense. But watch those levels!

ACOUSTICS AND ISOLATION ON LOCATION

Inside

If you are recording inside a building find a quiet location with carpets and comfortable chairs where you can still find a position to sit with the microphones well placed. The furnishings will take out the "bathroom" effect and make for a more comfortable, intimate recording. Clearly beware of hard surface corridors and rooms with large windows. Pulling blinds or curtains partially closed can help significantly. Stop to listen! Can you hear machinery or voices, the thunder of traffic outside or gardeners at work with lawn mowers or leaf blowers? It's amazing how many people are surprised at what's on the recording once they have got home, as their thoughts were only on asking the right questions.

Do not meet in a friendly communal place like a coffee bar, as apart from the masking noise of your neighbours, the banging out of used coffee granules and the roar of the milk steamer will win in the battle for the recorder's preferred capture.

Outside

The biggest factor outside is wind... You can't see it, you often can't hear it, but a breeze can obliterate your recording with ease. Fit your furry windjammer (see Chapter 4, "Microphones") and put your back to the wind to shield the microphone further. Try a test recording when time allows (a foam pop shield will not cut it!). Whilst it would be very unusual to have a totally silent situation, listen for aircraft, passing traffic, sirens, whistles and children at play; all can cut through to make a recording unlistenable. A passing motorbike is about the worst, but don't hesitate to re-ask the question once the annoying sound has completely passed.

Be wary of mounting a microphone (or two) on the table, as a nervous guest will tap the table leg with a foot, and this vibration will travel through the structure and add unexpected thumps on your recording.

Beware of auto-level controls; they track the loudest sound at any one time, and if that is our passing motorbike and not you, the recorder will cheerfully fade you out and replace your conversation with the louder engine noise until the motorbike is out of sight.

Making the Recording

Once you are recording, don't stop, as each stop and start will generate a new file which will need to be stitched to others. In detail, don't be tempted to pause the recorder for any reason; it's much better to have unwanted material to cut out in the edit than to find you didn't remember to restart the recording again after a pause or stop. Also, this guarantees one file number, rather than having several files to sort and stitch. You can always make vocal notes ("take three!") on the recording as you go, which will aid the edit later. It is wise to listen to the recording with headphones (one ear on, one off works for me), so you can hear distortion or extraneous sounds and repeat the conversation to repair the problem at the time.

As in all interviews don't be tempted to make vocal responses to show you are listening. OK, a-ha and hm-hm distract the end listener; use non-verbal clues such as nodding and smiling.

Record a longer duration than you want, assuming a loss of unwanted content in the quiet of the edit, where perspective will win over memory. Five minutes extra in 30 minutes, ten minutes in an hour.

Try to agree a method of communication with your guest, so you can keep in touch after you part company, even if it's simply a social media connection like LinkedIn.

When the interview has finished and the guest is gone, you can always stay behind and ask the questions again with the benefit of the authentic atmosphere behind. This focuses your read and ensures you are not talked over. *Record 30 seconds of pure atmosphere* to ensure you have something to ride your editing through bad cuts and disturbances.

Working with Video

The microphones can be in shot if, for instance, you are recording a vlog or a video podcast. However, it will be normal for the sound to play second fiddle to the pictures. Often in post-production, the actors will re-record their dialogue in a studio setting, and this will be dubbed over the location sound. Two points: although this possibility exists, always position your microphone(s) as though it doesn't. Do your level best to capture a clean and tight dialogue, either attaching the microphones just under the clothes (audibly transparent clothing, of course) or using a boom-placed microphone just out of shot, pointing reliably at the faces of the actors (obvious, but I have seen paid people too cautious, or just plain lazy, with boom positioning). Point two: as above for personal interviews, record a generous amount of local atmosphere to aid the post-production process; they will *love* you up the way!

If you are working in a team and make a small error such as the boom arm contacting with a doorframe, wait until "Cut" is called and tell the Director, so she/he can repair the problem whilst still on location. Of course, you will feel a little silly for damaging the routine of the shoot, but you will also be top dog for owning up and ensuring the day is won! That is the Sound Tech who will be hired again next time! Be prepared, as a Sound Tech, to be in the background, as video positions and lighting are in constant need of change to ensure consistency across the shoot. You can take solace in the knowledge that most film work is later sound led in the cut, and that the emotions felt by the audience are fired as much by the dramatic sound as by the stunning pictures.

BIBLIOGRAPHY

Parkin, P.H., Humphreys, H.R. and Cowell, J.R. (1979) *Acoustics, Noise and Buildings*, 4th edition. London and Boston: Faber and Faber.

The Health and Safety Executive (2025) *A Guide to Electrical Safety Checks at Work*. https://www.hse-network.com/a-guide-to-electrical-safety-checks-at-work/

14 Working in the Studio

ACING THE STUDENT PRACTICAL DAYS IN THE STUDIO

Before we get into the roles you will find in radio and music studios, I thought I would take a page to talk about the early days of practical training. Lecturers will assign roles, and these are laid out in depth below. If you are placed in a senior role for a practical day, it is essential to give thought to pre-session planning. There are several reasons for this; the first is the practical likelihood that the technical stores will have limited resources, so it pays to put your order in early, and if there are any problems, order the next best! A second reason is this process gives a good framework to the day, getting you thinking not only about the studio facilities, but items like sheet music or scripts and the softer people issues, security passes for guests, access for disabilities, refreshments, etc. It matters! I once arrived minutes late for an interview on national live radio because I hadn't remembered my guest, Steve Harley of band Cockney Rebel, had serious mobility problems. The longest walk of my life; he was great about it, but I should have thought whether he would have accepted a wheelchair!

Now our third and most obvious reason to pre-plan: a technical layout of people positions, microphone stands (see "Appendix 1 – Microphone Placement") and the best microphone type to use (see Chapter 4, "Microphones"). Each microphone lead will need a tidy, safe route to the wall box, where it will be assigned a desk channel number. Everyone works differently; one successful mixer channel allocation model is to place instruments to the left of the desk, vocal microphones in the middle and the drum microphones on the right. The sense behind this is that the balance engineer is most likely to ride the vocal microphones most, sensitively lifting them out of the mix, so having them front and centre makes sense! The instruments on the left will require elements of EQ, but once this is done, they take second fiddle to the vocals. The drum microphones are on the right, as they are likely to be grouped on a fader for an overall balance, which will be right next to the channels for more subtle single changes to the kit mix. Mark up the items on the input channels clearly with removable tape (masking tape is fine), so there is no doubt what each channel carries. The input chain secured, thought can be given to recording materials, vintage multitrack or a mix down to stereo two track tape, through HD drives, though patched memory; it must be set up in advance and ready to go! On multichannel, ensure you list clearly what each track contains and put a copy in the box of any removable device, also listing sampling rate, bit depth and peak level. The band will need to hear themselves; foldback speakers are unusual in sound studios as they colour the recorded sound. Closed-back headphones off

DOI: 10.4324/9781003717065-14

a distribution are the norm; the band members need to know how to alter their personal volume, so if applicable, allocate a channel number to each musician.

The pre-planning may seem like a bore, but with a team first thing eagerly awaiting instructions, it's great to hand them a sheet and let them get on with it. Too often, in my experience, training sessions have failed to complete the assigned objective because they have run out of time! Time spent running around getting equipment that had been forgotten, time rigging the studio without any plan, time spent configuring the recorder to work… And all this time the guests are sitting on their hands. As you go, keep detailed paperwork in case there is a second session. Durations, takes, whether publishing rights exist for the music inserts, anything to smooth the path of the next session. The desk should have been "normalised" at the end of the previous session, even if that is just a button push on an assignable desk. All the channels should be left with the same knob positions and switch positions and hopefully leaving it in a condition to work from the start. Too often that doesn't happen, causing confusion to the incoming crowd. Please, as your last act of the day, take pride in normalising the desk for the next group.

WHY IS RADIO DIFFERENT AS A MEDIUM?

I apologise in advance that this part of this chapter is UK-centric; it is borne out of personal experience, which can be insightful. However, it also narrows the field and can exclude readers unintentionally. I think the fact it happened in Britain doesn't mean it is not relevant to other countries. I'm sure this history has been repeated across continents and may well be very familiar. I hope you will let me know.

Since its development over 100 years ago, radio has held a special place in the hearts of its listeners. The entertainment of that time, variety shows in the great theatres, had been largely impersonal; the stage performers, Ladies and Gentlemen, communicating with the audience as a whole. Radio changed this form of communication, with the variety stars of the day finding their way into the BBC, realising they were talking to a wall, with an unseen family or perhaps just a single listener.

This early culture pervaded and enriched the stilted announcers; its need was felt particularly when World War II required light relief from the dark days of the threat of Nazi invasion. Comedians created scripts that painted vivid pictures in the mind, since the stage was unavailable to view. Characters and catchphrases were the height of conversation and the stars were household names – the first national celebrities.

At the height of radio's fame and penetration TV started in the UK. Two fuzzy black-and-white channels which only ran for a few hours a day. Early TV tried to capture the radio stars, creating on-screen versions of popular variety radio shows, but somehow these didn't gain traction with British audiences. I think this was because people had already formed their own pictures in their minds and the TV-generated ones just could not live up to individual dreams.

Because of these early failures, TV tended to take an impersonal route, and that broadly remains the case today – less of the "me to you" and more of the "us to them".

If we come to today, entertainment is very different from the dark and economically difficult times of the last war era, yet some truths persist. Dance clubs (call them what you will to fit the age) get good audiences, yet the stage seems distant to certain demographics and certainly costly. Television still tries to pitch a "one to one" philosophy with ever-changing presenter positioning and camera angles. Something in the human psyche does not buy it, and this medium broadly fails to capture the heart of the audience. On occasion individuals can get caught in the moment of a closely captured drama, a special dance or poignant soulful song, but can't feel "at the table" for a discussion.

Radio presenters equally have tried to address their audience as a group, perhaps used to stage performing, but this is often as foreign, even to young listeners, as a simulated closeness on TV. Radio, or podcasting, is at its best when the listener (each listener) is "in the room" with the presenter(s), feeling drawn into the conversation. The best presenters, the Wogans or O'Learys, have an intimacy that is infectious, capable of introducing guests to an audience, as though they have just walked through the listeners' front door. First-person phrases such as "I am delighted to introduce you to my next guest", or "I think you will love the next track", enhance the "one-to-one" environment that the medium thrives on. Using names and places as though they are seated with you makes them fellow travellers on the voyage that is today's radio adventure: "Jo, I just loved your story about the split onesie; to my friends in Solihull, stay away from any tiger-striped customers in the minimarket aisle!"

At the end of the day it's a mindset; success comes when each listener has at least equal value to the presenter's ego. It's an approach, but getting the illusion right makes you "our presenter" rather than "the presenter" and gives richness and longevity to a radio career.

ROLES IN RADIO

I have watched as a good number of younger UK friends have chosen media university courses, and whilst a number have made it into the industry, many have not. When I have looked at what went wrong, although the course content may have been poor on occasion, it was their *choice of course* that let them down. The single important driver is to find a way into the industry; once there, it is an individual's tenacity and ability to network and grow relationships with colleagues that will win through! This chapter sets out to dispel a few myths and to focus on the needs and skills of roles and how to get them.

The Producer

As a role in the larger stations and national stations, the producer will vary to a degree depending on the employer. In radio, the programme producer fulfils two roles that have been split out in TV and film. The first is the management of the programme budget, which includes paying the presenters. This budgetary ownership is split further in "pictures" where the action on the day falls to the Director. Radio is a simpler affair; both roles are mostly one, and it's arguable whether an in-studio presence is required at all in the case of a "Self-Op" presenter. Certainly, the original greats – Wogan, Moore, Blackburn, Edmonds and Jensen, to name a few – would be mostly pleased to see a rare studio visit by their producer; however, they would expect to run the programme unaided. To be clear, minute-to-minute editorial control sits with the presenter, where *technical* control belongs to a Studio Manager or Technical Operator (or "Engineer" in some companies) should the programme afford one. On large programmes one or two producers might work into an Editor who would carry the editorial can and hold the budget. The producer used to have programme secretaries before computerisation of many office functions; however, this has morphed into more of a production assistant, or perhaps a "Broadcast Assistant" (BA) who has technical and editorial skills to support the producer on day-to-day activities. It's only with added complexity that the role in studio production becomes useful and desirable.

The producer is responsible for the management of the guests, booking them to appear on a particular programme, getting them to the studio on the day and back to reception, management of audience feedback including phone-ins and social media in general, organising and editing the views to ensure they are clean, balanced and honest. Pre-recorded interviews need editing for transmission, and the producer would normally make decisions on what is kept and what is lost; and often today, they do the cuts themselves. News and sports have breaking stories, and the producer is responsible for picking priorities and lining up the inserts for transmission. In addition to this, it is the producer who is expected to understand the editorial rule book and to ensure all aspects of the programme conform to the expectations of the regulatory body; in the UK the body is Ofcom who can censure and fine broadcasters for breaches.

History has favoured people taking on the role who have previous experience of either BA or Studio Manager roles, where the grassroots skills have been honed over many transmissions, and so smart decisions are made from the experience of decisions by others. The role is clearly a leadership one, bringing together the varying contributions to be "on song" on the day. A simple wrong word on air from a presenter can end careers, and a smart producer will look to manage the situation at the time and the aftermath. As a BBC network producer you knew two immutable truths: you would be the "fall guy" if the presenter could be saved, and given a flawed DJ, no one is bigger than the station management, and if the presenter stands against them, it is invariably him or her who will walk! If this seems a harsh or negative stance, please see it as an iron underbelly to much lighter fun times that happen daily up top!

The producer is "content creator", "mother hen", "promoter" and editorial "quality control". It is a "buck stops here" role; however, it is creative, gives opportunity to shine and is rewarding when a good plan comes together. Either the computer picks the tunes for a show (boring), or the producer picks the music tracks for a show; partly because they are in tune with station policy, partly because it is tedious, but mostly to protect the DJ from claims of taking bribes ("payola").

Computer selection of the playlist is popular, particularly where resources are tight. The mix of music age, style, repetition rules and so on are set up in the digital playout software. More enlightened stations still choose the playlist for each programme by hand. A great music plot is very different from a club DJ set. Dance sets need to get people to the floor and build the momentum with subtle beat increments, which will top out in a crescendo and then start the build again. If it sounds easy, it's not, and the best in class have begrudging stories of the rare audiences who were impossible to move. A radio show must keep the listener guessing on pace, style and voice (slow/medium/bright – male/female – lyrical/raunchy); it must be constantly changing to keep the sound fresh and immediate. No matter the audience, the station headline, sequenced tracks which are clones quickly lose the listener's attention, and they'll reach for the dial.

Why is music programming tedious? All stations have repeat times on bands and their tracks to ensure the repetition is controlled and sensible. It always seemed, particularly with new releases, that my perfect build of a show was dented and reduced by substitutions, once we knew what the shows either side of us were playing and we had to change, or negotiate that they changed. The final point, and a thread running through the role of a producer, is that the DJ must be distanced from the music choices, or record companies and listeners will claim bias for or against certain artists. If allegations are made against the producer, it is easily rectified by moving the individual on. If it's against the DJ it has a high public profile and causes embarrassment to the station staff, because it smacks of mismanagement.

The Presenter/DJ

It's official: the profession of radio presenter is now over 100 years old! As you might expect, the role has changed with society, fashion and culture. The first "radio broadcaster", the BBC, even had its own version of the English language, "Received Pronunciation" (is that pro*nun*ciation or pro*noun*ciation?) and a Pronunciation Unit that guided the presenters on how to say words correctly. Up until the start of Radio 1 in 1967 all entertainment shows were scripted and rehearsed fully in real time before going on air with a live show. In addition to the programme producer, two Studio Managers would assist the presenter, one on the studio mixing desk and the other playing in "Grams" (as in gramophone records) and later "Tapes" of speech and music inserts. The battle for the presenter to operate their own equipment, "Self-Op", was a hard-fought one with new British DJs arriving from both the US format-based British "Pirates" who played their own records, and Luxembourg who were like the BBC and didn't.

The argument was won after historic talks between DJ Kenny Everett, Controller Robin Scott and BBC "Light Programme" producer Johnny Beerling who, incidentally, was the only BBC person to bother to visit Pirate Radio London to see Self-Op in action. The smart decision to use American jingle maker "PAMs" to make the station idents also came out of this meet. No surprise that Johnny also produced the first ever Radio 1 show (ex-Pirate Radio London DJ Tony Blackburn) on Saturday 30 September 1967 and Mr. J Beerling went on to be the Controller of the most successful radio station in the world. I have to say I wonder about how much Self-Op was about reducing the cost of the new networks (Radio 1 and Radio 2) to the bone? The battle isn't over yet; whilst many national stations still use Studio Managers for news and current affairs, most regional and local stations would be unable to afford the cost of the extra staff, so Self-Op is the norm.

In the 1970s and 1980s little changed, in that the presenters all spoke with slightly "posh" accents, wherever they came from in the nations, and as many now came up from smaller stations, they were able to Self-Op. The advent of independent radio in the UK did not really change this, with clear speaking being a requirement, set against a big personality and a confident demeanour. In this time, it was cool to be a DJ and a good number were seen to be bigger personalities than the artists whose records they played. It might then come as a surprise to realise that the finest presenters are, most often, former school class clowns because they craved the attention of their classmates; and as a result, sadly even the best presenters are surprisingly easy to wound! It's important to raise the point as this vulnerability is essential to the ad-lib style of the great presenters, the ability to wear their heart on their sleeve and to be self-deprecating. I have personally known self-assured national radio DJs and they either now work on the buses or came close to prison. It's a broader point about programme presenters in general, an inclination to being a willing communicator makes for a much more informed interviewer, and for better working in a team. You may say, "That's me!" and you may well be right, however you have been warned! Clearly showing sensitivity and support when working in with these genii, will be good for your career and theirs!

In the latter 2020s we are currently in the thrall of celebrity where dubious fame is seen as a suitable grounding over the efforts and passion of dedicated informed personalities. In my experience these careers are often short-lived, people pass through radio rather than stick and stay. There are notable exceptions in the UK, such as Michael Ball who made his name on stage, and ex-comedian Alexander Armstrong, both fabulous all-rounders, but then I'm struggling. You may think being an actor would in general be a suitable background, but in my experience, actors look to *play the part* of a presenter rather than be one! I guess that's why they went into acting in the first place, but I've witnessed it's so difficult to make a change in mindset. Newsreaders equally find radio presentation dodgy, used to the strictures of a script and the confines of practised intonation, the best can sound stilted and hesitant. I'm pleased to provide exceptions and John Suchet (brother of Poirot genius David) and more recently Radio 2's Tina Daheley show us the way!

The foregoing is to set you up for success in radio and podcast presenting. You need warmth, empathy, honesty, a natural curiosity, resilience for the knockbacks and a love for people so strong you are content talking to a wall for up to three hours a day (I couldn't do it!). The natural presenter sees beyond the wall, has the listeners in his/her heart and an earnest desire to make them happy. If self-promotion is your aim, or perhaps money, you will fail because you will not stand the course, much as a musician who strives to cultivate fame as the goal, over a passion to make pioneering music.

Sound Operations (Music Studio – See Engineer Below)

The defined roles of the last century have softened, yet one aspect remains; that sound operators/technicians can be drawn from both degree-holding backgrounds and non. Many national programmes are contracted out to production houses, each of which will have a view on the splitting of responsibilities and the depth of knowledge within each role. In the BBC traditionally a Studio Manager was degree entry, and a Tech-Operator was apprenticeship entry. Qualifications ebb and flow as the years pass, but the revival of apprenticeship schemes again gives a different route into sound work from a degree. If I take Radio Operations first, there are differences in responsibilities; however, whatever the role title the work is similar in many ways. In the BBC both degree entry and apprentice entry were technically responsible for the successful transmission of a programme and for taking charge to resolve technical issues. Both roles had to understand the format and run of the desk, being able to use monitoring points to track down, isolate and rectify faults, hopefully without the audience even noticing. Both require a solid knowledge of sound theory, which is more than covered in this book, and an appreciation of quality, which is as much experience as training.

The Technical Operator/Assistant skillset of today (no longer in National BBC) is still favoured for live working at the outputs of many stations across the world because of an in-depth technical knowledge of the entire broadcast chain; however, the professional Studio Manager is likely to be a specialist in a particular field of radio, where their skills often gain legendary status. The skills that come to mind in the BBC are (1) news and current affairs, (2) popular music, from DJ shows, through bands to orchestras, (3) classical music balancing, (4) drama including live effects and Foley and (5) daily features and sport programmes. On smaller stations the more experienced operators will be expected to manage across the range of skills and often do so with real flair!

Regardless, the skillset is to have an artistic ear to balance sound, to always be on top of the sound quality, to be able to work quickly and accurately and to be able to previously analyse and obtain the equipment to rig the studio on the day and derig and troubleshoot without recourse to others. The problem solving extends to sound effects, convincing listeners in dramatic scenes where both the actors and the background FX merge to create a realistic sound stage to mimic a scripted situation. The desk operator needs to be personable as many influential people will pass through the studio areas, and the ability to inform and often persuade is a necessary skill. Independence is key, as broadcast is fast moving, complex or both, and failing to keep up is never an option. Expect 24 hour working as that is the nature of the job. The difference between Operator and pure Engineer roles is clear. Operators don't fix faults! They rapidly and effectively route around them to get the transmission back up and running. The proper fix can wait for the engineer later.

Engineer

A description that is often, in radio, wrongly applied to desk operators. An engineer will normally have a degree in electronics or computing and will have extensive on-the-job training before they are allowed to work safely on broadcast equipment. They normally will not operate the desk in a broadcast situation, but they do need to understand how it works so they can make repairs. Whilst documentation should be available for all the equipment they might meet, a good Broadcast Engineer is expected to work from first principles, increasingly on firmware and software as much as hardware. An essential link in the chain, an engineer is science-based in background, highly patient in the rectification of challenging faults and, with experience, highly skilled. The engineer will also plan and execute technical revisions to broadcast equipment, ensuring that live broadcasting is unaffected and, where possible, there is a minimum impact on normal working.

Music Studio Engineer

In British music studios the role of engineer may well apply to the desk operator. The music studio engineer must place a high premium on sonic quality, as successful recordings are likely to be listened to thousands, if not millions of times. The engineer will have experience of many microphones and be able to select his/her choice for each voice or instrument, with a good feel for the sound before the mic is taken from the box. Their ears will be the final arbiter, and their knowledge of the studio acoustic will frame how the artists are placed. The engineer will decide on any outboard processing to modify the sound, with particular focus on the lead vocalist. Whilst not necessarily deeply technical, the engineer must evaluate sound problems and have strategies to solve them, as both studio and artist availability are at a premium. A studio's reputation can be made or broken with each session that is booked, and the buck stops with the engineer. A successful multitrack recording may be taken elsewhere for postproduction, however, mostly the final mixdown is carried out by the same engineer, as she/he knows the acoustic and monitoring of the control room and can get the best out of the studio sound. Just to note, the growing vogue to produce multi-channel versions is leading to new dedicated mixing rooms.

Assistant Engineer

On a meaningful music session, it is likely that there will be an Assistant Engineer. The role is to support the engineer in ensuring the smooth running of a booking from a technical direction. They will be familiar with the engineer's approach and know the technical facilities well. They will arrive early, power the studio (most desks sound better after a warm-up) and will rig microphones and stands as directed by the engineer. These will be patched through to the desk in readiness for the arrival of the artists. Once the session is active the assistant engineer will make small adjustments to microphone placement under direction, will swap out faulty or imperfect microphones and will deal with monitoring in the studio so all can hear their part clearly over headphones or personal earpieces. The assistant engineer will need to anticipate the needs of the engineer, as a theatre nurse handing instruments to a surgeon. She/he may well handle the multitrack recorder, particularly where it is analogue tape. In the routine of an ongoing session, the assistant engineer will have more downtime than many, so will be expected to get tea and coffee and take food and drink orders. Whilst this seems mundane, these smaller tasks are essential to the producer and artist's experience of the studio, and the personality and approach of a good assistant engineer will carry the day!

Broadcast Assistant

With a foot in the technical camp, and another in the creative, the BA is a great first role to learn the business of broadcasting from the bottom up! As the junior role in the team, the BA must turn a hand to the current priority (which might be a tea round) and use initiative from day one to tackle it. I like this entry, as it equips the BA with experience to apply for a range of jobs to progress in a direction of their choice. A concern is the obvious point of being a jack of all trades rather than a master of one. Time spent across departments sorts this out, but in my experience, it is not a position to stay in; it works best as a stepping stone, not as a destination.

Journalist

Journalist entry is by far the tidiest. The skill is taught at university, the direction chosen by the candidate after leaving school. The coursework is fairly well settled, meaning that unlike other broadcasting entries, there is a clear route into a post. Many broadcast journalists start in regional newspapers or internet news sites, gaining experience and making a name for themselves. If you are competitive, it's important to move out from regional work before it labels you as it gets increasingly difficult to move to a national or international career as the years pass. I say this in the obvious knowledge that many will love local work and the ability to get closer to the heart of stories within their community.

Newsreader

A successful journalist, the Newsreader is a skilled intonation authority, able to pick words off the page with ease and to apply the correct weight to the story, ranging from comedy through to tragedy. Light stories often end a bulletin to give a bridge back to entertainment programming, while the more serious stories, requiring authority and even pace, headline the News. The Newsreader is unflappable, and able to take on new stories with the red light on. The news will have a fixed duration; many to the second, as stations opt in and out by computerised switching. A skilled Newsreader will change pace to accommodate the unexpected and get back on time. It's a job for those who like to think on their feet. The Newsreader is often the emergency standby in case a presenter fails to show, or the studio goes down. The Self-Op news studio therefore has sources to play music as well as play in news inserts. If there are remote live inserts, it is more likely the Newsreader will have technical support from a Studio Manager.

GENERAL PRESENTER TIPS

A great presenter will keep up with the pronunciation of politicians' names and the positions they hold; and music acts, and even the venues where they are playing. The presenter will know the popular shows, the format and the characters who lead them. She/he will listen to the news and be aware of the sensitivities of the day; what shouldn't be said, and be on the lookout for the titles of music tracks which may be unfortunate, given the context of the day's stories: the track "Up, Up and Away" on the day a helicopter crashes or "I Don't Like Mondays" when there is a street shooting… You get the picture!

A good presenter will put her/his audience at the centre of all she/he does. They are your biggest friends, but also your harshest critics. Show empathy for the music and match the pace and energy with your voice to soften transitions between differing tracks.

A music playlist should be like a great meal out: full of variety, adventure and surprise; a rough feel for the balance of female–male–band with something for everyone on the menu. In the absence of a Playout System, when building the tracks for the show, your producer should mark whether a track fades or ends and add the track duration and length of the intro to your playlist (on paper or more likely on screen) so the presenter is confident and remains in control.

Pace your voice; gabbling may suit your flow, but it will not carry your audience with you. Take time to get the emphasis correct and to engender warmth in your delivery. To "Errrr" is human, but not for radio presenters. An ummm is not a springboard to the next phrase; practice using silence to fill the brain freezes and sound awesome!

MUSIC VERSUS SPEECH IN AN INTERVIEW

All stations have rules about music versus speech. Community-based radio can be more flexible, yet they still must attract and keep listeners with entertaining chats as well as decent music. Frankly the speech/music balance is about what feels right. It's good for both presenter and interviewee to take a break between questions. When those breaks occur is a matter of what feels right at the time… Is it a natural pause in the conversation, or are you deeply involved with fascinating stuff? Is the subject light-hearted or serious? Has the interviewee got a lot to say, or is questioning like pulling teeth? Is the interviewee itching to leave? Basically, the brain should be constantly evaluating "absorbing" or "tedious"? If tedious move on and play the music.

That said, in an "average" conversation with "average" content, breaking for music around every five minutes makes for a natural and balanced listen. As to total length, that must be judged by how stimulating the speaker is going to be for the listener. I'm sitting here smiling as there have been a few times when I've been directional over the talkback in the radio studio to a stand-in presenter who won't or just can't find a way to change the flow of the interview to introduce music. I suppose in a way it could be seen as an insult that the music becomes a priority over what the guest has to say, however, it is the listener who must come first, and so a way must be found.

You will not be surprised to find that there are no real rules as there is no "average" interview; however, the success in the radio industry of an individual presenter is often found in their sensitivity to the needs of their listeners and the skill they use to manage their interviewee to meet those needs.

TIMING IN BROADCASTING

This can sometimes jump out and surprise new people, but broadcasters often have to hit a mark with the accuracy of a second in an hour. Where items are pre-recorded, they still must fit the time between one news bulletin and another. TV is stricter than radio; however, the shows are more carefully laid out and rehearsed, with people in the gallery taking responsibility for getting the presenters to hit their marks with countdowns from ten in their earpiece. Where TV is a busy town, radio is a quiet village and often a live Self-Op radio presenter will find herself or himself completely alone for the duration of a show. Even if there is a "Studio Manager" on a national station to support the presenter, they will have their minds firmly on technical issues, leaving time keeping to the other side of the glass.

Where a programme is studio-based (the presenter is not operating their own equipment) there will often be two Studio Managers, one on the desk and the other providing the recorded inserts, whether they be music or speech. In this case it is normal to have a studio producer to keep everything together, and they will support the presenter with 30-second prompts from the end (music) or 10-second prompts (speech). They can also keep an eye as the programme nears the hour and swap items around to bring the projected total time back in the right ballpark. They may well ask the presenter to fill an underrun of, say, 30 seconds with little notice, so the pressure is not completely gone (and the ideas below come in handy!).

Being able to talk the hind leg off a donkey is a key requirement of live radio; however, there are a lot of tips and tricks to bring the timing under control, particularly for the less experienced. The first, and I personally believe essential, are "Time Sponges". These are from two sources: the Network and your own research. Network-based ones are information that the radio station wants you to read out, which might be promotions for other shows, upcoming events and, on commercial radio, spoken advertisements and endorsements. Often these can be verbally stretched by repeating important details or (where allowed) voluntarily reading out ones which weren't directly scheduled into your show. As a side point be sure to inject your own style and passion into these bland reads to make them your own. On talk radio the same salient facts can be repeated many times in an hour, particularly at the start of the hour to get an audience onboard with the topic of the day and getting involved with feedback to the show. In my own opinion some presenters are using the topic of the day as a crutch where the padding has worn thin and it's now sore to use!

Personal "Time Sponges" come in two flavours, those you research and those you use time after time. It's important to keep up with the news for several reasons, but I would suggest it is clearly in the presenter's interests to glean some fresh, fun stories of the day prior to going on air, which will enrich the programme with listener input and make great ways to bring timing back under control. The ones used time after time I call "Strings", and these can be practised and roughly timed as familiar buffers in unfamiliar situations. The neat trick is several can be sequenced in a tight spot to make any duration you require. Here are some simple examples that can be planned and even scripted before going on air:

String 1: Hi, this is Jane Smith on "Exceptional Radio" broadcasting to you online, 101.1 MHz FM and DAB+... Just ask your smart device for "Exceptional" to find us... The time is just after 8:35 and I know you are going to love...

String 2: I'm just looking out of the window, and I have to say this is a day... for ducks... for woolly hats... for sunscreen, etc.... High and low temperatures (by region if appropriate).

String 3: I'm with you till 10, and then (walk through times and presenters in today's schedule).

String 4: If you'd like to get in contact with me, I'd love to hear from you at Jane dot Smith at Exceptional Radio; you can text me on 3 double 5 double 6 and find me on WhatsApp as "Exceptional Jane".

String 5: We have got some great guests coming on the programme this (next) week, and I'd love to tell you about a couple of them...

NON-VERBAL TIME CHANGERS

The simplest, and perhaps most crude, time changer is to simply drop a music track or speech item to get back on time. Often it will take you the other way and you'll have to pad out your links to bring things right. I had a variation on this when programming daytime music on Radio 2. I would put an additional track in each hour, often of short duration, and put it next to a long one. That then gave the presenter the opportunity to choose the length that better suited the situation and drop the other one. It made the programme-as-broadcast (P-as-B) return (to pay copyright) a bit messy, but I think it did more good than harm.

Always back-time the last music track of each hour so you *know* when it needs to be started well in advance. In a music show you can lose a potential overrun by fading a track early, hiding the fade with a lively back announcement to bring you back on time to begin the last music track in the hour. A variation on this is to use the back-time of the last track of the hour and start it with the fader closed, fading it up under your voice when you are ready for it. Beware, it can sound awful if the track has strident vocals, and you are in the middle of a phrase. Best to bring it in so you can just hear it in the headphones behind your speech, and to talk until the end of a music phrase before bringing it into the foreground. For this reason, many presenters have known pretimed instrumentals which can be used as required, and often become part of the listener's day if they enjoy the tune. These must be reported in the "P-as-B" as a music track if you allow a significant section to play without a voiceover.

A talk show version of getting back on time is a "Pot Cut" where a recorded interview is smartly cut at the end of a sentence before it has run its course. The phrase comes from when there were large rotary faders, and that cut would be made with a swift hand movement. It is harder to do with modern faders, and worth practising. I use my thumb in a flick action to get the speed.

As a last point, at the end of a paragraph about waffle and timing I want to make the point that for intelligibility *less is more*, keep it simple and use plain simple English to get important points across. AI is not your friend here!

AUTOMATED SCHEDULING

Computer-based programming offers another approach; however, it can get glitchy. Times can be stretched or shortened within items in the hour, and it often goes unnoticed if the pitch of the song is brought back in line. If the pre-recorded presenter links are overly long for the hour, they can take on an unpleasant chipmunk tone, which is to be avoided at all costs! Matching the times of the music and speech roughly to the hour remains the best way of having a natural-sounding voice-tracked show.

SOUND EDITING FOR CREATIVE CONTENT

I have been a sound editor on national programming for decades, and I still get a kick out of making something formally raw, sound polished. At my start in the BBC the medium was 1/4-inch two-track magnetic stereo tape (there were still some mono one-track machines in the system for journalism). To make an edit the tape would be rocked past the replay head whilst listening on speakers, or more usually headphones, and the edit point found. The tape was marked with a chinagraph pencil on the replay head, at exactly the right point. We used a one-sided razor blade to cut the tape at the mark, place it in an aluminium block and cut at an angle at a precise point; the unwanted material would be pulled past the replay head, marked again and once the two wanted ends had been joined, the removed piece would be discarded. The join was completed with a special "splicing tape" which was a cross between parcel tape and masking tape. It was a glossy white tape, which could be peeled free if the edit failed to be exact to give another go. If this sounds cumbersome, skilled operators could manage six edits a minute! The reason for the angled cut in the block was that the blades became more magnetised with each pass through the tape, and in time this was enough to add a click or thump on the splice each time if the blade was perpendicular to the tape. Unbelievably this phenomenon has fed through to DAW editing (see Figures 14.1 and 14.2).

Now, many computers either come with sound editing software, or great multiplatform products like "Audacity" can be downloaded for free from the internet. These are called a DAW in the profession. In many ways, the easy availability of these sound tools has undermined the skillset of the people who did this year on year as a job. Something that used to be taught and tested is now assumed to be so trivial that anyone can pick it up in half a morning! The sad truth is it's easy to edit badly, but it still gets called an edit. That said, modern DAWs are capable of a fine result, given the appropriate skills of the user. The biggest blessing in the digital age is the "Undo" in the Edit menu, which is a great get out of jail free card. I'll talk more about some tricks and tips later.

So, what's wrong with many modern editors? Surely if stuff goes out on air it must be good enough? The throwaway nature of daily radio does mean a bad edit will be heard once and consigned to history. This is not about speed of working; it's about providing accuracy and pace in the same allotted time. It's an art like painting, pottery or flower arranging: we can all have a go at it; it's just that some are inspired to create to a higher level of perfection. The skill of a great editor is to render a result that tells a succinct story, which in turn is an easy and compelling listen.

Radio 4's quiz show "Just a Minute" has survived many decades of the BBC's history. Its start coincided with a determination by the BBC to be the best! In "Just a Minute" competitors must talk for 60 seconds on a given subject without breaking three rules: hesitation, repetition or deviation. I have often thought the programme idea came to a young producer undergoing editing training as these are the absolute goals of an inspired editor.

Hesitation: it's natural in speech to have times when the brain must catch up to the mouth. Some DJs never achieve it! The natural response is to fill the hole in conversation with an *umm*, and so where these are highlighted by their clear shape on the DAW screen, they are quickly excised. Many will repeat the end of the last sentence, or start a wrong sentence as a catch up, and whilst the other two derivations might claim it, to me it is a hesitation in the story which needs action. Decent removal of hesitations improves the flow of the work and particularly the clarity of comprehension by the listener. Big rule! *A breath is not a hesitation, it is a natural need, and cutting out all the breaths in a misguided attempt at pace is both amateur and unnatural.* Some *can be cut, if someone is short of breath, or perhaps nervous, they will breathe excessively, and up to half these extra breaths can be removed, leaving a loose rule of one per normal sentence, or two for a long one.*

Repetition: I'm not talking about a few words to fill a gap, but the need by the speaker to repeat a vague point to instil more clarity with a new version. Almost without fail, taking the second version and discarding the first will create a better flow, greater clarity and, of course, a more succinct piece. I might leave the repetition in if it is around a passionate core value of the interviewee, who is just hammering home their conviction, but that is a rare case.

Deviation: this is not perhaps what you think it might be... With a confident and experienced interviewer this shouldn't happen, but it takes all types! When an interview is planned, not only the person, but the *reason* for the interview should be pinned down. If a Liverpool centre forward scored a hat-trick there is a clear focus for the story, and all the questioning should support that end. Weepy footballers might want to talk about early life inspiration from their pigeon-keeping uncle, but it clouds the issue, and if it has made it past the interviewer, the distraction should be lost in the edit. Keep on track, keep the story simple and focused and you will bring the audience along with you.

TIPS AND TRICKS ON EDITING

If the item is unknown to you, or it's complex, and time allows; listen through once and take notes of fluffs, interruptions and retakes. Then the item can be assembled in the right order before you start to cut. This is also the time to create a cue sheet. The BBC way, and I think it works well, is to cut the first question and read it live; the playback then starts on the interviewee's answer. The cue sheet is necessary to write down that first question ready for the broadcast. Start moving through, leaving breaths, but tightening on pauses and mistakes. If you hear a decent breath, mark where it is because it's a great aid to repairing where one should exist but doesn't. The other important find is some clear background residual noise (atmosphere) which can be cut in to create pauses where they need to exist.

The first technical step is to try to uncover any flaws in the recording from the environment (although the smart operator will have detected and dealt with problems before the record button was pressed in earnest!). Wind noise, heavy traffic, induced hum from dimmers and structure-borne vibration all can be filtered to take back the level of annoyance, and this is the time to do it.

Next, versioning! When a recording is first made it will have an undecipherable number assigned by the recorder. When you copy it to the DAW, rename it immediately to something sensible; a footballer's score of three goals might be called "Hat-trick FC uncut". When you move it into the DAW rename it immediately to "Hat-trick FC edit" and save your work as you go. When you are happy the interview is ready for broadcast export the cut to a file named "Hat-trick FC final". I know it sounds like overkill, but when you want to use that recording two years later you will thank me! This process (*uncut, edit and final*) means whatever happens you have a recoverable situation. At worst, the recorder will still have the raw "uncut" version (if you can find it!), and you can start again.

Sometimes when you edit it will create a thump, splat or click at the join. This is sadly because the cut was not a clean one; it was made in the speech (or music), not in a gap. This happens because the DAW makes a 90-degree cut, like the magnetised razor blade in the section on "Sound Editing for Creative Content".

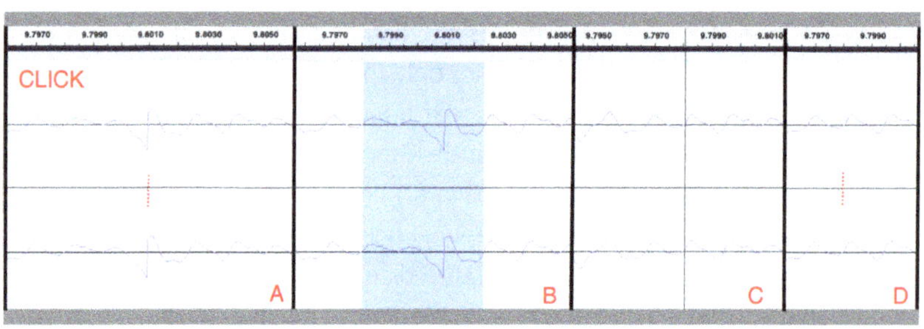

Figure 14.1 Method to edit out a click

The reason for a click is that the edit has inadvertently introduced a hard vertical edge in the waveform (Figure 14.1).

A second reason (more a thump or splat than a click) is that a cut has introduced a DC offset (Figure 14.2). In plain speak, the waveform on average spends the same time in positive volts as it does negative. If this pattern is disrupted (instead of the waveform going positive negative positive; it goes positive, positive or negative, negative), a DC offset is created. The solution in both cases is simple. Find the problem by listening, put it in the middle of the edit window. Repeatedly magnify the waveform until each wave is visible and look for either the sharp edge of the join or time spent too long north or south of the 0 V line.

The bad edit discovered, join a nearby positive to a negative at 0 V (the line crossover point). In the diagrams, the red dotted line in each "A" shows the position of the fault, "B" shows the selection to cut across the damage, "C" the result, and "D" the final waveform (red line shows edit point). This surgical removal will not alter the flow of the recording, and all should be fine now.

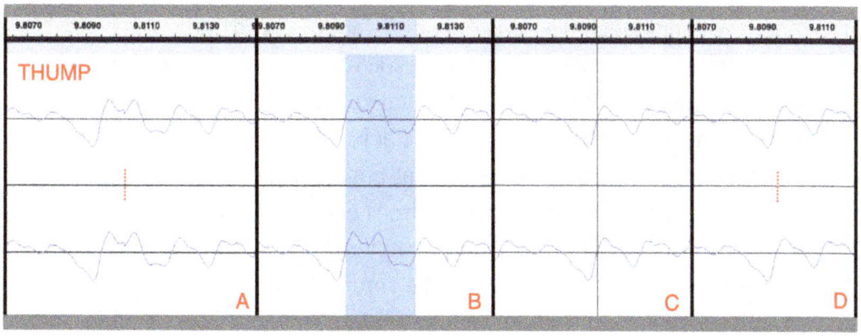

Figure 14.2 Method to edit out a DC thump

While editing, look to cut keeping a natural rhythm, particularly to gauge the size of pauses between words. If you can cut on a hard clean word such as *c*at, *d*og, *g*et, *k*ite, *p*et or *t*ap. What they share is that you can't slur into them from a previous word; you must change the shape of your mouth which gives a distinct edge to the sound and the waveform on the DAW display, meaning the cut is clean. I'd carry that forward; if there is a retake with a soft word at the start, look to cut out to a retake in the middle of the word when there is a hard sound there. Phan*t*om, wire*l*ess and ban*k*ing are examples. I've already said take the second version on repeats unless they have an obvious flaw, and major savings on time can be made by slicing out the content which is not relevant to the subject of the discussion. Sometimes there will be an interruption by extraneous noise that was not noted at the time of the recording. If the noise is sharp and brief the words can often be taken from elsewhere in the interview to replace affected ones. Yes, it does work, and there is a certain smugness about achieving a repair like this against the odds!

Once you are happy with the edited content look at levels. I've known people start sentences loud and drop to 18 dB lower to a whisper as their breath constricts. Look to even up any clearly noticeable level problems, as otherwise an audience may struggle to hear. Next, if the recording is an insert to a music programme, it will need an element of the DAW essential, limiting (see Chapter 12, "Outboard Devices") to bring the average volume up to the rest of the programme. Limiting reduces the dynamic range of the piece, making it more of a match to processed music, but be aware it also raises the background sounds, namely the recording's ambience and the technical noise floor. Finally select and amplify the whole interview to near peak, *ensuring of course that the highest sounds are still under MSB (max modulation)*, so it is in line with the presenter and other inserts into a programme.

MUSIC EDITING

Music editing, like speech editing, is great fun and a rewarding occupation. It might be that there is a simple need to take several takes of a song and construct a single good one. Another popular requirement is to take a song and edit it to match with a speech dialogue of known length to run under the script. This is called a "bed". It is likely that the bed will have a catchy start and a distinctive end for impact, and may feature several items of music or just instrumental cuts (like the middle eight) from a single track.

The technique remains the same in all cases, and that is to find the *underlying beat* of the piece. Success comes from cutting out and in on the beat, or by cutting a known distance between beats on the out, and carefully matching the distance to the beat for the in on the return. To give an example, a singer will often sing over the first beat of a new bar at the start of the song, so the cut needs to come early to remove the voice, but that must be matched by an early in before the bar in which you return. I hope that makes sense; practice will bring you many examples to test this out. In the days of tape editing an editor could be left with the problem of a piece of tape a quarter of an inch long, so make use of the brilliant "undo" feature that the tape editors didn't have, and follow the adage of measure twice, cut once. If dealing with drop-ins on multitrack, the issue of following the beat is still a golden tip, even if to subconsciously ignore it whilst slipping a track or two for dramatic effect on occasion.

ARTIFICIAL INTELLIGENCE

A symbol sits at the start of this line, waiting to enhance my writing skills, the same for PowerPoint and many phone apps. This is exciting and not a little scary. The photo apps can now easily remove that annoying hand from the corner of your selfie, and much more, but what does it mean for "Sound"? You may ask, is it a threat or an opportunity?

The answer, of course, is both! AI has hit the workplace; sound technology is not an exception. Already an essential tool in room analysis when setting up multichannel sound, this goes beyond speaker numbers and positioning. AI-generated surround sound creates a simulated soundscape which optimises the system, scene by scene, to provide a mesmeric experience.

It will be a while before we fully understand how day-to-day operations will be hit, but of course the simpler repetitive activities are the most prone to a form of automation. I firmly believe the subtleness of music balance will elude AI for some time to come, as this is more about artistry than mechanics. AI will not be able to unload lorries to carry equipment on to site; however, it may make the list of what is required and might even assemble items to pack the lorry. It will not position microphones nor rig them into a mixer; however, it might make a fist of a PA balance, able to keep the mix out of howl round. AI may well monitor radio programmes effectively, able to spot problems with a passing programme feed, but for a good while, a professional will need to step in to remediate. There will be a time of error and frustration, as AI programs learn what sound is required and what is not, but it is likely the church PA and conferences will be overtaken by AI mixes. Multichannel band mixes are very different; whether for PA or recording, the subtle interplay between the instruments and vocals will keep AI on its toes for some time to come. I can see the industry is already affected; the top postproduction studios have seen a serious downturn in work, but I also believe there will be gains. The growth of the games market and the needs of epic audio tracks has proved that as some elements diminish, others will thrive.

It is over 20 years since machines were routinely able to beat humans at chess, because the game is a logical matrix of options and outcomes. It has certainties. Computers to date cannot step outside a logical rulebook; they cannot see connections which are outside their experience; they will not value and tally emotional responses by humans which will drive an unpredictable decision-making process. Humanity is the key. Business leaders have asked, "What worked last time?" This is where computers have an edge. If it has been done before, it can be in the arsenal of the bot! It is up to us to employ positive new thinking, but that is just where artists, musicians and poets soar above bot creativity, because new is what we do!

Each time a new wave of music hits, it breaks the rules of those that went before it. Motown, Brill Building and the Mersey sound in the 1960s; Disco, Philadelphia, Punk in the 1970s; New Wave, New Romantics, new everything! It's hard to imagine a world without Prince's production values, or Quincy Jones or James Brown for that matter! Breaking the mould is our normal, and whilst AI will be quick to create clones of a style, it will not be the instigator. Artists are fighting to protect their styles and their voices from AI cloning, and I wish them well. It is their very humanity they fight for, and they deserve to win!

If this seems a little bleak, let's look at when AI is our friend. I'm sure many readers will have come across the impossibility of Johnny Cash singing "I'm a Barbie girl". Maybe you have heard the Beatles singing Beach Boy songs; this is all possible with AI and frankly is worrying if presented as fact, rather than clever AI fakery. However, the technology brings an amazing opportunity which is just starting to be harvested! In the early days of CD remastering, popular music albums became a big industry, using groundbreaking tools such as the Cambridge University inspired Computer Enhanced Digital Audio Restoration (CEDAR), looking to recreate the "sound of vinyl" on the new medium. However, the engineers could only go back (if they were lucky) to the multitrack tapes to rebuild the mix with new EQ, limiting and use of the burgeoning digital effects. CEDAR is still great and doing good work, but there is a new kid on the block!

AI offers a new perspective. Your sound system, whether it is part of the room or your mobile phone, can personalise the listening experience, recognising moods at varying times of day to make recommendations.

Work with mono Buddy Holly tracks (which were beautifully recorded in the first place) has shown where this new technology can take us, and I would recommend reading up Peter Jackson's successful trawl of the Beatles achieve held at Abbey Road for the film *Let It Be*. Machine Assisted Learning (their "friend", MAL) offered the ability to teach the computer an instrument, and with knowledge gained, choose that instrument in a mix, isolate it and completely remove it from the track! Source by source, a mixed music track can be deconstructed removing each component until only the studio background noise exists (and yes, it is that good!). At this point, what might have been a mono track can be rebuilt in stereo or better Dolby Atmos surround, ensuring each component is pristine and placed perfectly in the new sound field. This is not just audio separation; it allows the brain to access the material in new ways, to gain a perception that simply was not possible to hear from the linear mono tape. I predict a growth in the reimagining of classic soundtracks using software suites such as "Demux" and "Spectral Layers", creating a product that was a distant dream when the spools were turning on the studio corner multitrack.

As ever in the evolution of people, passion and creativity will conquer. There is no doubt that those who learn to work well alongside AI will win out over those who ignore it. Further, the more "emotional intelligence" you can project into your choice of career, the more protected you will be from a bot takeover. Working as a member of an inventive team, where the synergy of the individuals creates the artistic outcome; be it in a theatre, on a film set or a radio, TV or music studio; your future will be assured.

Acknowledgement – My thanks for the insight and counsel of audio industry prime mover Peter Thomas in the paragraphs on AI.

BIBLIOGRAPHY

Shah, H (2025) The Symphony of AI: How Artificial Intelligence Is Changing Audio Forever. *Forbes*. https://www.forbes.com/councils/forbestechcouncil/2025/03/26/the-symphony-of-ai-how-artificial-intelligence-is-changing-audio-forever/

15 Balancing Down to a Mix

This chapter is not easy for me to write, as programme balance, dynamic range and the result of a final mix are not just personal credos but are the subject of fashion. Accepted practices get thrown out of the window when the new kid on the block rewrites the rules. Technology improvements drove early changes as quality improvements in both the recording equipment and home reproduction improvements created new possibilities in creativity. I'll tell the story and then offer an outline of support for you to colour in!

POPULAR MUSIC IS BORN

I don't propose to go back as far as wax cylinders, but it is worth saying that well into the 1950s levels were always kept high, as the background noise generated by the playback medium of 78 revolutions per minute (rpm) acetate records and the damage done by each performance of it with (quickly worn) steel needles, meant that skirts had to be constantly lifted clear of the mud. Loud was simply better!

The advent of microgroove vinyl 45 rpm singles in the middle of the 1950s brought with it much-improved cartridges and pliable diamond stylus mountings which were far kinder to the records. Stereo kicked off in America and in 1959 the Wurlitzer 2300 was the first stereo jukebox. It wasn't stereo (unless you bought a separate external speaker) but it enabled the company to break an old contract for obsolete mono pick-up cartridges and equip their jukebox with a new stereo pick-up which could track at a much lighter (and so less wearing) tracking weight. In the 1960s, the mono two-sided "long player" now allowed producers to experiment with stereo sound and the seventies gave us a much wider dynamic range with Ray Dolby's "Dolby A" and David Blackmer's "DBX" modules making possible exciting leaps of volume which were breathtaking to a new generation.

RAY DOLBY – AUDIO LEGEND

This revolution could not have been achieved without a true audio legend, Ray Dolby. The microgroove vinyl disc allowed for noise values far better than available from the best tape recorders; tape hiss was a problem at source, but each time a tape was copied, the level of hiss increased. The idea of bouncing tracks to build the complexity available on so few tracks was very clever; however, the residual tape hiss grew quickly to unacceptable levels. Recording engineers of the day were at a peak of genius in finding ways to control and manage the tracks to ensure a clean result, but it was all a compromise. In 1965 Dolby

DOI: 10.4324/9781003717065-15

Laboratories introduced the "Sound Compander". The Dolby A™ system split sound into bands and compressed the signals fed onto the tape up to 15 dB, which was backed out in replay, taking down the noise floor of the tape by 10 dB or more at higher frequencies.

This freedom to further use the dynamic range in production at the turn of the seventies pioneered "high fidelity" sound. On mass, out went the all-in-one radiograms to be replaced by Hi-Fi separates of amplifier, turntable, radio and later cassette and in the 1980s CD, feeding a new breed of stand-alone quality stereo loudspeakers. Bass became extended down to 50 Hz or so in top-flight systems, and people craved clean "open" top frequencies. It was a new religion!

It must be said that way back in the 1950s, the competitive singles market wanted their new track to sound better than any others on the only mass distribution medium of the day, the AM transistor radio. To sell, the sound needed to stick out, to sound more strident than competitors. AM radios were distinctly middle frequencies only, with small single speakers. No hope of bass, and a top frequency of 7 kHz on a good day. At the start of the 1960s the Brill Building in New York was the hit factory. People such as Neil Sedaka, Paul Simon, Carole King and Neil Diamond worked in tiny offices with pressure to create a new hit song a day (really!), and the singers and band were expected to record the song down to a single-track tape recorder in one take! The amazing Motown sound in Detroit was little different, and let's not forget the soul sound of the Stax label. These were a perfect prelude and inspiration for British pop with the Rolling Stones, Who, Kinks, Searchers, Hollies, obviously the Beatles and the other Mersey Sound genii.

THE WALL OF SOUND

Back to the Brill Building. At the start of the 1960s one person at the Brill Building got away with spending weeks in the studio to get each song right. Phil Spector created his "Wall of Sound" as a breakaway from traditional techniques. Across the water, George Martin was put in charge of producing the Beatles, using one of the first Studer J37 four-track recorders, which supplied an element of mixdown after the studio session but also allowed for tracks to be "bounced" to allow further vocals/instruments to be added. To come full circle Phil Spector produced the final album by the Beatles, *Let It Be*, in 1970.

In 1967 *Sgt. Pepper's Lonely Hearts Club Band* became the first Beatles album to be available in both mono and stereo, so I think that's a great yardstick in the introduction of new technologies. Interesting that the mono version is now worth a great deal more than the stereo! Albums became as important as the single around 1970 as vinyl became the dominant home entertainment, and along with the development of high-fidelity sound in equipment, record producers wanted to exploit these amazing home improvements, although studio equipment had some catching up to do. But of course, 12-inch vinyl records could not be played in the car!

MUSIC ON THE MOVE

Cue the American eight-track cartridge, which was developed as a mobile medium from the start. First known as "Stereo 8" (stereo being the new "cool"), it was developed by Bill Lear (yes, he of Lear Jets!), together with tape machine maker Ampex, car makers Ford and General Motors, car radio manufacturer Motorola and music publisher RCA-Victor. This elite group took an existing idea of the Fidelipac DJ (jingle) cartridge with a neat trick of one endless loop on one spool, the tape feed being pulled from the centre and wound around the outside on its return. From 1965, it was a premium purchase for the car, and so a status symbol at its start, like the car phone. The slightly dodgy quality meant it suited "wall of sound" music best, although classical albums were released. In a time before quality headphones in the home, it was an amazing freedom to have your own music playing in the car, rather than the radio DJ's choice. For young people this was their territory, away from the parents, and it gave the opportunity for new expression without risking being judged! This was an opening for American bands like The Eagles, Santana and Allman Brothers and British bands like Black Sabbath, Uriah Heep and Deep Purple; loud and soulful was in!

THE COMPACT CASSETTE

Another home reproduction product made an appearance in 1964. Philips of Eindhoven created the "Compact Cassette". The Cassette player took the idea of an analogue reel-to-reel tape player but made it much smaller and encapsulated the magnetic tape on two tiny reels in a mess-free hand-sized housing. At first, they were mono sound and poor quality, invented for office dictation; it took time to become a music rival against the eighth track. The bulky reel-to-reel tape machine died out over ten years to be replaced by this new medium; many came with a crude Automatic Gain Control, which gave confidence to people to record and play sound at home who otherwise would never have expected to own a tape recorder. Better still, in time manufacturers started producing pre-recorded cassette music tapes featuring the best albums of the day and so found yet another market to sell an album.

The early tape machines tended to be portable, like the transistor radios, with small speakers, and that was fine as they would sound awful on quality systems. In time, the mechanism was compact enough to sit inside the car radio chassis. Now, unlike vinyl records, compact cassette tapes would work well in cars, and the compact players were quickly incorporated into European car radios for music on the move! The cassette tapes were half the width of traditional tapes and ran at a very low speed of 1 7/8 inches a second, half that of the average domestic tape recorders. On top of this each channel utilised just a quarter of the tape to provide stereo recording on two sides. In the early days these underwhelming features coupled with rudimentary tape formulations meant a high tape hiss level, and poor high frequency response.

Step in a much better tape formulation and a domestic version of Ray Dolby's noise reduction system, "Dolby B" (and later "Dolby C") and suddenly Compact Cassette could also be "HiFi". By the mid-1970s, all the major manufacturers produced stack-mounted stand-alone cassette players for the home systems, and of course these improvements filtered their way into car radio systems. Add to this new commercial radio stations coming on-line with super quality FM radio transmission, and the market for quality in-car entertainment (ICE) was born. Multiple high-quality speakers with decent stereo amplifiers echoed the developments in lounge listening with the experience heard in the car.

DISCOVERING HOME PRODUCTION VALUES

The music producers in the seventies were quick to grasp the ability to provide huge quality improvements in the Studio Masters they created, knowing the home listener would, and could, enjoy them. The impetus for thrusting middle frequencies stepped back, replaced with a driving bass and a clinical sharpness of zingy top, often with a wistful quiet introduction as a prelude to the storm to come. Classic releases, at last, could represent the full dynamic range of the orchestra, from solo violin to a crescendo of the entire ensemble. The popular music scene was also unshackled. New production values shone with the Sound of Philadelphia, particularly Barry White and separately Maurice White of Earth, Wind and Fire. A special mention to Giorgio Moroder and Prince, who broke so many rules and created new sounds for a new generation. Of course, there are many more examples of rule breakers; each sound engineer will have affirmed favourites, but I guess I want to point out that each technical revolution created its own ripples of production evolution that changed techniques and reinvented the rule book for a new generation of sound balancers.

In terms of recording technology, the four-track analogue mastering tape machine turned to 8-, 16-, 24- and 32-track recording during the 1970s. The 1980s were the decade of the widespread introduction of digital audio, first as a mastering medium and later as a mechanics-free multichannel recorder. This in turn lent itself to further automation and a cheaper route to greater track channel numbers. As the electronics improved, 16-bit professional digital systems were replaced by 24-bit, which substantially increased the dynamic range, allowing either brilliant low noise, or at the other end, a generous headroom or perhaps a bit of both! Greater sampling rates allowed the recording of frequencies well beyond human hearing, and I guess a debate has raged since as to the benefit of that to mere mortals.

ELECTRO POP

Analogue Synthesisers had remained a tool for the rich and famous (Chapter 11, "Introduction to Sound Synthesis"), but in 1983 the Digital Yamaha DX7 put powerful production tools in the hands of jobbing musicians, and they were quick to bring a new zingy sound to the charts, Electro Pop. In the mid-1980s a synthesised bass line was heard on half the new releases, often with synthesised strings and brass to accompany. New "instruments" were created that humans could never fabricate. DSP allowed enhancement of the audio through manipulation of the signal in its digital number state, allowing it to be arithmetically manipulated to do what was impossible in the analogue domain. A finished sound was made more exciting and louder than ever before and much loved old analogue devices were successfully aped in the digital domain for a fraction of the cost, and in the privacy of a bedroom. In an unexpected twist, people who couldn't sing could be brought spot on key and sell records! Sadly, the loudness wars returned in the 1990s, and I've not seen much technology change in the recent decades as modern sound design is simply better than our ears and so change no longer seems worth the research. Whilst it is sad to see we are largely past the point of fervent experiment in sound, I guess what there is, is largely available to the home user, and the bays of humming hot racks are gone in favour of an affordable downloaded computer programme.

THE DIGITAL AGE

I've not mentioned "Compact Disc", introduced in 1982 (I was at the London launch – it didn't go well!). The medium is much maligned, and I think unfairly. Bear with me! The uncompressed 16-bit coding of native "Red Book" CD allows for an amazing theoretical dynamic range of 96 dB (90 dB aurally adjusted) and a flat frequency response to 20 kHz. That is more than good enough for a home medium. Higher quality is available from specialist companies at mastering quality or close, but most people will be listening off far lower quality formed from lossy streaming services. So, I think in the age of music the "CD" neither contributed to, nor adversely affected the progress of quality music. It is true the medium should have been the most transparent, as the duplicating process could directly call on the quality of the mastering process, so why do so many people call CD a "music filter" or worse? The answer lies in the early critic's comparisons of the same album on CD and vinyl; to the ears of the HiFi glitterati, the vinyl always sounded better.

I was very fortunate to meet and watch the work of Melvyn Abrahams, an exceptional "Disc Cutter" in the age of 7-inch and 12-inch vinyl. A valued and much sought-after skill, the Disc Cutter created a disc pressing master from the analogue studio master tape. The reason for their existence was to make the best use of the space available on the record. Bass frequencies created large excursions of the cutter head quickly using up the time available, so the Disc Cutter had to plan these just right: (1) duration set against (2) volume, to get the maximum bass into the tracks that needed it. Over time the equipment rack of tools available to support the role grew; a great cutter could not just create great bass but clean up the dynamics and provide a clean crisp top that the studio had not dreamt of. The Disc Cutter was the stand-alone critic of the studio session; however, this critic had the tools to enhance the sound to new levels. Studios turned a blind eye; they were not going to admit an outsider nailed what they had missed, and so an uneasy acceptance of the end-to-end process became normal.

The advent of CD had a major selling point! The recent adoption of 44.1 kHz digital studio mastering could be used in the CD production process without the detriment of rescaling or analogue artefacts; the studio produced digital files were directly used at the CD pressing plant without recourse to an intermediary. In earlier times the master tape was analogue, I have to say the copies supplied to many of the CD pressing plants were not given the priority of care they should have had with bias and azimuth adjustment, it became a bit pedestrian and half-hearted. Unfortunately, the Disc Cutter and his/her toys were now out of the loop, and out of a job! Comparison with disc cutter inspired "vinyl" in the home against CDs mastered from the untreated analogue tapes in the CD duplication process proved an unexpected and unfair start for CD. Interestingly it took a matter of a few years for the studios to catch on and produce new augmented versions of the albums that resembled the sound of the disc cutter modified vinyl records Those who were early CD adopters were expected to buy their collection all over again! These updated versions were called "remastered", and often had a member of the band, or the producer involved, to give the new work authenticity. Of course, some of the most popular albums have truly been reworked from the old analogue multichannel tapes, the best by the artists who originally created them. I will leave you to be the judge of whether the reissues do show a clear level of improvement over the original.

In balance of the above, I have to say I agree with several career sound professionals, one might say the caretakers of high-quality reproduction. There is growing concern that generations of digital copies are not as proof against deterioration as we originally thought, and that we simply have not invented appropriate tools to test this, as the problem is yet to be universally acknowledged. From the first, people talked of a digital dryness, a loss of musicality; and experts have noted that whilst, in a digital product, an image of stereo width between loudspeakers is maintained, the depth and height of this image can flatten to a flaccid two-dimensional. This is not a pop at CD; the medium is capable of great results with the right source material and manufacture. It must be noted though that the source file used for the CD master, the pit's lasered on the master stamper and the quality of the plastic chosen to reproduce the stamper accurately, can all undermine listener enjoyment of a brand new disc, often sold at a premium price by a reputable label! It is a point of concern that with widespread sharing of music files in many differing formats between countless online providers, the count of the number of generations away from the studio master is ignored. Can any digital sound format be universally trusted?

Some have suggested that it is not in the best interests of music companies to let their pristine material out into the public domain, that they increasingly need to keep their works safe to the point where the raw studio quality is never allowed to be accessed on any home medium. This stands up to the point where iconic 50-year-old material is sold as renovated in £80 sets, yet many enthusiasts would assert better session sound can be found from the time the works were first created. If access to quality sound is denied, how do budding sound engineers learn to differentiate and discern best practice, and do we really want to dumb down the expectations and acuity of the buying public? Would Stradivarius have bothered making violins if his colleagues "improved" each one before they were allowed out of the workshop?

SOME POINTS ON HOME PRODUCTION

I hope that is not a depressing scenario for you, as the creative aspect is still in your hands and new talent can use social media to find a platform with the right promotion, although that does not gaurantee a willing audience. The need for being loud to the point of "compression to death" has been neatly castrated by the algorithms of the modern internet streaming providers. I recommend the 2019 New York Times article by Greg Milner entitled "They Really Don't Make Music Like They Used To" as he is as incisive as he is brilliantly descriptive. The point he makes is that streaming services "normalise" the level of tracks; all the headbanging effort in the studio is strained through a grey algorithm, so the listener is now tapped politely on the shoulder by a mildly disturbing volume change.

So, I entreat you to be one of the enlightened sound balancers. The best-selling album of all time is the surprisingly dynamic *The Eagles Greatest Hits (1971–1975)*, with a release date of February 1976, plumb in the centre of the HiFi revolution! In your own endeavours, I encourage you to set out a dynamic range that surprises and delights the ear, more a shaggy hedgerow of sound than a wall. Individual instruments are traditionally tailored in frequency response to be "tighter", and I encourage this, although I accept there will be artistic exceptions. Some instruments will benefit from light compression to punch their weight against the others, and drums (if not electronic from the off) can be limited and gated, to provide stout cleanliness in the beat. The instruments need to be placed across the sound stage, you can go natural (laid out as the band sited on stage), or choose a dynamic variant, but avoid the extremes as they are wearing, unless for short duration effects. To balance music is a gift; however, like singing, it can be learnt, and the more practise, the better.

In my experience the ear suffers from fatigue, particularly at louder levels, so take breaks to regain a sense of perspective. Do enjoy the energy and detail loud listening gives you, but check your work at low levels to ensure none of the soundtrack gets buried and you still have the buzz you enjoyed at the louder volumes (check out "Using Monitors in the Studio" in Chapter 5, "Loudspeakers"). Particularly in the early days, seek advice from colleagues and friends; they may just have a perspective you have missed. Remember, you are only as good as your last work. You never stop learning in the field of sound recording and balancing.

PRODUCING A RECORDING WITH CLIENTS

What does a producer do at a session. Well, it varies between organisations, but as said elsewhere it is where the buck stops; yet it is also where the energy and cohesion in the session are formed. The producer is responsible for the outcome of the time spent in an expensive studio, but more than that, she/he is judged by perhaps millions who will hear the result. So, how do you optimise the opportunity and minimise the snags? Rule 1 – communication, communication, communication. Once your precious guests are snared, you need to keep them that way! You need to use absolute clarity to ensure they understand where to come, when to come, how long they will be detained, what to expect from the session (or interviewer), any taboo subjects that should be avoided and, of course, favourite snacks! On the day arrive early to ensure the engineers are in and correct microphone placements are set up. During the day ensure someone (probably you!) looks to get drinks and food at the right moments. That's the housekeeping done; now as the session starts put the guests at ease, use comments/compliments to show you are on their side and avidly listen.

Keep a track of each instrument/singer, for quality; look for small mistakes that you can retake, and ensure your focus is absolute. It will be noted, and it will be welcomed, that you are kind, but a hard arse! I'll give you a real example. I was producing an 84-piece orchestra featuring world-renowned jazz singer Cleo Laine. As we got into the run of the morning session, I was surprised to hear the trumpet solos were just a little off; slight warbling on the high notes and occasionally missing ones in the more complex parts. Heart in throat it dawned on me that the player had probably got a nasty problem called "trumpeter's lip". Sadly, it can end careers! These sessions were for national broadcast, not an album, but I still wanted the best sound possible. We were close to the morning break, so when I called it, I asked him politely to stay behind. I told him I knew, and seeing his hurtful expression, I said how sorry I was; but I firmly informed him he was to give his "second" the lead for the remainder of this session. I was very careful to say that he was to present this as an opportunity to shine, for the lead trumpet to give his deputy a generous favour to gain experience. The session went very well after that; we did not talk again, but I was aware the man's career was in jeopardy, and I knew he would have some difficult, painful decisions to make. I left it entirely in his hands; I said nothing to my peers or management that might affect him; his dignity was at stake, and that I would not wound! I'm sure readers will draw different conclusions from my actions, positive and negative, and I'll take that on the chin. I was responsible for the quality of the session; I saw something wrong, and I put it right with the minimum fuss. That's the job!

LEARNING HOW TO LISTEN

The initial BBC training was very hands-on, always with the threat of termination if you didn't perform to expectations. As future continuity operators we had several sessions where we learnt to recognise both analogue and digital faults, comparing different forms of speech and music and hearing subtle faults introduced by the instructor. We were individually tested on what we learnt, naming the problems as we heard them and praying we got them all, as there were consequences if you didn't! You could say I was led by the nose to gain these skills, and in part that is true. However, a footballer instinctively knows by the feel of his/her boot on the ball that the aim is true and knows it's going into the corner of the net. I think hearing sound is like that; if it's something you love, you will grow instincts to know when the "boot" is millimetres off target! Listen a lot, love the sound, but listen for how you could improve it.

As I write we have just had a major UK outdoor concert, with American funk/disco heroes of mine on the main stage. It was screened live, and within seconds I found myself shouting at the TV. It took three whole songs for the music balancer to realise there was no bass guitar in the mix... In funk numbers!!! *If you forget something live, don't creep it in... Bang it in on the next bar*! My film colleagues say they can't watch a movie without seeing grain, poorly framed shots, bad edits and lighting faults. They say it's ruined watching films for them. Luckily, I remain largely immune to these problems, and still enjoy a decent movie – until they mess with the sound! The thread of this book is we can all do this; it takes practise and not a little dedication, but I think cloth ears are more a life choice than an affliction! Critique audio as a group and compare notes afterwards, before listening a second time to see whether they had a point. Again, I say trust your personal instincts over a cocksure adversary!

MONITORING AND LISTENING

If you are in technical charge of a mixdown, a knowledge of Chapter 8, "Acoustics", will help to optimise the placement of loudspeakers to use the available space to create the best sound.

In Chapter 6, "Sound Levels and Connections", we found that sound level metering had gone in two directions, the European adopted BBC PPM ("Peak Programme Meter") and the American VU ("Volume Unit") meter. The popular mixer LED strip meters ape either one standard or the other (maybe even switchable), or perhaps a compromise to overcome the differences between them. There is currently a lack of a universal standard pertaining to these, so tread carefully. The PPM is an expensive and accurate tool to stop our sound engineer from missing peaks in sound which would otherwise overload the mixer. It's been a great tool for 90 years but is being overtaken by LUFS loudness metering.

A "real" VU Meter is expensive to make, so few are authentic. The meter is accurate at capturing a static level like line-up tone, but is sluggish and fooled by sudden peaks, so the meter now normally has the addition of a flashing red "peak" light in the corner, so you know when to turn the signal down. Whilst the meter movement is not reliable on peaks, its champions say it gives a better idea of the loudness of the sound. Like all tools, experience with each piece of sound equipment will teach the limits of metering as well as the other variables like sound quality and headroom through the chain. This meter does not suit digital recording.

Winning through in a digital world is "loudness"-based metering with standards agreed by world-leading bodies including the EBU R128 and ITU 1770. These are available to meet legal requirements in certain countries and are now available cheaply for the rest of us, routinely installed in software. The peak level to correctly modulate tape recording equipment, as for distribution, is PPM 6 (+8 dBu). How does this equate to the digital world? In both Chapter 9, "Digital Audio", and Chapter 6, "Sound Levels and Connections", we learn that frankly it doesn't fit well. The better way is to reference the metered extreme peak level of the system and back off to a protective lower level. This is to create necessary headroom, as digital systems, unless very well protected, will fall over with the briefest of overloads.

Loudness metering should not be confused with the metering of digital levels in a sound desk, from both the initial channel levels and then balancing a mixdown for a master of a music track. The component tracks of a digital mix are best set up in rehearsal by using the PFL or solo button to isolate the signals so they can be adjusted to a correct nominal level of the loudest point of the performance, checked on the metering (which should switch to follow the solo) and adjusted by changing the input gain. This should be a maximum level of −6 dBFS or better in my view −9 dBFS on each channel (particularly if the system is 24-bit), to allow for changes made at a final mixdown by digital production tools.

There is also the concern over "Inter-Sample Peaks" which can create unexpected gain in the digital to analogue converter (DAC) process, creating distortion of the "true peak" of the recreated analogue waveform. It is a problem at lower bit rates, and so cheaper setups, but it makes the point that *caution is always better than bravado.* I note there are now software metering tools that can measure true peaks and those which can effectively limit them. The final mix is also open to inter-sample peak issues; however, that is where we will use tools to set a final commercial loudness referenced in LUFS against 0 dBTP. Some editors, such as Protools, allow for off-line loudness measurement. The programme must be integrated into a single file and then imported as a Protools session. In the "AudioSuite" menu choose the "loudness Meter plugin". Highlight the entire audio stream within the session file and click "Analyze".

LOUDNESS NORMALISATION

Loudness normalisation is a curse and a blessing in one. Used carefully it can take your completed file, ensure the content is as loud as allowed (by LUFS) throughout the duration and prepare it for different distributions. For instance, your –24 LUFS file can be repurposed as a –18 LUFS for use in a podcast. We have already discussed that audio peak values are a poor indicator of loudness; moving into the digital domain and using psychoacoustic research, audio is adjusted to provide an average loudness and then moved up or down to suit the required level of output. The curse is a steep jump in level will not be sorted; it will remain in the mix, so production faults of level problems will need to be corrected before the process takes place. Further, (already) loud sounds on high LUFS levels can be driven into clipping by the process. A peak limiter may be present to prevent failure, but it can degrade your intended mix. Different plugins take differing approaches; read reviews with the knowledge gained here, make your choice and live with the good and bad, as no system is perfect.

Above all, in mastering use your ears, keep on top of the loudness meter (short-term) changes as the programme develops, and keep loudness normalisation as a tool for the end, when the level of the entire project is modified for distribution. At the time of writing, I am yet to find a field recorder fitted with a loudness meter, however, you are most likely to have a bar graph meter. As stated before leave a decent headroom; a practical way is to keep the signal in the green and use the yellow segment as "red" (i.e. don't go into frequent yellow peaks). Adjustment to an appropriate LUFS can be made back in the studio.

Stereo content assumes a wide stage of coherent and non-coherent material. If you are looking to create a mono programme on a stereo system, the content will be entirely coherent and will fool a loudness meter which samples each channel in isolation. There is no simple solution; I would seek advice from the publisher; however, as a rule of thumb, this coherent material would work better if you chose a level –3 LUFS lower, i.e. –26 LUFS rather than –23 LUFS for a broadcast programme, to better match the loudness of surrounding stereo content.

AURAL MONITORING

Visual monitoring is a great aid to having confidence that the sound is present and being fed to the destination at a sensible level. It cannot tell you if the sound is distorted or has an imbalance in frequency or level problems compared with the other channels at the exit of a mix. There are three choices of quality aural listening: main monitoring loudspeakers, nearfield loudspeakers and headphones.

Only one of these is not radically affected by the surrounding acoustics of the room, and that is the headphones. Do not underrate a quality pair of headphones, a portable reference a sound engineer can carry from place to place. At the risk of repeating myself, like loudspeakers, many manufacturers will make headphones that sound great on most sounds, but in doing so they are inaccurate. Be careful and read reviews before auditioning several pairs to find a personal favourite. The inaccurate headphones have their place with broadcasters as they can sound superb on your own voice, but never if the content is to be later adjusted in a mix and tonal adjustments made, as the material will sound dull and lifeless when played back on an accurate system elsewhere.

I've indicated the importance of a flat or "linear" response across the human range of hearing; the other is transient response. Quality headphones handle delicate higher frequencies well, producing upper harmonics with precision without introducing harshness. Headphones come in two sorts: open back and closed. I believe, with long experience, open back have an edge in quality over closed – an airiness; however, they can't go on the studio side of the glass as the spill from the headphones colours the sound picked up by the microphones and, at worst, can lead to a howl round. If you are involved with personal content creation, as well as managing that content, stick to closed headphones; frankly, you will lose very little.

A little on leads: great headphones will often come with flimsy leads which are long and thin and will easily trap under the castor of a studio chair and break. Some headphones may be OK quality but come with a robust coiled lead. To have the quality and the robust coiled lead is a high-cost option. The optimum is to have headphones with an owner-replaceable lead (available as a component from the manufacturer), but if you can't afford it, get the thin-wired quality pair, check out the sound and put them safely back in their case before the session.

In Chapter 3, "Introduction to Electronics", we introduced the concept of impedance, and you will find quality headphone models can be bought in a range of impedance values. The headphone amplifier that you will plug into (whether that be your mobile phone or a 64-channel sound desk) will have its own source impedance, which will sit in series with the load of the headphones (Bet you wish you had read the "Introduction to Electronics" in Chapter 3 now? You still can!). Basically, the higher the headphone impedance, the more threat there is that the amplifier will top out, and you may find you can't get enough loudness to hear and do the job. So why not always go for low impedance headphones? The higher stated impedance on many stereo headphones comes from the transducers employing finer coils of wire (each ear), but many more turns. Many would argue this system produces a greater nuance, as the finer coils can be lighter and whilst that might in theory be true, I don't have the ears to hear it. With headphone impedances ranging from 8 Ω to 400 Ω, *if* you wish to take your headphones to different venues, I choose a compromise of around 80 Ω, so I know I will have enough drive to create the level needed in the ears to make a decent mix, yet have the quality gained from finer coil windings.

Eye-watering amounts of money are spent on making the internal acoustics of a high-end studio control room neutral so when the speakers are placed in, the sound is as intended by the speaker manufacturer. Trust me, very few studios can either afford the acoustics or the speakers (US$60,000 plus a pair) we are talking of, and so the use of high-end main monitor loudspeakers, whether it be in a studio or a lounge at home, is fraught with compromise. That does not mean that we should not try to get as good as possible. I've put forward ideas and concepts in "Appendix 2 – Typical Acoustic Environments and Their Design" on acoustic treatments, which will get those who care a long way. My reason for making the point is that the price paid for quality speakers is just half the story.

It's a matter of interest that American sound and British sound took two differing traditional routes in the heyday of "HiFi" in the 1970s. The British preferred an extended and flat bass response, and the Americans an emphasis on upper bass with a small boost. Top frequencies on American systems were also often boosted to create an initially pleasingly crisp top, and even on high-end monitors this could catch out the unwary. Both had their markets, and of course music is readily exchanged across the Atlantic; these days a flat extended response is normal across the globe in professional monitor speakers. If your purpose is to solely listen in the home, buy what you love; but if you want to analyse and adjust sound to be loved in all homes, you do need to look for a neutral wide and flat response monitor speaker. It may well sound "dry" and unexciting on certain material, but at least it is honest, and the sound (acoustics allowing) is as the producer intended.

"Nearfield" monitors are one or more small speakers placed near the sound engineer, normally placed on or near the top of the mixing desk. The small size does mean the bass is bound to be compromised; however, they have an important role to play in modern studios as the acoustics of the room play far less of a role in the "near field", so imaging is naturally more accurate and room effects on the mix are reduced at the ears. Further, although our means of consumption have improved with high-quality Bluetooth headphones and multi-speaker in-car entertainment, a content creator still wants to have their work cut through better than the opposition and that is best judged on small or cheaper home reproducers.

I'm sure you can find a familiar thread to this chapter: to get the sound monitoring as good as you can, but accept there will be compromises and learn those compromises so you can adjust for them. Great engineers see through the imperfections to the sound beneath the monitoring and can feel what is right. Work daily with a system you know; try the recordings out in as many other places as you can to gain confidence as to how your vibes will transfer outside the studio. The brain is amazing in this; trust yourself with the kit you have, and take your trusty headphones on the road, so in unfamiliar places you can use them to zero in and find the sound you know and love...

WHAT TO LISTEN FOR?

Great question, difficult answer! For instance, how much is nature, how much is nurture? Many people have perfect pitch from birth and just know when something is wrong. I was part of the investigation when an Oxbridge music professor wrote into the BBC to say the hourly "pips" (Greenwich Time Signal) had moved frequency. At first it seemed laughable, but on investigation he was proved right; the oscillator had an age-related fault and was off by just a few cycles. That said, the great news is we can all improve with practise!

Let's talk about what's in a mix! I'm a great believer in the final sound being "open", in physical terms the difference between a blocked nose and easy breathing. Bass should be deep and firm without being overly invasive or softened by distortion. Top frequencies should be clean, extended and bright without being brittle and hard. Middle frequencies should be warm and rounded (without woolliness), lifted with a purity and cleanliness in the diction of a sung or spoken voice. The instruments should vie with each other, each taking an opportunity to soar in places before returning to the maelstrom. Vocals should swim smoothly on the sea of instruments, never drowned nor flailing in the air.

I want to go back before digital manipulation to find a balance "masterclass" for you. In the early 1970s the popular balances of famous artists were virtually impeccable. Simon and Garfunkel, Elton John, the Eagles, Cat Stevens, Genesis, etc. all follow faultless production values. It comes as a surprise that one of the top albums of the period, the 1972 "Transformer" by American Lou Reed, threw away the balance hymn sheet. The producer was no less than David Bowie, whose British album masterpieces were on the ascendancy. The album was a huge success, particularly beloved by fellow musicians. I'd like you to take a listen to the entire album (preferably with others) and analyse the music balance, particularly the vocal distance perspective and level on different tracks. Why did David recess Lou on certain songs and bring him in your face on others? I have my views, but the value is in you forming your own and how well this style sits with each track. Ask yourself, would David do the same again? Should others rethink the status quo and push their vocal boundaries further? Happy listening! I'd love it if you suggested other albums which push the balance envelope!

Levels aside, many shape each instrument with equalisation to capture the essence of it. I think that's fine provided the harmonics are not curtailed on sensitive orchestral pieces. Also jazz quartets, I've found, are another arena where instrument shaping should be used with caution. Vocals should be bright, rising out cleanly from the rest of the mix, meaning you should be riding the fader through the track, keeping your vocal surfer on the crest of the wave. The vocals must be compressed to compete with an instrumental background, and the more electronic instruments used, the more compression.

There is an intentionally wide margin in my descriptions; you must find your own unique path. My confidence comes from the constant surprises found in literally thousands of hours spent critically monitoring commercially produced sound, and trust me it's a good job if you can get it! Apart from, of course, arriving in BBC Radio 3 Continuity on a Saturday at 9 a.m. to Wagner's "Der Ring des Nibelungen", learning that the 14-hour production would run beyond my 12-hour day! That was demoralising to say the least. Billed as a "moral, intellectual and emotional challenge", I can truly say it lived up to, and beyond, the hype...

In closing, I firmly believe that creative skills can be learnt by all; it simply takes passion and perseverance. No time is wasted in experimentation, none in examination; all listening is good listening, and the barely discernible will be increasingly obvious. It just needs to be done with an open and questioning mind! Whilst being open to the thoughts of others, thoroughly and completely believe in yourself, and it will happen!

BIBLIOGRAPHY

Alton, F. (1984) *Acoustic Techniques for Home & Studio*, 2nd edition, 4th printing. United States of America: Tab Books Inc.

Architectural and Civil Engineering Department BBC Engineering (1980) *Guide to Acoustic Practice*. BBC Engineering Training Department.

Beyer Dynamic (2023) *What Impedance Should I Choose for My Headphones?* https://blog-en.beyerdynamic.com/what-impedance-should-i-choose-for-my-headphones/.

Milner, G (2019) They Really Don't Make Music Like They Used to. *New York Times*. https://www.nytimes.com/2019/02/07/opinion/what-these-grammy-songs-tell-us-about-the-loudness-wars.html.

Watkinson, J. (n.d.) *Time Domain Accuracy in Human Hearing and Loudspeaker Design*. https://www.thebroadcastbridge.com/content/entry/7066/loudspeaker-technology-part-1.

The Audio Engineering Society (n.d.) *K-weighting and Other Frequency Weighting Schemes*. https://aes2.org/resources/audio-topics/loudness-project/learn-more/#true-peaks.

16 Real-World Problems with Audio Systems

So, the theory is buttoned down, but what is waiting for the sound practitioner in the workplace? I'm going to break down this section into two halves: problems that are operational and those that are physical.

OPERATIONAL PROBLEMS

The world of the audio engineer is beset with possibilities of errors, many of which are obvious and those that are less so. I have said in Chapter 13, "Planning for a Location Recording", if going out on site, set up the kit at home to check everything that should be packed has been! This obviously includes equipment, but what about venue maps, event ID, bad weather protection, spare (fresh) batteries, car parking options and running orders or scripts? If you've not made a planning visit, arrive early at the venue (whether location or traditional studio) so you can sniff out prospective problems such as the siting of suitable power outlets and safe routing to get the power to you without making a trip hazard.

Famously reported in the national press, a live BBC Radio 4 political programme was taken off air by an unimpressed audience member who idly flipped off the supply power to the mixer on the switch at the wall... Yes, you do have to think of everything! Paul McCartney's vocal microphone failed to an audience of billions across the world at Live Aid, and it took minutes to get a spare out to him (why?). It's not realised by most that he later dubbed his voice for the video release to limit the damage. An XLR cable plugged into a spare mixer channel, ready to be deployed, is clearly a "sound" investment.

As pointed out in Location Recording, one year the "Brit Awards" came from a packed huge arena, produced by an independent production company, with many performers and star presenters (I went to the first one in a small central London venue where seventies band Hot Chocolate were the only entertainment!). The prestigious BRITs were due to go out live on primetime BBC television, but I was present in Broadcasting House London, when the venue belatedly realised they had forgotten to book British Telecom lines to get the broadcast back to us... Nothing could be done. No transmission! My advice is to plan from the first microphone to the last destination, leaving nothing to chance; then you truly are a Sound Director!

CABLE RUNS

On the day, the running of cabling is important. Don't run signal cables next to data cables (such as lighting DMX) as the square wave data has harmonics

DOI: 10.4324/9781003717065-16

at many frequencies and will find any chinks in the armour of the audio wiring. Long runs of signal cabling (analogue or digital, input or output more than 5 m) should be balanced. Keep sound and lighting apart, as dimmed lighting produces huge amounts of radio frequency interference (RFI), that again will blight audio cables and certain radio microphones. You will only know if there is a problem if you sit through a full rehearsal. Radio microphone frequency bands and licensing vary by country; however, if you are a visitor to a venue, you will need to ensure the frequencies you wish to use are not the same ones that they already have in use. Take additional microphones in case of failure, and ideally a compact sound desk that can replace your state-of-the-art unit if it fails to power. That might be because you left the separate box with the power supply behind, so ensure all is packed! Monitoring can get broken on the road as there are several parts to assemble, so take your favourite headphones as a backstop!

MICROPHONE PLACING

When setting out the microphones use the right sort of polar pattern and technology (see "Appendix 1 – Microphone Placement"). Beware of hard surfaces behind the microphone that will reflect the PA and create a howl round at comparatively low levels. Equally beware of the pickup each stage microphone position has off the audience, as this much louder "line source" of the people can overpower your mix. Use directional microphones carefully to isolate each microphone from extraneous noises. On that point, microphones pointing out to capture the audience applause will be in reverse phase to the stage microphones (as they will be pointing in the other direction). Beware of phasing between the audience and stage microphones; it sounds weird, and it's more common than you think! Consider a separate PA mix to the main mix to help with acoustic compromises.

TRACKING DOWN A FAULT

Tracking faults is practically a science on its own! There are various ideas for finding out why something doesn't work, and each has its advocates. As elsewhere in this book, I'm going to tell you what works for me. The BBC taught troubleshooting skills to its operators and tested them by introducing live faults to equipment whilst they were using it. Failure to spot and fix errors fast was essential to stay the course. Hopefully your employer will be more relaxed, but it doesn't hurt to be well prepared. Here are some ideas…

- If the initial rig doesn't work when first powered up, check each unit is powered; it's amazing how often one can be missed! If its stereo (or multichannel), is just one channel faulty? If so, you can quickly track down the fault by swapping leads between units to see if the fault swaps too! The "Half-Split" technique works well here. You start around halfway in and swap leads, testing each half to identify which unit contains the fault. You can then investigate the faulty half by further dividing and testing until the exact location of the fault is found.

- If you are live "on air" you will be delighted that you spent a previous tea break well, looking at where monitoring points exist in the chain and memorising them! I use the "Input/Output" technique here. When the output is lost or damaged (in radio we used to listen to a local receiver to hear our output back from the transmitter). The next point back is clearly your last checking point, which is your desk output. If that's OK (phew) it's someone else's problem to fix. If not, check the source into you is OK. If that is damaged (phew), contact the stage before you. Sadly, on this fictitious occasion it is fine incoming and bad outgoing. Check forwards using listening points such as pre-fade listens, auxiliary sends and returns. If this is not conclusive try plugging across to an adjacent input, so you are using a different channel in the desk.
- If it was something like the Queen's Speech at Christmas (obviously King's Speech now), I would have a complete parallel chain set up and tested so I could go over to it in seconds.
- If you have more time, and the system is not multichannel, you will have to go down the road of "Unit Substitution" where each block in the chain is replaced one by one, until the fault clears. The Half-Split technique works well here where the chain is complex.

Let's try a simple everyday fault diagnosis to try and pin down some ideas. My bedside light doesn't work. Is there another one the other side of the bed? Does that work? If so, then use it until I have time to investigate further. (Later) Take the faulty light and plug it into its working twin's socket. Still not working? Exchange bulbs with the other light. Still not working? Exchange fuses (if fitted). Still not working? It must be the lamp cabling, which often includes a switch. Put to one side and use the working lamp whilst waiting for a repair.

This example introduces two points I want to expand on. The first is the logical nature of fault finding, excluding the more obvious answers first. The bulb is a consumable, and often the bulb will take a fuse out with it, so needs checking too. The second is a pure operational matter. The operator's role is not to repair, but to restore! The operator must resume the programme quickly so the audience can continue to enjoy it. The faulty equipment is not a concern for the moment; it must be taken out of use and stored until it can be repaired or replaced. That is what separates a professional operator from an engineer.

CONTRIBUTORS

There are two ways of sending a listening (cue) feed to headphones, whether the contributor is local or remote. The obvious way is to send the desk output so they can hear the programme they are contributing to, and this is called "Cue Feed". The other was developed for old analogue phone lines where the incoming and outgoing feeds sometimes had very little sound separation and could easily howl round. In this, the cue feed to the contributor has all the sounds from the desk mixed, apart from their own contribution. This is called "Clean Feed" or "Mix Minus" and is very effective in difficult technical situations. Note, it won't stop a howl round if the contributor has a radio on to hear themselves in the room, as the other sounds of your output are still being played down the line. Talkback can be added to Cue or Clean feed so you can keep the contributor happy with reassuring chat. Just note it is worth telling the person that they won't hear themselves (Clean Feed) as when they start to speak and don't hear their voice, they may panic and stop… Not great broadcasting!

PHYSICAL PROBLEMS

Our look at both analogue and digital sound systems has largely assumed a perfect world where an outside influence on our pristine audio circuits is not considered. We did touch on balanced line working to protect low level signals from induced noise, and the use of a series capacitor in a low pass filter circuit to block DC and a parallel capacitor (low pass) to roll off unwanted AC frequencies above the range of human hearing. This chapter will expand on the everyday problems in the practical application of sound systems, how to discern what the problem is and how to tackle it. The advent of multichannel sound is useful in fault finding as parts of the bad channel can be swapped out with the good, to tie a problem down to a particular unit or board, and although it may sound obvious, it's always worth checking all the units in the sound chain are powered. Here are the basic problems you may encounter:

- Dirty connectors or switches
- Aging components
- Poor earthing
- Fuses
- Hum loops (coils – humbucker)
- Power supply smoothing (transformer-based designs and switch mode)
- RFI (cable-borne and air-borne)
- Dry joints
- Component failure

I think it's fair to say that if you don't follow the science, you will still learn a lot from this chapter about matching an unwanted sound (or lack of it) to a probable cause. That said, there are combinations of problems that we can't hope to solve in a simple constructive overview, and new technologies will bring new problems that are not covered in this book.

DODGY CONNECTIONS

The first I am going to tackle is the most common: dirty connectors, switches and potentiometers ("knobs" like gain, bass, treble, etc.). A dirty connection obviously crackles, but at low levels it can also form a partial connection which acts as a semi-conductor where low-level sound is peculiarly distorted, a fizzing "Dalek" effect created by rectification across the contacts. I've heard this a lot with balanced lines where either the live or return is lost, the resultant low level distorted whisper is called "one legged". The initial solution is to work the junctions, twisting jacks or wiggling other connectors, working faders and switches back and forth, until the dirt is cleared back to bare contact faces. It is worth noting that quality contacts including relays and switches are routinely plated with a precious or semi-precious metal such as nickel, silver, palladium or gold to improve conductivity. The fact that dust could enter suggests that a better cure is available through a "Switch Cleaner" aerosol, which can pump in cleaner where the dust previously found a home. A "must" in an audio engineer's toolkit, this spray cleans and lubricates the contacts for months to come. If a cure is not immediately apparent on a potentiometer, it is likely that the carbon track is worn through to the backing (often cardboard on cheaper units), and replacement is the only option.

AGING COMPONENTS

If crackling remains it is likely there is a stray DC voltage on the contacts, which points to an ailing (leaking) capacitor which will require replacement. As well as crackling, there may well be a deep thump as you plug and unplug the unit from other equipment. In passing, the only modern basic electronic circuit component to age to the point of failure is the electrolytic capacitor (the polarised one marked with + and – terminals). Most contain a simple gel between the two plates, which dries out over time, particularly when sited close to hot components such as power transistors, and so changes the value of capacitance, perhaps to the extreme of a short or open circuit. These ailing components can sometimes show no outward signs, but fortunately most will exhibit slight bulging of the can or staining on the body, which can flow to deposit hard brown residue onto the circuit board. Replacement is the only option. If exact values are hard to find, go higher in value in both "working voltage" and capacitance, if necessary. Any replacement will "work" but if the equipment is special to you, look for a quality component with a low internal resistance. This is the "ESR" value in ohms, and lower is better. Leakage values improve with price, as does the ability to work at higher surrounding temperatures. Again, use the internet to take your knowledge to the next level.

Whilst resistors don't age, placement near overly hot components (as for capacitors!) such as power transistors or their heatsinks can cook a component, perhaps to the point of failure where it can either go partially short or open circuit. A similar problem can happen if the component was specified without enough wattage to do the job. Look for a blackened exterior to the resistor which may well have faded the ring colours to the point where it's hard to see the value. Some valve equipment treated a resistor as a fuse, as it was better to sacrifice this cheap component rather than the expensive ones. In this unusual case the replacement should be the same wattage as the failed one, once the fault is cleared.

POOR EARTHING (AN OLDY BUT GOODY!)

Poor earthing is often as much a matter of design as ageing. Bear in mind that the AC signal earth is often also the DC power supply earth, so problems here can be complex. Areas which are tied well to earth reject RFI, however should that connection become dodgy, it's open season! The design rule is to tie each section or board to earth once, and strictly once only. Where there are multiple boards in one unit (casing), this needs care as the input to the board will bring the earth from the previous board and will pass this to the next board, and in this daisy chain, problems are easily exposed. As mentioned earlier, a DC current on an AC sound wave is easily detected as a distinct "thump" is heard as contact is made or broken.

"Hum loops" are a direct carry over from the previous paragraph, a situation where the earth wiring finds two distinct paths, forming a loop which can induce stray currents. In vintage equipment this was often the case when the earth of the electronics was shared with the mains earth, both being tied to the metal chassis to shield the electronics from RFI and to protect the user, as should a live mains feed touch the case it would blow the fuse.

Imagine two boxes such as, say, a mixer and a power amplifier. They are connected using a signal cable to carry the sound feed, and this links the earths of the two boxes together. They are also both plugged into the mains to power them, and so the plug earth pins are joined through the mains wiring; therefore, there are two earth paths between the equipment. The "loop" formed by the two paths induces local interference, which is predominately from the AC of the mains itself. 50 Hz (60 Hz in the US) is often too low a frequency to register on the equipment, but the harmonics at 100 Hz and 150 Hz (120 Hz and 180 Hz) create an overpowering "buzzy hum" which, once heard, is never forgotten. This problem is rare now as most equipment power supplies are "double-insulated", which means a mains earth wire is no longer needed, but with a trend of new love for the sound of vintage equipment, you may well encounter it. Bizarrely, I am aware of a more recent similar problem where the screen of a USB cable is connected to two units, which terminate it in different ways. Whilst not dangerous in the least, a loop can be formed, and interference produced.

FUSES

Many people reading this book will be aware of fuses but may not have needed to get to grips with them, so it's worth a paragraph or two to get everyone up to speed. The fuse is a consumable item (made for replacing), yet the vast majority never need replacement, and most go unnoticed. It is a protection device, sitting in an electrical or electronic circuit, to ensure the circuit does not draw too much current. This protection is twofold. (1) To protect the supply by partitioning off a faulty circuit from other good ones, so maintaining power to the rest. (2) To protect components in the fused circuit from damage caused by a fault, either in the circuit or caused by an external factor connected directly to the circuit. The fuse is a very old style of component and has been replaced in many situations with modern resettable devices such as micro circuit breakers but the humble fuse does remain a huge participant in audio equipment and, for that matter, technology like your car (Figure 16.1).

Figure 16.1 A selection of fuses
Photo by the author

The fuse is simply a piece of wire that burns out if the current is too high for it to take (measured in amps). Fuses are available in many sizes and differing current ratings. Rare, but there are also thermal fuses which blow when there is excess temperature rather than current. Thicker fuse wire means a higher current rating before it will blow. It is possible, but unlikely, that in an old system a dead fuse has simply fatigued and failed. However, a blown fuse should be seen as a warning that something is amiss and needs investigation. When the problem is resolved, a new fuse of the right size and current rating can be inserted, and the unit repowered. Sometimes (but not frequently) specialist fuses will be employed. Some can resist short surges in current, like at switch on. These are "Anti Surge". Fuses can also be "Quick Blow" in cases, and where these are fitted, the replacement should be the same.

POWER SUPPLIES

Unless your equipment is battery powered, the following is important to know: from previous sections, the reader will understand that the public power supply across the world is AC as this is much easier to generate and distribute from the power stations. The electronics of our sound system will require a DC supply so the local power supply will need to lower the voltage to approximately the value required by the electronics and to change it from AC to DC. Power supplies were traditionally transformer-based, but there is a well-established move to "switched mode" which uses far less copper, so is both lighter and cheaper. Both types of supply are not perfect in the change from AC to DC, leaving behind an unwanted undulation in the DC which is referred to as "ripple". The traditional transformer circuit will use large capacitors strapped across the output of the supply which are a "short" to AC but allow the DC to pass and this is normally adequate. Further sophistication is to use transistor regulation; series chokes and parallel small (but fast) capacitors remove any remaining high-frequency components. These power supplies are bulky and heavy, but the quality and consistency have been long assured.

Switched mode supplies (often configured "Buck Converters") are increasingly popular as the cost savings are very tempting, particularly with high current activities like large power amplifiers for PA uses. The saving on weight is amazing, and cost/watt is decimated; however, components are more heavily stressed, and the long-term reliability is not as assured. The ubiquitous 5 V USB supply is now always switched mode, and these are most often used for charging batteries in equipment, rather than powering it. However most custom supplies for mass produced semi-professional sound equipment also use a switched mode supply. As the title implies the output is created by a high-speed oscillator working at around 150 kHz. This is a rough square wave and so creates harmonics at many times the base frequency. At the time of writing, DAB radio signals in cars (~200 MHz) are being destroyed by dash cam switched mode supplies running near the antenna lead. This causes the DAB signal to mute on some stations, so it's either dashcam or radio, but not both!

The solution to hang large capacitors off the output of the supply, as in traditional transformer supplies, will not work as the circuit operation is fundamentally changed and ceases to work. The answer is to use inductors called ferrite beads to remove the troublesome RFI and electromagnetic fields (EMF). Frankly all audio devices, in both digital and analogue designs, should ensure that stray high-frequency signals cannot destabilise the actions of our equipment; however, an unusual combination of units may well not have been considered.

Figure 16.2
Suppression using a ferrite ring
Photo by the author

A more effective solution is to purchase large ferrite rings, an inch or more across, and to pass the power lead through the ring tightly (secure with tie-wraps if necessary) several times (Figure 16.2). Between five and seven times has been found to be effective in most circumstances; however, note that the bulk of the ring must be hidden in a well of the car somewhere, hopefully where it will not rattle around!

The easiest way to protect equipment is to encase (shield) it in metal. Tin and copper are the best, and to ground the case by connecting it to the 0 V power supply rail. The same can be said for the sources of noise; the power supply should be separately metal cased and earthed, even if the outer is a plastic casing to allow it to meet double insulation rules. Allow for ventilation of heat.

Clip-on commercial ferrite beads can be added to signal wires at their ends (where the induced signal is at its strongest) to lessen the vulnerability to RFI and EMF. The only weakness with these measures is where the inductor works in parallel with existing capacitance to form a band pass filter (see Chapter 3, "Introduction to Electronics"), which can exacerbate the problem rather than remove it!

Common Mode Choke – our solutions so far have not involved breaking into the power supply wiring, but I'm sad to say, they may not have got your precious signal back! So where do we go from here? We have previously said traditional transformer power supplies also have an AC component on their output from the rectifiers. This is easily filtered for most purposes by strapping large capacitors across the supply, putting inductive chokes in line with the supply or a combination of both, to achieve a steeper slope of high frequency attenuation. Unfortunately, this approach cannot be used with the buck converter, as I mentioned earlier; it changes the parameters of the circuit design.

Fortunately, a variation of the simple choke does suit high frequency filtering on the output of a switched mode supply. The Common Mode Choke uses a ferrite ring on which two mirrored, yet otherwise identical windings are made. DC volts for the audio equipment pass straight through the negligible resistance of a few turns of thick copper wire, however, the AC content is increasingly out of phase with itself (differential) with rising frequency and so is attenuated. Example – a 1 millihenry choke (6-amp winding) is the size of a thumbnail. With a typical dash cam loading, this gives a −3 dB at 300 Hz, which is already attenuating our supply switch-mode frequency by over 50 dB. As frequencies rise to DAB multiplexes, the loss exceeds measurable limits. The dots on the diagram in Figure 16.3a push home that the coils must be wired in phase for the choke to work, and Figure 16.3b shows a typical choke the size of a thumbnail.

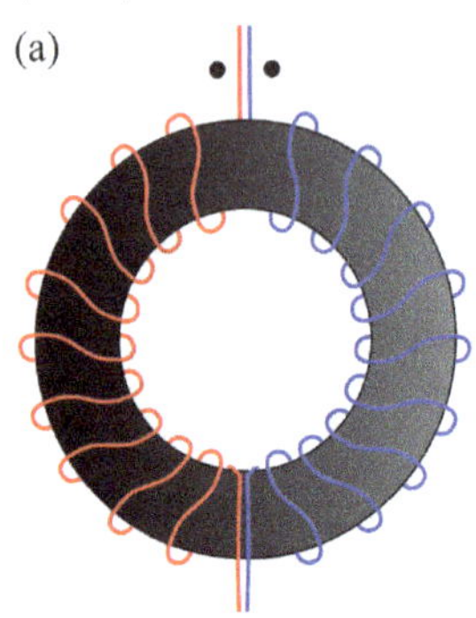

Figure 16.3a A common mode choke diagram

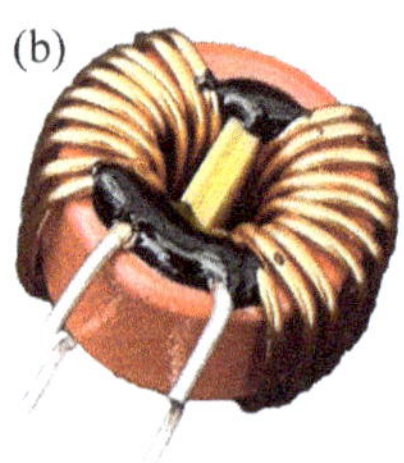

Figure 16.3b Common mode choke
Photo by author

The rarest faults, especially these days, are dry solder joints and component failure. A dry joint is often created in manufacture, where the components are slightly soiled and so the solder can't adhere properly. Although the unit may work well at first, if the joint moves, perhaps through slight expansion by heating up in use, the effect is the same as a dirty switch, with failure or distortion through rectification across the joint. Component failure is mercifully rare; however, electrolytic capacitors are the most likely to break down partially or totally fail to open or short circuit (see above). If you are unlucky the latter can take out semiconductors, particularly in discrete power amplifiers, which are "direct coupled" (DC can pass through uninhibited) for higher quality. Capacitor failure will either stop the circuit working, may cause volume changes, create pops, clicks and hash, or occasionally make an audio channel cut in and out in a spurious manner. As mentioned above, a faulty capacitor will often leak gunge that looks like glue. They are easily replaced, bearing in mind the maximum working voltage, as well as the value, measured in fractions of a farad. Note that electrolytic capacitors are polarised, meaning they must be put in the right way around.

DIGITAL FAULTS

This chapter is mainly an introduction to analogue fault finding, to build the confidence of the reader to help themselves, rather than to have to rely on others. In general, digital circuits either work or they don't, and provided the interfaces utilise the same protocols, there is not much that can be done, other than swapping complete units out, or using specific software containing digital investigation tools. In my experience the most common problem is with a damaged coaxial or optical link. A simple kink in the optical lead or a dirty end is enough to stop the high-speed transmission working. Replacement is the easy way to check if the lead is the problem (Make sure you carry a spare!).

On a rare occasion you can be unlucky with lead length; even with clean connections there will be some reflection back from the end, providing an out-of-sync echo of the original. In my experience, this echo will change dramatically with the lead length, just like standing waves and room dimensions, and, as above, a simple change of lead length may well cure the loss.

Modern digital electronic circuits are very sophisticated; the designs have been improved over half a century and are largely self-healing. The phenomenon of jitter (timing inaccuracies of the regeneration clock) in the DAC is well sorted, and whilst we can artificially produce problems to hear what huge amounts of jitter can do, the likelihood of encountering them is now very small. Again, early problems arising with the need to "brick wall" the highest audio frequency have been mitigated with higher sampling rates and oversampling converters. However, more is not necessarily better with sampling rates, and thought-provoking research has pointed to problems of meeting or exceeding 192 kHz as your chosen rate. 48 kHz is likely all you will need, but 96 kHz will do no harm. It's difficult to see the point in going above 24-bit, as the four least significant bits will be lost under a blanket of thermal noise (see the section of Chapter 6, "Sound Levels and Connections", entitled "More About Noise"). A reminder that a doubling of either sampling rate or bits doubles the amount of storage capacity required. In the role of wet blanket, I believe it is unlikely anyone will hear a difference beyond a well specified 48 kHz and 24-bit linear system. 32-bit Float is in its early days and has extraordinary on-paper advantages in multi-channel session recording, but I am yet to work with it.

BIBLIOGRAPHY

Baker, B. (2016) *How Delta-sigma ADCs Work, Part 1.* Texas Instruments Incorporated. https://www.ti.com/lit/an/slyt423a/slyt423a.pdf?ts=1758953962563.

Coates, E. (2020) *Module 3.0 Switched Mode Power Supplies.* Learn About Electronics-Power Supplies. https://learnabout-electronics.org/PSU/psu30.php.

Duncan, B. (1984) *Studio Earthing Techniques.* Home and Studio Recording- September 84. Music Maker Publications (UK), Future Publishing.

Eco, J. and Limjoco, A. (2016) Ferrite Beads Demystified. *Analog Dialogue*, 50(2), February.

Engdahl, T. (2013) *Ground Loop Problems and How to Get Rid of Them.* https://www.epanorama.net/documents/groundloop/index.html.

Appendix 1 – Microphone Placement

In Chapter 4, "Microphones", we introduced the concept of a sound pressure to electricity transducer called a microphone. We examined dynamic (moving coil), ribbon and capacitor (condenser) types. Being careful to keep polar patterns distinct from transducers, we looked at the three most likely patterns in a studio: dynamic (omnidirectional), figure of eight (bidirectional) and cardioid (unidirectional). The chapter went on to discuss stereo recording techniques and in part tried to match a need for a microphone to the type best suited for it. For instance, dynamic cardioid for stage vocalists and a ribbon figure of eight for a studio violinist.

In "Appendix 1 – Microphone Placement" I want to expand on Chapter 4, "Microphones", further matching microphones and their positioning to instruments you would expect to find in bands and orchestras. I'll state from the start that this is just one interpretation; music engineers spend a lifetime perfecting positioning and types to get that golden sound! I hope to at least provide a starting point for further experimentation and to give you the confidence to take on more complex challenges. I need to make the distinction between live working and studio working. On stage there is often more freedom on space; however, there is always concern around a howl round off the PA. An orchestra is likely to be packed in at a studio! Whilst it's true, the artists will be listening on headphones, there remains concern about achieving enough separation from the sound made by other musicians. The drums will require isolation from the other instruments as they are so much louder. Drum acoustic isolation surrounds called "tents" are now routinely used on stage as well as in the studio, and it's common to have the expense of a separate room just for drums in a studio complex! Thought must be given to keeping the drummer communicating well with the rest of the musicians, both visually and audibly. Each situation will require a unique version of a general solution, watching and listening for the interaction of a microphone picking up what it shouldn't!

When people are gathered in a small space, the likelihood of tripping over cables increases. Of course, there is the issue of safety and falls; however, further, there is concern for the thinner cables fracturing if dragged by feet or chair legs. These are likely to be the headphone cables, or miniature capacitor or dynamic microphones, which are clipped to instruments to get a close sound and only transfer from a thin wire to a robust XLR on the floor or taped to a convenient object like a chair.

Sensitive instruments, and by that, I mean the lower output ones like violins, are exposed during a solo. Used sparingly neoprene or rubber kneelers or pads, placed under the mic stand, will ensure structure-borne noise does not travel up the stand to the microphone and be amplified with the soft violin. Structural noise can be produced by clumsy close neighbours, or from the outside from juggernauts or, in London, the &!@@?% underground!

I'm using clip-on microphones sparingly in the 20 microphone placement pages, not because they are poor, but I find the sound front directly at the mouth of the instrument is yet to coalesce, so like placing microphones down the larger horns, it's not for me. That said, clip-ons are fine if you are looking for the effect rather than correctness and some might say doing it my way, I run the risk of players going off mic in exuberant moments. I've tried to push the boundaries to give newcomers some insights into methodology, and over the course of 20 instruments I hope a pattern will emerge. There are many ways to cut this cookie, and most of them will taste just fine!

USING MULTIPLE MICROPHONES

Chapter 4, "Microphones", covered stereo microphone pairing and positioning and touched on the importance of absolute microphone phase as otherwise it creates sound cancellation across all frequencies. But what of the unintentional interaction between adjacent microphones on stage and in the studio? If they point directly at each other, with the sound source between them, that is a similar situation to out-of-phase, but there are many situations where the microphones are physically close but miking different people or instruments. Perhaps it is a single microphone near a reflective screen or wall, so it receives the direct sound and the longer path length of the reflection. Depending on the difference in path length between the sources, multiple audio frequencies are cancelled out in an aberration called a comb filter; it's the same problem as a destructive first reflection in stereo monitoring (Chapter 5, "Loudspeakers"). In this appendix we will take a closer look.

The effect of the comb filter can be highly useful in video and in cosmic technologies, in interpolation and investigation. In audio, it's a pain! In our introduction in Chapter 1, "Introduction to Sound", we introduced linear plots of frequency response but noted we generally use logarithmic ones as that is how the ear hears. Figure A1.1c shows the effect where a second path length roughly twice the original microphone distance is added to it, and you can clearly see the "comb" distortion in this linear plot. The "comb" may be created because a second path has bounced off a nearby flat as in Figure A1.1a, or simply as shown in Figure A1.1b when a sound source is picked up by more than one microphone. You can see there has been a "constructive" gain of around 3 dB, but this is offset by the destructive "notch filtering" of many frequencies. You'll remember we don't model with linear frequency plots, so Figure A1.1d shows the damage in the context of what the ear will hear.

How can we avoid comb filtering? Clearly, by being aware of the problem, steps can be taken. Close miking, where possible with a directional pattern, is common sense, and carefully setting out a microphone plot is useful, but is there any industry guidance?

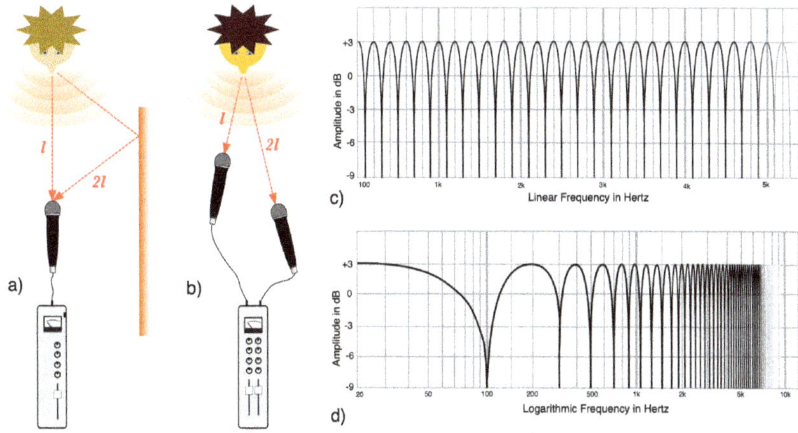

Figure A1.1 Microphone phase problems cause comb filtering

THE 3:1 RULE

Decades ago, the phasing effects of comb filtering were first noted when using multiple conventional omnidirectional microphones and there was much experimentation. This led to the "3:1 rule". As the years have passed different interpretations have sprung up, many seeking to clarify, some missing the point! I look at it like the specified braking distance to a car in front; you get asked questions on the minimum distance on the driving test and probably forget the next day. But I ask what car, what tyres, is it raining? 3:1 is a simple practicality based on old technology, where a meagre choice of microphones had roughly similar sensitivities and directional microphones were rare. Mixing desks were also basic.

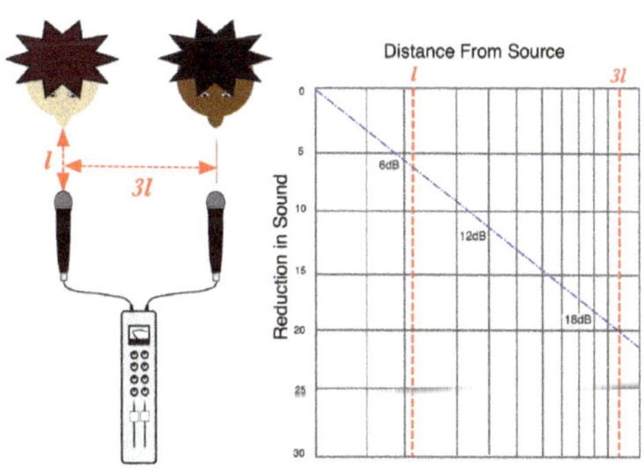

Figure A1.2 The "3:1 Rule"

Given these restrictions engineers found combing effects could be minimised where the omnidirectional microphones were placed *more than three times further apart than the distance from the sound source to its microphone*. This is shown in Figure A1.2. The graph is borrowed from Figure 1.7, the point source inverse square law diagram in Chapter 1, "Introduction to Sound". The 3:1 rule simply seeks to have *at least* 12 dB level advantage between the wanted microphone and its unwanted neighbour, which greatly minimises (not removes!) comb effects. In general, a modern cardioid microphone is 6 dB less sensitive at 90 degrees, which is already half the battle! Is it still relevant? Yes! Just like you don't tailgate the car in front of you because you don't want to crash, use the process to consider the microphone pattern, where it's pointing, how close it is to the wanted source, and does it also point at unwanted sources? Look to make these checks a habit.

Extra gain on a quiet source will make a microphone more sensitive to surrounding sounds, so get as close and directional as you can. Consider near reflective surfaces like a drum tent wall, and if you rig a reserve microphone, don't make it active till you need it. Multi-miking drums will always be a compromise, but the lessons learnt in contemplation of the 3:1 Rule still hold true. The rule is based on a coherent point sound source; if it's not, like say a 4 × 12 speaker, that wide sound face can comb filter even where there is a single microphone, so a solution needs common sense. The next section places microphones in likely acoustic positions, but it is also about the fidelity of the sound and managing the sound pressure level from the instrument. It is not a rule book; it is a reasoned start to get you thinking about techniques to experiment with and improve on your own journey. I start with the simplest techniques and build the complexity as I go. Final words: when you come to mike your own band, put breaking the rules at the centre of your path of discovery!

MICROPHONE PLACEMENT – A MICROPHONE SELECTION

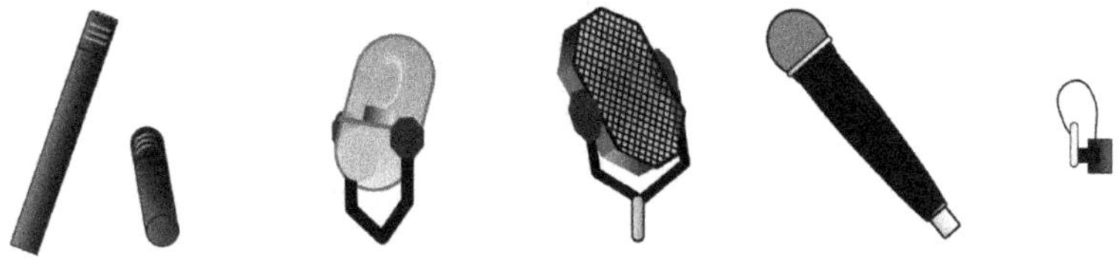

Figure A1.3 Microphone choices for use in the following placement examples

TWENTY DETAILED MICROPHONE PLACEMENTS

1. Vocalist
2. Electric Guitar
3. Trumpet
4. Trombone
5. Double Bass
6. Cello
7. Viola
8. Violin
9. Grand Piano
10. Acoustic Guitar
11. Clarinet
12. Alto Saxophone
13. Tuba
14. Congas
15. Xylophone
16. Harp
17. Flute
18. Accordion
19. French Horn
20. Drum Kit

Vocalist (Studio)

No problems with placement, a simple microphone stand with an obligatory mesh pop shield. A suspension mount is always a good idea. The difficulties come in microphone choice, which traditionally has been practically anything that "sounds right" on the day. With male voice frequencies down to 100Hz a large diaphragm condensor makes sense with cardioid pickup, but don't be afraid to experiment.

Use a 80Hz bass cut, or put the stand on neoprene matting to suppress structural breakthrough from instruments and location. The microphone diaphragm should point directly at the mouth, and I usually have it looking slightly up at the face as instinctively many singers "emotionally hug" the mike looking down on it in sensitive lyrics.

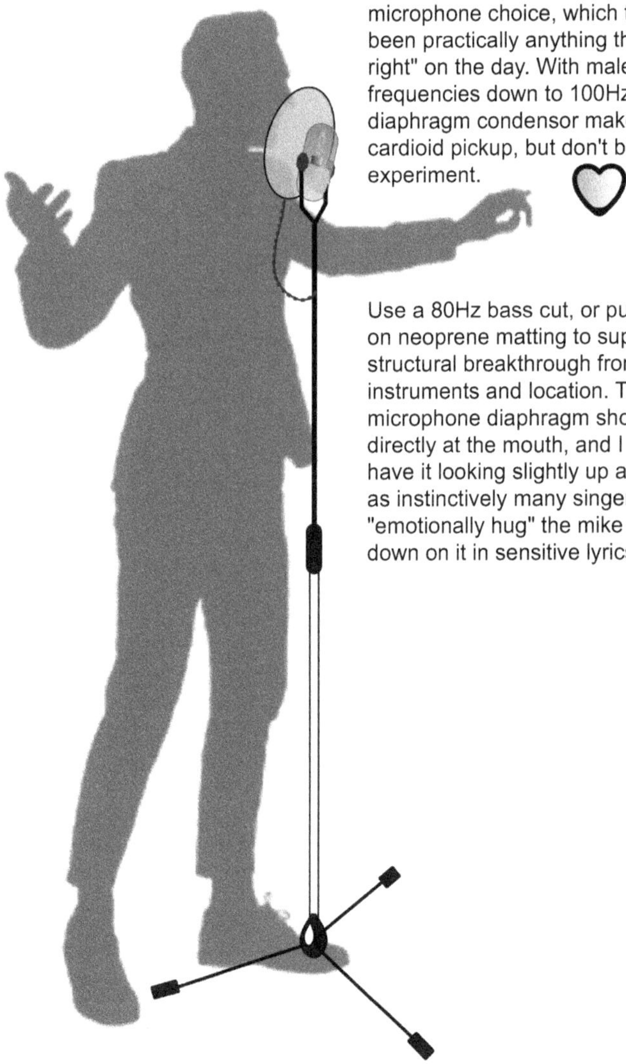

Figure A1.4 Vocalist

Trombone

The trombone has a high acoustic output, yet needs direct on bell pickup to sound natural, as it is directional. This combination suits a fine mesh pop screen on a cardioid condenser or a foam pop shield on a clip on. The large diaphragm microphone should be 30-40cm (12-15 inches) from the bell. Use the microphone switched pad.

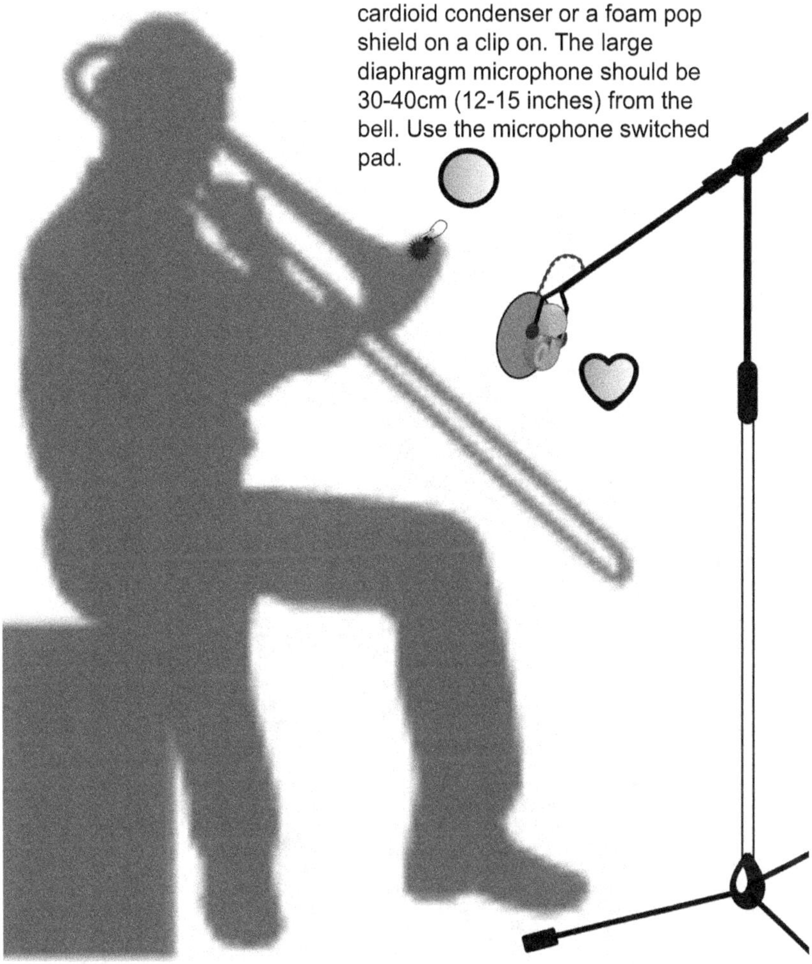

Figure A1.5 Trombone

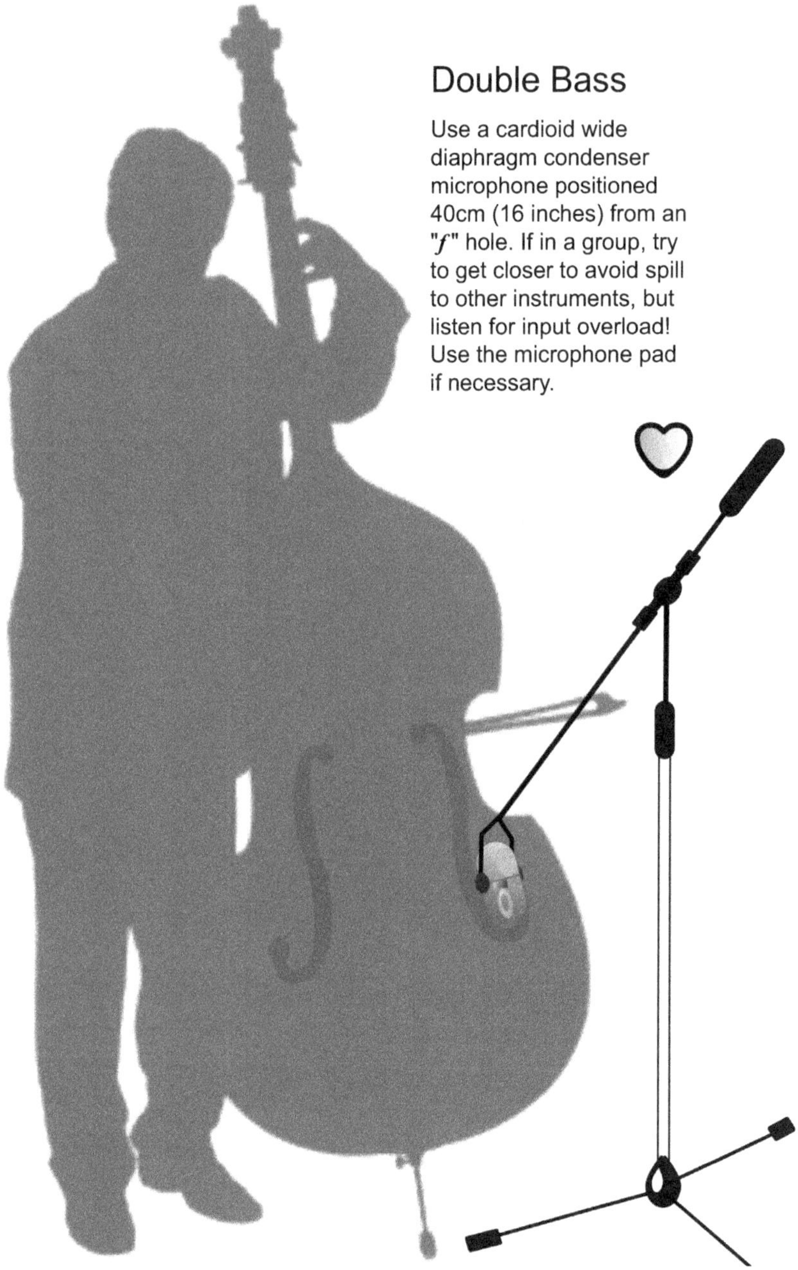

Figure A1.6 Double bass

Cello

Position a large diaphragm condenser microphone 30cm (12 inches) away between the bridge and and right "*f*" hole. That may be too close for an exuberant player, but it is a decent starting position.

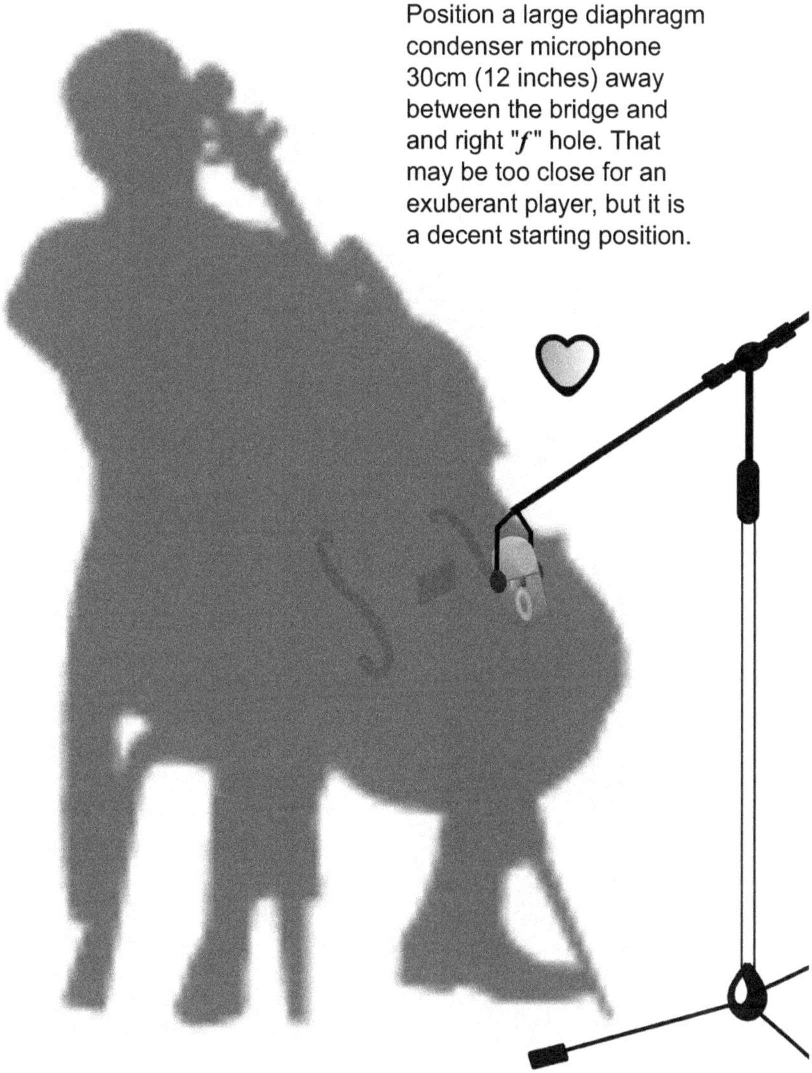

Figure A1.7 Cello

Viola (Session)

Use a short-form small diaphragm condenser on a body clip, placed above, looking down on an area between the fine tuners and the centre of the *f* holes (listen to check no resonance from the mount). Gain an intimate sound at a distance of 10cm (4 inches above) and move the microphone perpendicularly along the strings to achieve the right sound between body resonance, bowing and tonal brightness.

Figure A1.8 Viola

Violin (Orchestral)

If there is a decent ceiling height a ribbon microphone is the perfect partner to the smaller string instruments both in concert hall and studio. Placed above, looking down on the bridge between the *f* holes. Try just above head height for the violin, and 25cm (10 inches) higher for the viola. If not a ribbon, try a super cardioid. An alternative is to mount a clip-on on the strings behind the bridge (listen to check no resonance from the mount).

Figure A1.9 Violin

Grand Piano

This instrument should routinely be recorded in stereo, and I prefer the splayed XY technique (120°) of two large diaphragm condenser cardioids mounted above the middle strings (the alternative is two microphones mounted at the bass and treble ends which can have mono compatibility problems). 30cm to 40cm (12 to 15 inches) above strings gives less hammer and less boomy, but remains bright and well balanced.

Figure A1.10 Grand piano

Guitar

A surprisingly difficult instrument. This is just one way, for an open, intimate and balanced delivery. Place a small diaphragm (large often overwhelming) condenser microphone perpendicular to the frets, and experiment in a line between the edge of the body (upper bout) and the 12th fret. You will be amazed at the tonal change in this short movement. Distance 10cm (4inches).

Figure A1.11 Acoustic guitar

Clarinet

Place the microphone slightly off to one side, pointing at the lower end (chalumeau register) 30cm (12 inches away). The ribbon microphone is fabulous with this instrument, revealing the range of rich yet harmonically complex notes with warmth and precision. Obviously ensure the other live side is not in line to pick up stray sound.

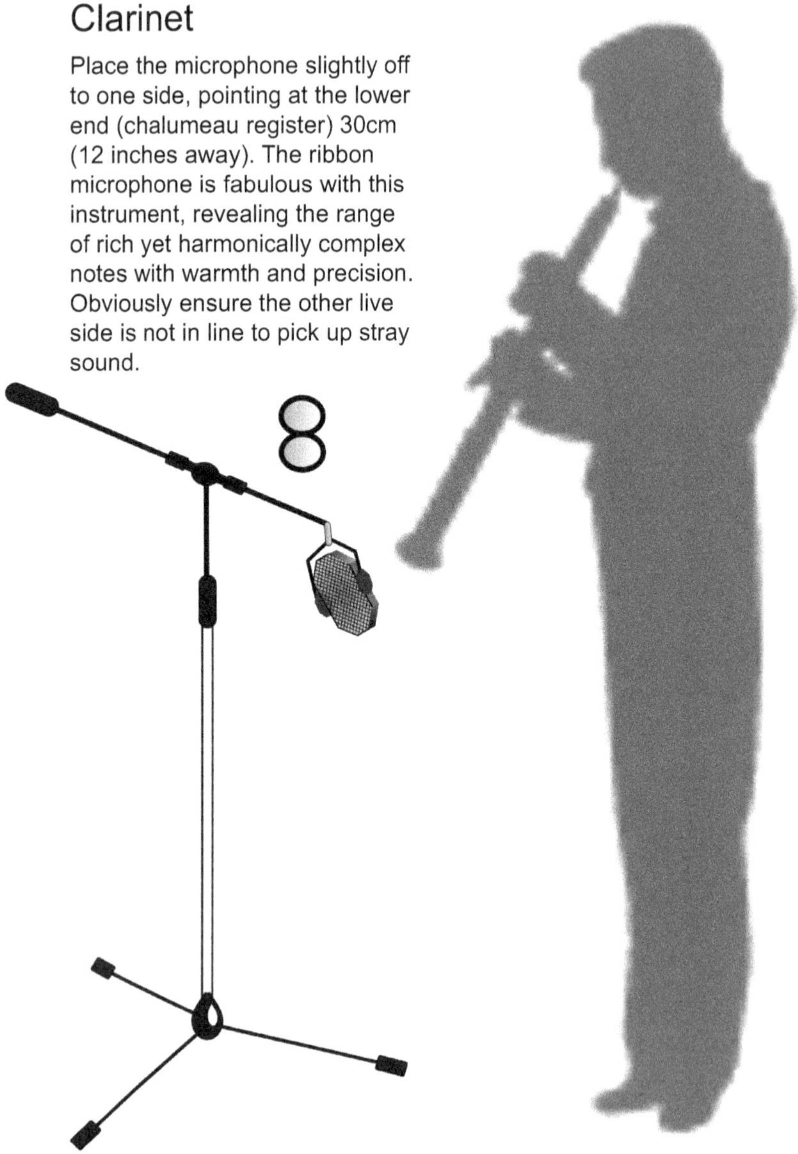

Figure A1.12 Clarinet

Alto Saxophone

The Saxophone creates sound from both the key holes and the bell. *The most natural contains both*, with a vocal mike suiting the edgy response well from 50cm (20 inches) above the instrument. An alternative, particularly on stage, is a clip-on to the bell. Less atmospheric, but punchy, open and good isolation.

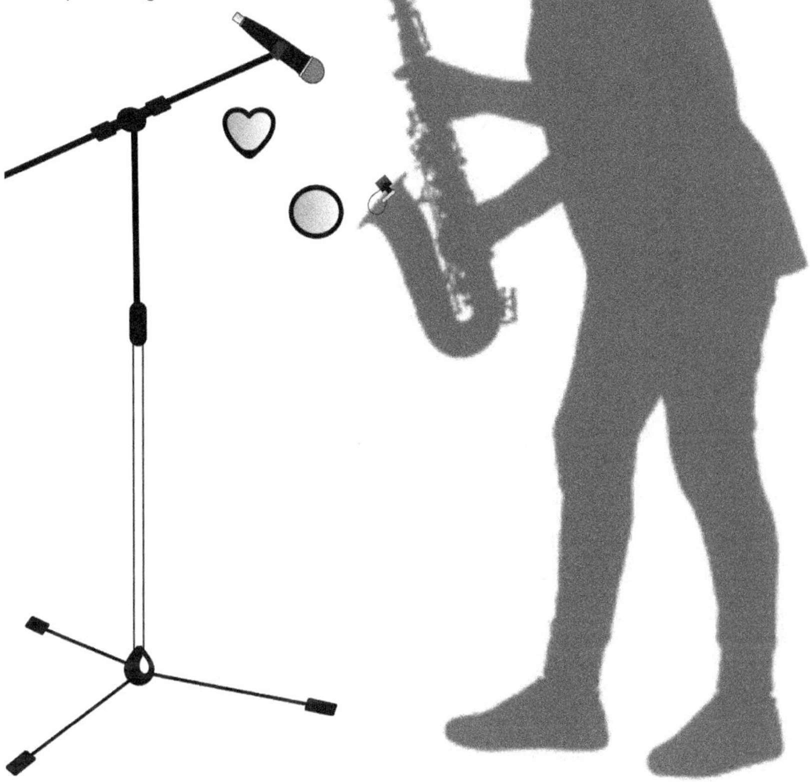

Figure A1.13 Alto saxophone

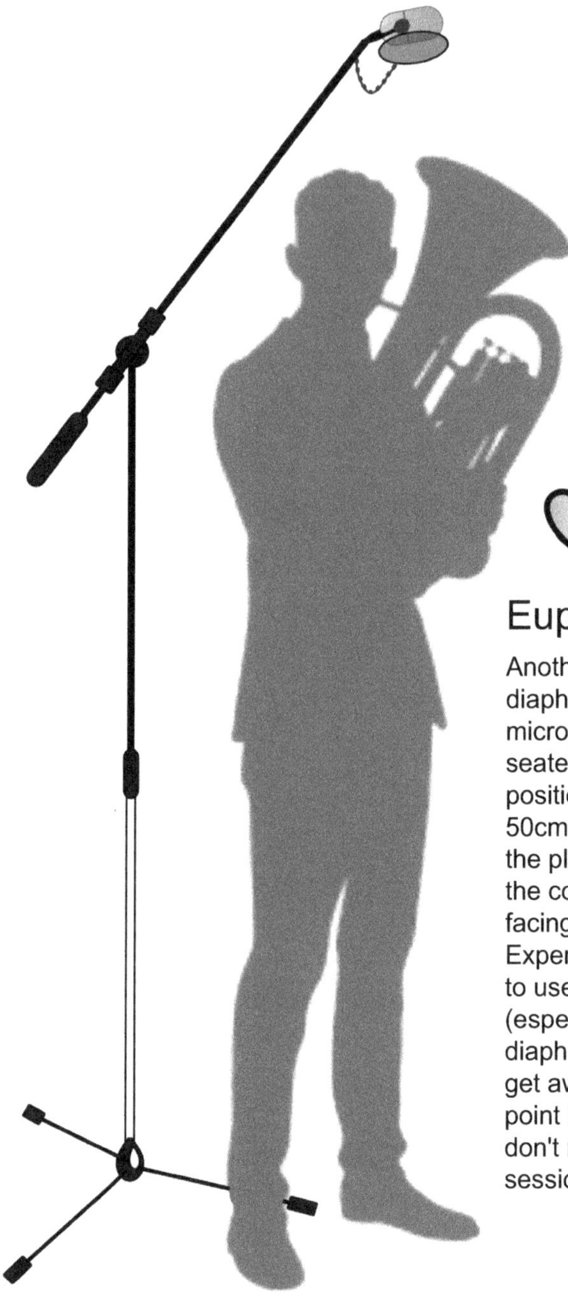

Euphonium/Tuba

Another job for the large diaphragm condenser microphone. Whether seated or standing, position the microphone 50cm (20 inches) above the player's instrument at the corner of the bell, facing the centre. Experience has taught me to use a pop shield (especially with wide diaphragms), but you may get away without it. The point is, if it's there you don't need to stop the session to fit it!

Figure A1.14 Euphonium and tuba

Conga

This instrument has three sizes, quinto, tres dos and tumba, all of which can be tuned to different tension, bringing different notes. There are three styles of hand hit: bass, tone and slap. These variables make for an extremely versatile instrument, from deep and resonant to taught and tight. Because of this, microphone choice on the day will change with the style of the piece and the player. Pop favours "tight" so small diaphragm cardioid condensers will work fine, but more traditional players will work with the larger variety. My player has two identical congas, but will tune them apart. I've chosen a stereo mount, but note the sound doesn't have to be in stereo, it's just handy! Place 5cm (2inches) over the rim, at a height of 8cm (3 inches) at an inclination of 45°.

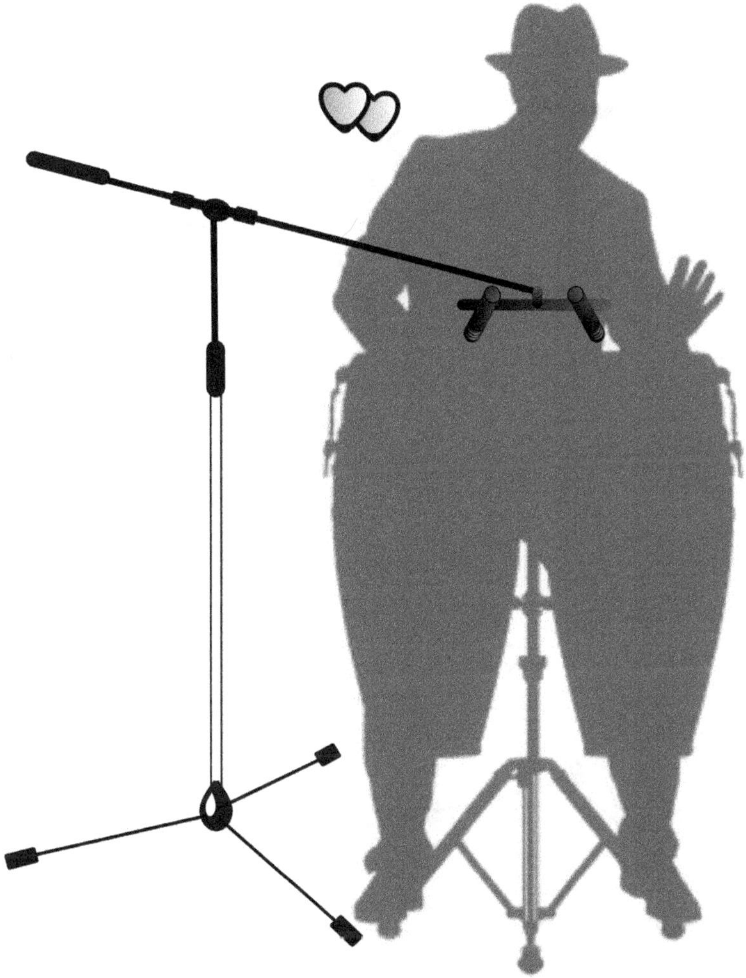

Figure A1.15 Congas

Xylophone/Marimba

An excellent use for a coincident pair of small diaphragm condenser microphones, 130° spread with capsules adjacent, positioned 30-40cm (12-16 inches) above the playing surface. Work with the artist on the effect of differing mallet finishes.

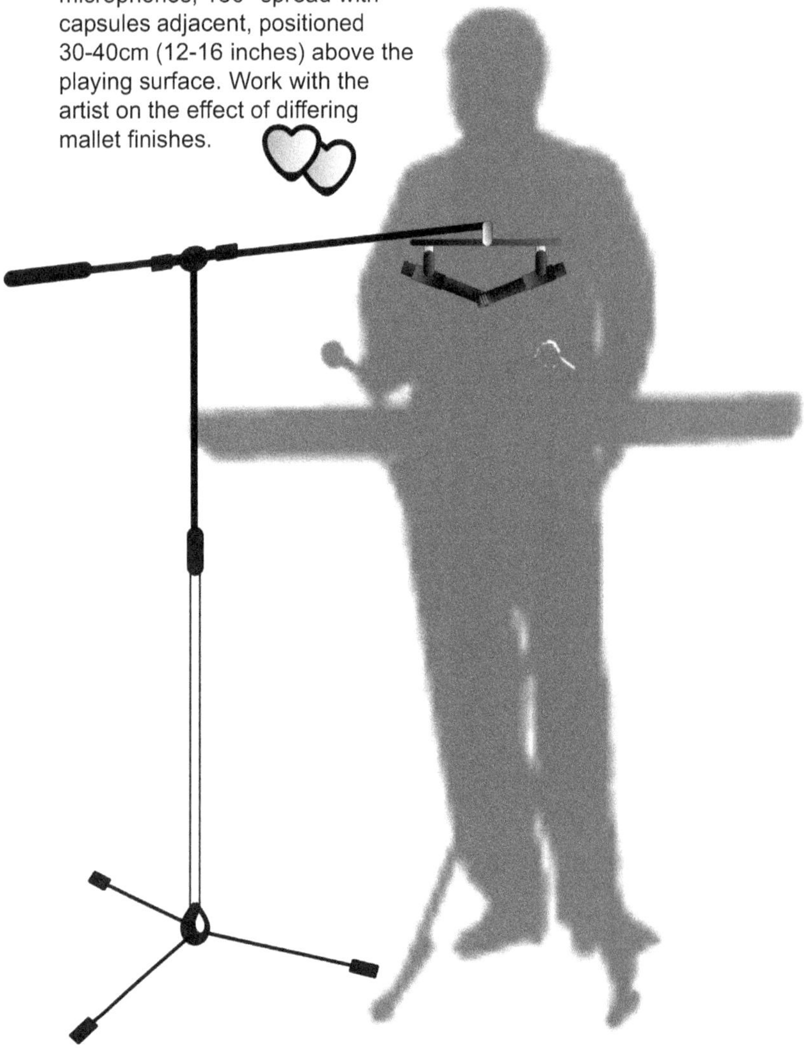

Figure A1.16 Xylophone

Harp

The harp is an instrument designed to fill a whole room (like the piano does). A traditional Blumlein ribbon bidirectional XY pair (90°) is focussed on the middle of the instrument at a distance of 125cm (50 inches), roughly 20° off vertical. The pair's direct face supplies the pluck of the strings and provides a sound balance across the range of strings. The rear faces of the pair provides a coherent instrument acoustic and essential sustain. Mono compatibility should be checked with fine tuning of the placement.

Figure A1.17 Harp

Flute

A flute's major output is from the blow hole at the artists mouth. A small diaphragm condenser does not have the range so point a large diaphragm microphone (position 1) to the nose (so slightly above) at a distance of 15cm (6inches) for breathy jazz, and 25cm (10 inches) for classical. Add in position 2 for atmosphere, at 120cm (4 feet) *pointing half way down* the instrument.

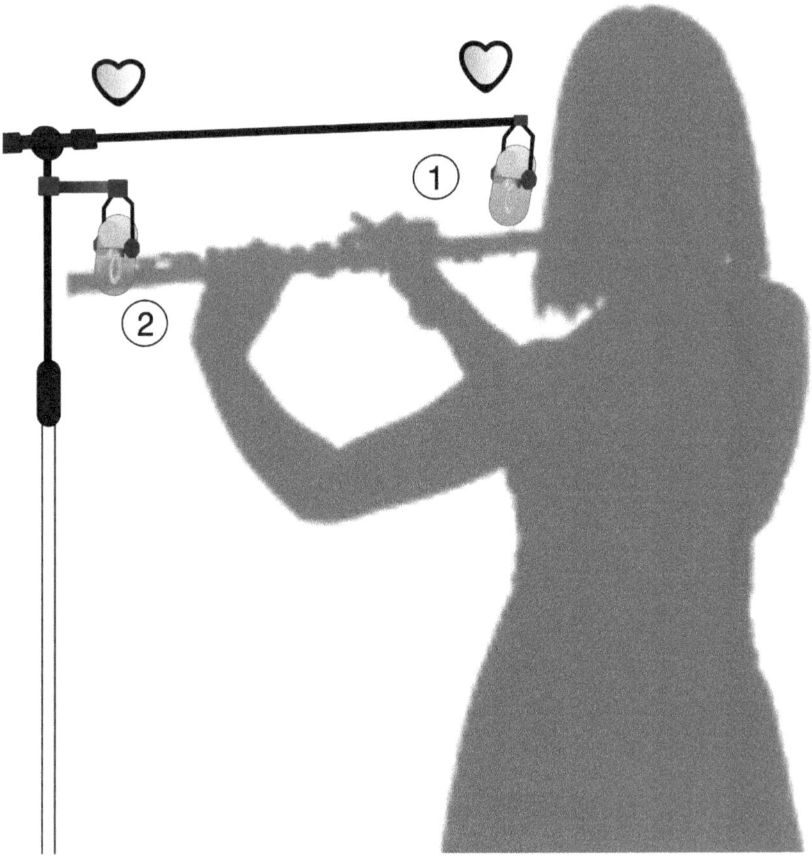

Figure A1.18 Flute

Accordion

This one *has* to be split between studio and live performance. Studio - The accordion produces sound from both ends, the bass (dependant on historic region) has different key placements, however bass is generally from the left hand. Use two cardioid's placed 40cm (16 inches) further apart when the accordion is extended, and forward of the front by 50cm (20 inches). Pan these mono sources to the same image position (as required) and then pan one mic 10° further to create a narrow stereo image. There often will be mono compatibility issues, so this is the time check and adjust! A quilt hung in line with the mic fronts between the stereo pair will improve the sound and the image compatibility.

Band Stage Performance - (Black microphone stand). Use a cardioid dynamic vocal mic for punch and volume handling. Position 30cm (12 inches) forward and to the side pointing at the centre of the right hand instrument grille. Other members of the band will favour the bass line in this scenario, creating a balanced sound.

Figure A1.19 Accordion

French Horn

My earlier point concerning brass instruments and allowing for space so the sound can coalesce outside the bell is repeated here. The microphone should be mounted on a level with the top of the bell, slanting down to capture the centre of the bell. Distance 50cm (20 inches). Watch positioning, the microphone is pointing in reverse to many other members of the band, try phase reverse on this channel and mono compatibility tests when listening to the whole mix.

Figure A1.20 French horn

Drum Kit

Specialist drum microphone kits are legion, this is a guide to typical mic's available to a student thrown into a session. Several microphones inside a hard surfaced drum tent are a disaster waiting to happen. Phasing issues are common, so listen out! The snare is loudest, so position the "catch all" stereo coincident condenser pair (1) equidistant to it (120°). The pair are positioned 30cm (12 inches) roughly above the drummer's head to capture the kit, however spot mics are still useful. The "Snare" (2) is a problem; high SPL, yet delicate high frequencies. Try a dynamic vocal mic, the natural 5kHz boost is useful. Put a small diaphragm condenser mic, between and just above the two "Toms" (3- short form rim-mounted). Point away from the snare! Same again for the Floor Tom (4). Mike the Hi Hat (5) and the Crash/.Ride (6-Cymbal) with small diaphragm condenser microphones. The Kick Drum mic (7) is often placed inside, try a robust vocal mic with the top rolled steeply off from 1kHz. Damp down the resonance with a folded bed sheet laid across the bottom of the drum, and place the mic on top of it.

All microphones are unidirectional for decent separation.

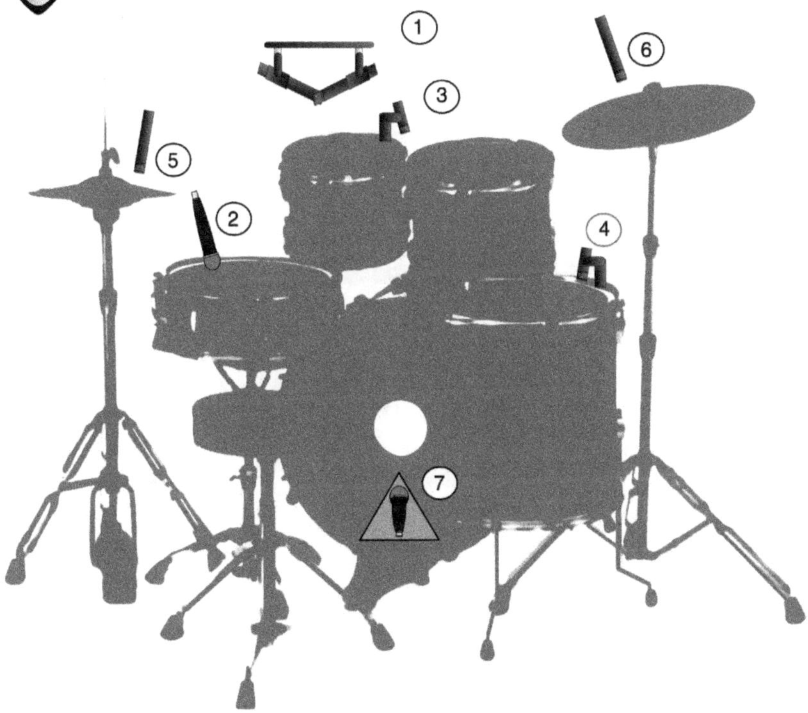

Figure A1.21 Drum kit

BIBLIOGRAPHY

Beyer Dynamic (2024) *Microphones: How Each Different Type Works.* https://blog-en.beyerdynamic.com/microphones-how-each-different-type-works/

Jones, S. (2025) *How to Choose the Right Microphone to Record Anything.* https://www.akg.com/choose-the-right-mic.html

Shure (2014) *Microphone Techniques for Recording.* https://www.shure.com/damfiles/default/global/documents/publications/en/performance-production/microphone_techniques_for_recording_english.pdf-bb0469316afdb6118691d2f3f5e3ff01.pdf

Appendix 2 – Typical Acoustic Environments and Their Design

OVERVIEW

I think we can make a couple of assumptive divides to start the conversation. The first is whether a sound session takes place in a controlled environment or an unknown venue. The second is whether the sound is the sole capture, or whether we require pictures as well. In the following paragraphs I want to dip a toe into varying situations to give an insight into the world of professional sound. I hope you find it useful, but I equally want to say that each location style has a level of expertise that is above the remit of this book, which may I remind you is to give a solid, well-rounded foundation in the world of sound, so you, as an enthusiast, can confidently take the next steps wherever your career takes you.

I am sure a space which is dedicated to recording the voice must be the most popular at the time of publication. Since the earliest home tape recorders of the 1950s people have sought to make their own masterpieces in their bedroom; the Pirate broadcasters of the 1960s and mature DJs born out of the 1940s will without fail have started with home recording equipment. Time has gone by, and the industry has settled to roughly three levels of studio: (1) the professional sound studio, (2) the dedicated home studio and (3) the temporary home studio. The professional sound studio, whether it be radio or voice production, uses expensive levels of acoustic isolation to ensure it can work through the day without interruption from the outside. The internal structure will have a separate cubicle acoustically optimised to monitor the sound from the studio, which will have its own neutral acoustic to neither add nor subtract from the timbre of the artist. It will have well-designed silent air conditioning to ensure you can work safely for long periods without fatigue. A common factor in all the rooms is the small size, which is close to the wavelength of the low-end register of frequencies. More critical than the larger spaces, the studios and their Control Areas (cubicles) will suffer from standing waves and, in some cases, structure-borne noise created by the activities in the studio resonating the wall and ceiling panels.

HOME STUDIOS

Dedicated Home Studio

Many professional voice artists will now give over a room in the house as a working studio. This genre is also likely to include well-known podcasters and radio presenters who routinely work from home "voice tracking" their broadcast links and uploading them to the central station server. In fact, some stations no longer have a studio, relying entirely on home studios and location recordings (see Chapter 13, "Planning for Location Recording") for their content. It is unlikely that the finances will stretch to a professional acoustic design; however, much can be achieved by trial and error. Obvious parts of the room will be reflective, the windows and doors in particular, and these must be prominent in the design.

The door should be heavy and windowless and should be carefully sealed with draft excluder, making sure even the smallest holes are taped (don't forget under the door with rubber stripping). It makes sense that secondary double glazing is considered in town settings, another barrier to gaps, but also to damp resonance from the existing panes of glass. Structure-borne noise is best tackled at the microphone, using an Anglepoise stand with a shock mount. This also allows the microphone to be nearer the artist, which better excludes exterior sound and the effect of the acoustics of the room on the sound. If an Anglepoise is not used, a table desk stand should be heavy and supported on slightly compressed foam (neoprene is excellent) to mechanically isolate it from the table.

So, what can be done with the room acoustics? Keep the rear of the microphone away from hard objects, particularly (1) a tabletop, (2) a window and (3) a parallel computer screen. Consider taping around the floorboards between the skirting and floorboards to trap noise before adding a decent underlay and a quality carpet. Break up the sound (diffuse the reflections) with bookcases containing many sizes of books and air gaps behind. Hang tapestries on the walls (spaced 40 mm away with battens) opposite hard surfaces such as glass and bare walls. Consider soft furnishings if the space will take it, and heavy lined curtains as an "adjustable absorber". If after this "trial-and-error" problems persist, make simple absorbers to the modular designs in "Appendix 3 – A Guide to Home-Built Absorbers". Consider middle frequencies first, and when happy, see whether bass absorption is required in addition. It is unlikely high frequencies will need much taming as the other absorbers achieve this, particularly those already in the room like carpets. If sound from above is a problem (regular overflying aircraft) a plasterboard ceiling can be sound insulated by applying a layer of kiln dried sand directly on top of it. Clearly the sand has weight *and so be careful not to bring the ceiling down*! I have found 1 inch (25 mm) evenly distributed to work well, but I can't know how well your ceiling is fixed! Common sense rules!

Resonance from corners will be highly influenced by microphone positioning, and even a move of a couple of inches can make a difference. You will be monitoring live work by headphones, as any speakers will colour the sound and, at worst, "ring" and howl round whilst you make the recording. The headphones must be *closed* back and should be neutral through the frequency range. Some popular "monitor" headphones sound great working in the studio; however, they are not authentic and may make you make adjustments that will unbalance the recording when heard elsewhere. As a professional, *please* be careful in your selection.

A final point is fresh air. Although I have seen it in cost-sensitive installations, air-conditioning is unlikely to form part of a home studio design, as it is expensive, complex to design and the fan noise can be the most intrusive sound after other treatments have been made. It is common sense that if working in a sealed box, regular breaks must be taken to open the windows and doors to replenish the air.

It was not originally to be in the remit of this book, but I have thought carefully about providing basic ideas on acoustic treatment, and being clear on what is not appropriate, so you will find a guide in "Appendix 3 – A Guide to Home-Built Absorbers" to support you in building your own absorbers in the home using readily available cheap DIY materials, which will, in my humble opinion, achieve an excellent result for the money.

Temporary Home Studio

Every part of the dedicated studio that went before applies here; however, clearly the dual needs of an area mean compromises need to be made, so I will highlight the important aspects. Choose a quiet room in the house on the rear side (most often), one with carpets and soft furnishings. Work as near the centre of the room as possible to avoid boundary effects and get the microphone as near your mouth as possible (remembering bass tip-up!). This is best done with an Anglepoise *with shock mount*, which is a sensible investment, as you can push it out of the way when not using it. This also gets the microphone off the table, so structure-borne noise is largely avoided. Be careful not to use furniture like a dressing table which has a hard surface to the rear of the microphone (the mirror), and don't back the microphone onto a window. Try using the curtains to see whether the sound is improved. It's equally important to keep hard flat objects from being right behind you.

Be careful of setting up in a "box" of surrounding furniture where nearby hard surfaces reflect and bounce your voice around like a pinball. Bring in winter dressing gowns and duvets to drape across hard furniture, the dressing gown as a high frequency absorber and the duvet as a mid-band absorber. Use the best closed back headphones you can afford and play back the result in other rooms to hear how natural it sounds. Get to know what is right and be careful not to rely only on what you hear as you record without sample checking elsewhere. As above, you are likely to close windows and doors to keep out extraneous sound; remember to take breaks to keep the air fresh.

THE PROFESSIONAL TALKS SUITE

The Studio

Well, in theory the easiest acoustic to construct and control, but many fail, even in "expert" hands. There is a "kings' new clothes" element to criticism of such "simple" spaces, but to be fair, they are not simple! This is because many will size the "box" by the number of people who will sit in it. Often built for one, these small "broom cupboard" spaces typically suffer from a rise in bass response making the sound woolly and indistinct. Close working on the microphone will mask the acoustic of the studio area to a degree, but that creates a further bass tip-up (see Microphones). The upside is that we only need to control the range of frequencies the voice can make, and a dry sound is preferable meaning we have an opportunity to use a decent amount of tailored absorption. An RT_{60} of around 0.3 is almost universal (not more than 0.5); however, we must size the room for bass absorption (deep membrane panels) to ensure the tip-up does not exceed around 3 dB at 100 Hz, as it will noticeably colour the sound. This problem rapidly reduces as we increase the size of the studio until a relaxed talks studio for four guests becomes far more routine in acoustic terms. It is unwise to seat a guest with their back to the glass, as of course it is difficult to communicate, but more so the reflective glass facing the microphone will play havoc with the sound. It's no better to have hard surfaces behind the microphone either! The obvious solution is to sit guests facing in from the sides of the studio, so they can glance through to the cubicle, yet can hold an easy eye to eye conversation.

The Control Room or Cubicle

The control room should ideally have a slightly less live acoustic than the studio to be able to hear the reverberation of the studio. That's no mean feat, as the hardware of the desk and ancillaries will be packed in hard reflective casings, and the ubiquitous glass panel to communicate with the artist is perfect to generate unwanted standing waves if mistakenly mounted parallel to the opposite wall. That will normally be the only window to improve the chances of decent acoustic design, so such boxes are often in the centre of buildings away from the noise of the street and therefore daylight. In over half a century of stereo control area design certain norms have stuck. The control desk faces the glass through to the studio area and the two speakers are placed either side, either wall mounted or on stands to raise the monitors above the plane of the desk. The speakers *must* be angled in to face the desk operator and must not be in a corner or too near the wall (particularly rear ported types) as the bass will be artificially enhanced.

In the debate on "absorb or diffuse", it is general to favour absorption behind the speakers and diffusion on the opposite wall behind the control desk. The side walls should dampen and break up reflections and be designed to tame the acoustic down to the required reverberation level (as they have the largest unimpeded surface). At these short RT_{60} times, calculations using Earing modified Sabine will be tricky, particularly at low frequencies, given the presence of the equipment and the usual small space normally allotted to the function. The smart money is on a level of trial and error, and of course we should acknowledge the expertise of world-class acousticians who travel from continent to continent, able to repeat well-proven designs in which they (and we) can have confidence. On the upside, the lack of low frequencies in the human voice means smaller monitors can be used, and the problems of low-frequency standing waves below 100 Hz or so are therefore abated. Larger speakers will be needed if the studio is for music-based shows where the only sound recorded is the presenters' voice and their guests; however, the output of linking music from the studio desk will need to be balanced against the studio voices on decent monitors. The quality of the speakers in the given acoustic can be checked by neutral headphones, and *open* backed provides a small but tangible advantage over closed in this situation.

THE PROFESSIONAL MUSIC SUITE

The Studio

Wow, well whole books have been written about this subject alone, so I'm not sure where to go with a write-up here. I guess the first point is what is professional? Music studio areas have been built in barns and churches as well as basements and offices. A thatched roof will suck the soul out of the sound, a church will be massively over-reverberant, yet great results can come out of these repurposed buildings. The answer lies in the passion of the owners and engineers who use trial and error to make the best of the space they own and love and grow to know just what works and where within the space. Again, a matter of scale, the better studios of my personal experience, like Abbey Road (built in the back garden) and Maida Vale (a former roller-skating rink), are tacked on to existing sites, where decades of experiment and generations of kit have honed the environment to manage the need.

Panel absorbers will deck the walls and ceilings, both velocity and pressure-based designs to tame the structure, yet local climates are made up on the floor of wheel mounted screens with hard (possibly diffuse) and soft sides to (semi) isolate the louder from the softer and to provide local absorption where needed. I have produced the BBC Radio Orchestra in Maida Vale studio 3 and a sectioned-off five-piece band. Great results come from the skills of the people, the Abbey Road Engineers and the BBC Studio Managers. Reverberation times are designed to be longer than talk studios, at between 1 second and 1.5 seconds to add life to the instruments and allow the players to hear each other.

The extremes will need to be tamed, the drum kit will have to go behind screens so it doesn't swamp the other instruments, and non-electric string and reed instruments will need close "miking", so they are not coloured in the mix. In the more professional environments, every performer will hear "the mix" on personal closed back headphones, with key performers able to request their performance to be varied in the foldback mix to get their participation just right.

The Music Control Room

The Music Control Room will be used to record the instruments and voices to a multitrack recorder, but most will have the secondary function of doing the final mixdown to stereo. Multichannel mastering needs a dedicated area (see "Multichannel Mastering and Listening"). The basics of "Control Room" design criteria are the same as the professional Talks studio, save the need for the acoustics to allow for accurate positional imaging from two (or more) speakers and the need to allow for larger speakers than "Talks" for a flat bass response down to around 30 Hz to capture all the instruments accurately. The space is going to be larger anyway to take the multichannel desk, the room dimensions will be skewed to ensure there are no parallel walls and the desk will need to be kept away from the rear wall and the glass in front, so the cubicle is likely to be deep in dimension. Eigentones become far more critical with the requirement to produce lower frequencies, and it is likely acoustic designers will absorb a good amount of the building budget as a necessity to get the sound neutral at the desk engineer's position.

Modern techniques have allowed diffusion to grow over absorption, bringing a livelier natural sound to the modern mix, but it does take a lot of space (deep voids) to make that happen with "large absorption mass". I can't pass on this subject without talking about Live End, Dead End™ (LEDE™) construction first mooted by Don Davis in 1978. Don carried out a great deal of work examining the wave patterns close to the monitoring speakers using the emerging technique of time delay spectrometry. He demonstrated a comb filter effect in the middle frequencies created by the early reflections off the side walls and ceiling of the control room mixing with the direct path of sound created by the stereo speakers.

This had not been considered before as the long-known Haas effect suggested that the ear was not sensitive to time difference effects until the path length time delay exceeded 35 mS (11 m). As frequency rises, wavelength shortens, and the delayed reflections set against the higher frequencies created the anomalies. This cancellation affected frequencies in the first 15 mS and was shown to be worst between 1 kHz and 8 kHz. His bold solution, which has been successfully tested in many locations, was to make the front half of the control area (back as far as the ears of the music balancer) as dead as possible with deep absorbers and leave the rear of the room bare and fully reflective. That means the reflections returning from the rear wall (in a normal sized music control room) are around 40 mS and upwards, where the ears can clearly disassociate the return from the direct sound, and discount it in the mix.

The cherry on top was the rise of mathematically designed 2D and 3D diffusers in the next decade by Schroeder, again acting from around 1 kHz, which allow the rear wall to remain reflective but break up the reflections to get an open acoustic expected from a much larger room. In the chapter on loudspeakers, we learnt that monitors become 1) more directional with higher frequency, and 2) the longer path lengths of the reflections lose energy striking against the boundary and 3) have inverse square losses. The three together mean our brilliant ears and brain can already interpret and reject the direct sound from the delayed reflection (law of the first wave front) with cancellation in the Dorsal Nucleus of the Lateral Lemniscus. That's one to use in the pub tonight!

However, please be aware of the subtle losses of some higher frequencies through the comb filter effect, and to credit the work of these pioneers in creating superlative listening areas. Note the high precision architecture attending to successful LEDE™ installations and that it cannot be simply crudely copied as an idea that will work in your space without access to specialist resources.

Multichannel Mastering and Listening

This book so far has concentrated on stereophonic or two-dimensional sound utilising two channels. The acoustic layouts to control reverberation and time domain cancellation effects are well researched (see "The Music Control Room"), and for such areas, the acoustic treatments are understood. The use of more speakers to create the illusion of complete sound surround (including sounds above the listener) is a more recent science, and I would suggest one where debate is yet to settle; yet the need for multichannel listening spaces is growing and cannot be ignored. The focus in the book has been for the sound creative rather than the listener, so I am more interested in acoustics for the mastering process than the end listener, although frankly they are technologically aligned. Let's break this down further.

I'm sure I can convince you to dismiss the acoustical isolation as it is the same science and solutions as for traditional mastering areas. Damped (resonance suppressing) high mass materials to control the invasion of low frequency noise, and the sealing of gaps to suppress air-borne sound at high frequencies. Where air must pass, ventilation being the obvious one, porous damping along the ducts can make the suppression level of unwanted sound acceptable.

Reverberation Times can also be borrowed from traditional mastering rooms; an RT_{60} of a minimum of 0.2 to 0.6 seconds, maintained across the sound frequency spectrum, provides much-needed life to the session. Too much damping (anechoic) has been postured, but I think that is a mistake and ignores the Stokowski ideal.

Clearly, LEDE™ is an excellent stereo (two speaker) mastering room tool, yet left as is, does not translate to a multiple speaker surround installation. The secondary question is whether first reflections from the front speakers should be suppressed by porous absorption on the adjacent side walls to stop comb filtering of the midband, as in stereo. My answer is that this has only proved critical in narrow listening spaces, and precision multichannel listening will not work in this layout anyway.

The goal of acoustical control of any area must be two-fold: to damp down reflections by absorption to dampen the area to a known reverberation time (RT_{60}), and secondly to disperse sound reflecting from the remaining surfaces, so excessive comb filtering across the frequency range does not occur. In a multichannel mastering room, speakers are increasingly sited on all room walls and often the ceiling. Unlike single wall reproduction (stereo) it makes sense to disperse the absorption in approximately equal amounts between the speakers, so the sound is uniformly modified on each flank. Ideally this would gain more precision by placing wideband absorption in each position (see "A Wide-Band Studio Absorber Solution" in "Appendix 3 – A Guide to Home-Built Absorbers"), rather than a patchwork of membrane and porous boxes across the walls. Ceiling treatments can easily be modified from more traditional rooms.

I'm sure readers will try differing speaker positions to gain the best spatial 3D outcome. The same must be said for the wideband absorption panels placed between the speakers, the installation of dispersive structures and finally ceiling treatments. Into this three-ring circus enters the subwoofer(s). Placement of this (these) must avoid obvious eigentones (standing waves) and have a coherent phase relationship with the other speakers. Computer analysis (including that within encoders/decoders) will help, but smart initial positioning is crucial to ensure the alignment software does not have to work beyond its competence.

CONCERT HALLS AND THEATRES

A reminder from Chapter 8, "Introduction to Acoustics", that auditorium seat bottoms are often padded to absorb the same sound as an "average" person in the seat. Long before electronic means to make the stage performance sound louder, smart people were creating spaces for many people to hear the actors and singers clearly. The earliest roofed structure I can find is the Italian Opera House which relied on multiple tiers of seats to bring people as close to the stage as possible and with a surprisingly short RT because the many segregated boxes acted like mid-band diffusers. Wagner created the first modern purpose acoustically designed building at Bayreuth in 1876. Buildings started to utilise the simple concept of reflecting sound off surfaces to carry it to the back seats.

Imagine a torch pointed at an angle to a white ceiling. Light will reflect back downwards and disperse to cover the floor on the other side of the room. The ceiling could equally be made to tilt to get an exact reflection to the right place, or even concave or convex to focus or spread the light out. Equally a reflector artfully erected at the back of the stage, like the throat of a horn, can push sound forward. The secret is to avoid unwanted reflections which could muddy or cancel out the sounds we want to hear.

A reminder that sound energy follows the inverse square law and so the volume falls by half for each doubling of distance. In addition, the sound loses energy "grazing" over the heads of the audience, particularly given the shallow rake of the seats in the stalls, so often by the back the volume is much lower than that experienced by the front seats. We need help! In Figure A2.1 is an imaginary diagram of a large purpose-designed theatre, I've used three reflectors. The ceiling is sculptured to reflect the sound to the circle at the rear, a reflector is positioned above where the orchestra plays to push sound forward beyond the level of front rows, and in addition the void below the circle has a sloping roof to bring the sound to the rows at the back of the stalls.

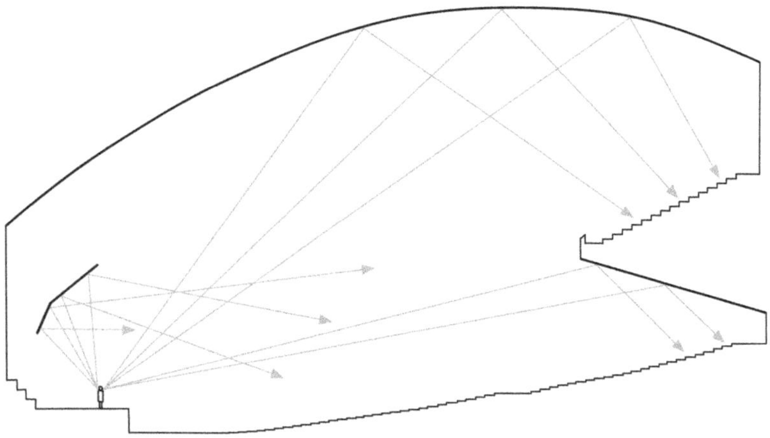

Figure A2.1 Reflections create a natural sound spread in a large building

The golden rule of Haas must be followed; with a point source multiple audible sound paths should not be separated by more than a maximum of 11 m path difference. It's worth noting that by raking the seats at 20 degrees and above, it's been found the audience grazing effect is all but eliminated.

CINEMAS/SMALL THEATRES

The origins of public performance of film came from Paris at around 1900, but the addition of a soundtrack to the formally "silent film" sound had a difficult beginning, only available on "shorts" until the feature length *The Jazz Singer* in 1927. The "talking pictures" or "talkies" took off globally in the 1930s, although many filmmakers felt sound subverted the "unique aesthetic virtues of silent cinema". The various ways in which experimenters tried in vain to synchronise the sound with the picture make great reading if you love your history (hence keeping early "talkies" short), but it's not for this book.

Of course, like the music of the day, early films were "monophonic" a single channel. Whilst the optical soundtracks of the day were poor quality by modern standards, the cost of the loudspeaker system was small in comparison to the structure, and so the installed systems were invariably very high quality and capable of large volumes! As there was only one channel the speakers were placed behind the screen. The auditorium had plush seating to absorb reflections, and an absorbent rear wall. Many early venues had an upper circle to bring more people closer to the stage, and no special design criteria were generally required to get the loudspeaker system sound to the back row.

Stereo sound in cinema started in 1941 with the release of Walt Disney's *Fantasia*. Since this was 1) an add-on to existing picture houses, 2) needed modified sound installations, 3) had poor quality of the "optical" soundtrack (two wiggly translucent lines embedded at the side of the picture) and 4) a failure to update the acoustics for multi-channel, resulted in cinemas falling out of favour, just as TV rose in prominence in the 1960s, and many traditional cinemas closed. There was still a demand, and this was met with poor quality small "boxes" in new shopping centre schemes called a "Multiplex".

In 1974 Universal launched the film *Earthquake* with heavy bass "Sensurround" and Dolby Cinema noise reduction came a couple of years later with *Star Wars* (1977) and *Saturday Night Fever* (1977). By the mid 1980s state-of-the-art film sound such as Lucasfilm's THX and Sony's "Dynamic Digital Sound" was incorporated into the new multiple screen sites, with the multiple sound channel "Megaplex".

Where the early cinema buildings tended to house a single screen or perhaps two at most, the arrival of the "Megaplex" meant that designers went back to the drawing board! Multi-channel sound required new bottom-up thinking, ensuring lateral reflections didn't destroy the sound imaging by applying porous absorbers mixed with diffusers to all the cinema surfaces. Isolation became a huge headache with the advent of the multi-screen cinemas where several venues showing different films at the same time were architecturally positioned adjacent and on top of each other, each with its own distinct and often loud soundtrack.

Table A2.1 Typical noise criteria for a multi-channel cinema

100 dBA	The maximum sound level at 1 kHz in a cinema hall equipped with Dolby Atmos
120 dBA	The maximum sound pressure in the 63 Hz octave band in a cinema hall
30 dBA	The maximum permissible level of penetrating noise at 1 kHz as per NR 30 (noise reduction contour)
72 dBA	The air-borne noise insulation index between cinema halls (Acoustics Group)

Stopping the acoustic transmission of noise between areas is distinctly tricky, but the structural isolation required between spaces of modern "action" soundtracks with deep explosions shaking the building is phenomenal. Add in the vibration of air conditioning, chillers, catering and generators, and it's amazing the standards are met! In Table A2.1 are typical figures suggested by the industry.

An acoustic technology developed and finding favour in the 1980s was the use of large mathematically designed diffusers to break up sound waves, rather than absorbing them with the porous mats and the crude dispersion planes of the earlier cinemas. The size of these buildings lent themselves to the new technology and keeping the sound live rather than damping it right down fits the style of the cinema experience well! These diffusers will have differing sizes and shapes of profile to ensure a wide range of frequencies are equally dispersed (Figure A2.2).

Figure A2.2 Auditorium diffusers, in position and a profile close-up
Photo by the author

OPEN AIR CONCERTS

Again, we have a split in usage between "Amphitheatres" and field events like Glastonbury. The Romans were of course the first to exploit permanent open-air semi-circular public seating (although many contend the Greeks invented the concept), banked high to bring people as near the stage as possible. When Frank Lloyd Wright designed the open-air Hollywood Bowl stage, he could not imagine that Rolling Stone Magazine would name it one of the best ten live music venues in the world! The design follows the Romans in that the Los Angeles landscape formed a natural bowl shape, formally called Daisy Dell, which could be further refined to produce the highly raked seating, a natural Italian opera house, if you like.

The 1919 crude design was dramatically improved by Frank Lloyd Wright in 1927 with the addition of the "band shells" on the stage to magnify the sound made by the performers and acoustically project it forward to the audience. Permanent versions of Lloyd Wright's weathered structure were installed, and modifications made over the years to this brilliant natural amplifier, yet many argue the Lloyd Wright original was the best. These days an inbuilt powered sound reinforcement system uses a pair of columns, a line-array configuration of multiple loudspeaker enclosures hung vertically in the now very familiar curved "tusks", with the lower enclosures facing the front sections and the upper enclosures angled towards the rear sections.

Field style "ad-hoc" venues tend to use the same style of sound reinforcement systems, even though they start out with an empty canvas. Professional systems have modern sophisticated audio processing, allowing each loudspeaker enclosure in the "Line Array" stack to be "tuned" and directed towards the location of the listener, regardless of where in the venue they are sitting. The electronic processing includes sound level, frequency equalization, and most importantly time delay. As we have discussed sound passes through electronics much faster than through air; therefore, the sound coming from the speakers must be delayed as the distance from the stage increases, allowing the actual sound from the stage to "catch up" so both sources reach the listeners' ears simultaneously. Similar systems, although much simplified in small venues, are in our theatres and concert halls to ensure the audience in the rear seats hears the same audio, at the same level, as in the front seats. It bears repeating that a path length difference of more than 11 m between sources of the same sound muddies the reception and confuses the listener, so the smart delays introduced to the sound are crucial to a professional performance.

COMMUNITY HALLS

Most towns and villages are still fortunate to have a community space, the village or town hall. Many will have been adjunct to a church, and the older ones are likely to have a stage area for local plays or for leadership at meetings. The window curtains and stage carpets are long gone to be replaced by hygienic washable surfaces such as laminate or wooden floors. The need to find external funding in more austere times has introduced yoga, exercise and dance classes, nurseries and fairs. The remaining hard surfaces such as plasterboard, glass, steel and concrete offer little absorption and reverberation times can be found of 5 seconds and more! In addition, and crucially, most sound sources now come from the floor of the hall itself, rather than as intended from the stage. I have covered these problems in a separate section at the end of "Appendix 3 – A Guide to Home-Built Absorbers", and also that of a similar concern in restaurants.

LOCATION RECORDINGS

So hidden at the back of the expression of the greatest structures built in the science of sound we come to the low-budget, suck-it-and-see, communal garden location recording. Gone are the architect-tuned acoustics, the finely tailored sound systems and the knowledge that you have control of the environment you now find yourself in. Many professional sound people will baulk at a situation that has so many variables, yet this is one of the most common needs in the industry, and one of the largest employers. Add in that the location is for film or television and the need to keep the microphones out of shot annoyingly becomes a bigger problem for the director than the intrinsic quality of the sound – sad but too often true! This is why this subject is broken out as a separate chapter in the book: Chapter 13, "Planning for Location Recording".

BIBLIOGRAPHY

Parkin, P.H., Humphreys, H.R. and Cowell, J.R. (1979) *Acoustics, Noise and Buildings*, 4th edition. London and Boston: Faber and Faber.

Alton, F. (1984) *Acoustic Techniques for Home & Studio*, 2nd edition, 4th printing. United States of America: Tab Books Inc.

Burd, A.P., Gilford, C. and Spring, N. (1966) *Data for the Acoustic Design of Studios*. BBC Engineering Monograph No. 64. London: BBC Engineering Division.

Davis, D. and Davis, C. (1980) The LEDE™ Concept for the Control of Acoustic and Psychoacoustic Parameters in Recording Control Rooms. *Journal of the Audio Engineering Society*, 28(9).

Mackenzie, G.W. (1964) *Acoustics*. Bungay, Suffolk: Focal Press Limited, Richard Clay and Company Ltd.

THX Home (2025) *The Entertainment Experience You Want Wherever You Want It*. https://www.thx.com.

THX Ltd (2025) *Cinema Certification*. https://www.thx.com/blog/faq_category/cinema-certification/.

Appendix 3 – A Guide to Home-Built Absorbers

INTRODUCTION

A thread through the chapters of this book has been the woes of having amazing equipment placed in an untreated or poorly treated acoustic. The value of room improvement in professional settings, in both recording and listening areas, is paramount; otherwise, it is a lottery and is unlikely to yield good results, unless by beginner's luck!

In Chapter 8, "Introduction to Acoustics", I suggested the type of practical improvements that can be made, but in the interests of brevity, and the introductory nature of this book, I was short on detail and advice, and I do recognise and encourage further reading of several brilliant books on the subject (see each chapter's Bibliography). From personal experience I acknowledge that these specialist books can be scary as the science of acoustics is complex and wide-ranging, with some hefty formulae. Therefore, I also recommend to readers a suck-it-and-see philosophy with acoustic treatment; but I acknowledge it does need some form of foundation. My approach also means I can free you from the maths that acousticians fondly use to model designs, particularly in small rooms with difficult, short reverberation times; again, I salute them!

This appendix is an attempt to help small start-ups and keen amateurs who are short of an acoustician's fee to find a path that through an element of trial and error should give predictable pleasing results. Whilst I can show handy designs of sound absorbers, the amount of each type (by square metre) will vary on each application, as it will change the RT_{60} and this is where local knowledge and testing are key to get the sound that works for you. You may be surprised to find I have not utilised Schroeder's brilliant invention, "Diffusers" (originally called "Diffusors") because "proper size" ones do not function predictably in smaller spaces like living rooms. Why? They don't achieve specification in the near-space to sound sources, they are an unknown off-axis to the direct sound, mostly they tackle limited mid-band frequencies (which are easily in the grasp of crafty porous absorption in small spaces), if poorly built they can cause colouration and unwanted frequency response anomalies, they change characteristics when placed in with other clone panels and lastly, they are heavy, complex and expensive to make properly. I am disappointed with a number in the prosumer marketplace who offer inaccurate, scaled down designs, to be used in inappropriate settings, and claim to operate at much lower frequencies than the science deems is possible. The best that can be said is that many of these unproven structures are an attractive addition as a fashion statement; "Wood Art" if you like, however appropriate they may be in the reality of tailored sound control. Frankly at these sizes they do little harm, but it's a bit like putting throaty exhausts on your motor so you get noticed on the street, but not bothering to up the power.

DIFFUSION IN LARGER AREAS

Schroeder's pioneering work on diffusion in the 1980s, and subsequent modelled and tested designs, however, do remain a key tool in the arsenal of the professional acoustic designer, particularly in large bespoke structures such as concert halls where they enhance the purity of the music and at the same time keep the acoustic delightfully live! They can also play a key part on the rear wall of a larger Control Room, designed and modelled to work with the space. I have also seen them used to good effect mounted on screens in the studio area to modify the acoustic; but again, I would point out that performance is unpredictable in the near field! The alternative of acres of absorption is fortunately dying out; many spaces were routinely over-damped as people assume adding more vanilla panels would create more control and therefore a better result. Don't do it, whatever the space! Furthermore, a small change of position of speakers or the listening/measurement position will create a different set of results, so please don't get too bogged down in the crutch of cheap computer-based measurement kit. Use your ears! Expect surprising results; a BBC paper on a favourite topic, Control Rooms noted the high-end reverb time went *up* when the carpet went down. The more you are fixated on a microphone calculated sweet spot, the more volatile the space becomes for other users and uses. The same applies for the desk monitoring position. At the end of the day, despite the lively rhetoric, the treatment of room acoustics is a challenging compromise between the acceptable and the practical, and don't let any snake oil salesman tell you otherwise.

DISPERSION

A very early technique for breaking up sound wave fronts is often overlooked in modern texts, maybe because of its sheer simplicity and cheapness. Dispersing the sound through simple flat plate diffraction was widely used in early concert halls, and then cinemas in the last century. The major advantage is that angled planks or sheets attached to the walls or ceiling will work to disperse frequencies from a dimension of just 1/10 of the wavelength, making the structures much shallower than the other techniques of disrupting problematic sound waves, and so much more suitable as a technique for use in smaller rooms (Figure A3.1).

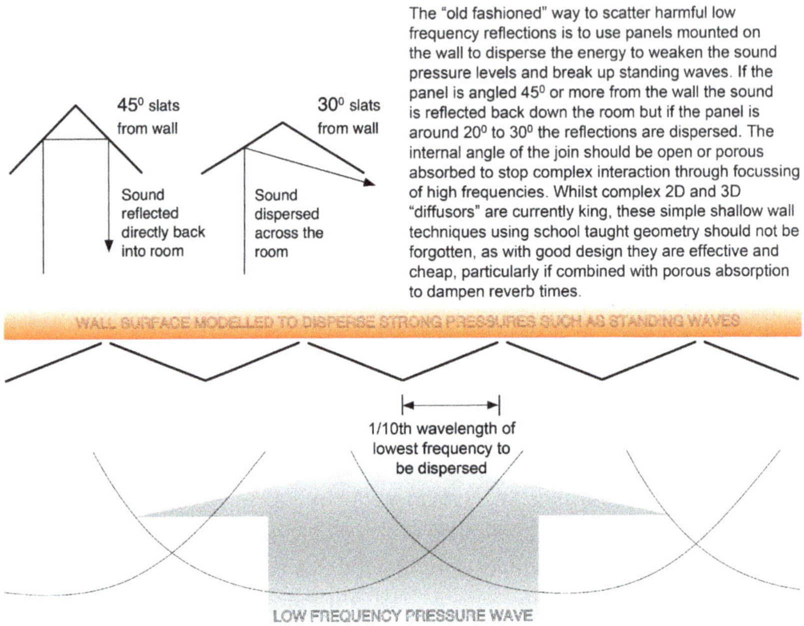

Figure A3.1 Wall mounted dispersion

ROOM MODES

It is always best to avoid parallel walls before ever contemplating compensating for them. The next concern is the room dimensions. We start with a problem of three obvious modes of standing wave in the room: its width, depth and height. As well as these "Axial" modes, there are Tangential and Oblique modes between the corners. Resonances in the room of just $1/2\lambda$ (and its multiples of up to 300 Hz or so) will produce unwanted peaks approaching 10 dB in the low frequency response. These modes can be problematic alone, but architects creating "neat" room dimensions that have mathematical common roots to the height, width and length compound the problem in that the same "standing waves" are generated in more than one plane and add together to create super waves!

STANDING WAVES

Earlier, in Chapter 8, "Introduction to Acoustics", we touched on standing waves, also known as eigentones. To understand what is happening in the room, we will introduce another level of detail. Acousticians have several descriptive models to try and illustrate sound in pictorial ways we can easily understand. We can point to the "ray" approach, which shows sound bouncing off structures like a laser off a mirror (suits higher frequencies), or the "wave" approach where we have fronts of sound pressure powering over the room like waves on a beach (suits lower frequencies). Frankly no simple model stands up to the reality. We have touched on two models of acoustic control, one which looks at the continuously changing squash and expansion of particles of air that in turn create sound waves. This looks at the velocity of movement and is the "Particle Velocity" model. The other measures the amount of force required to move the ear drum (or microphone diaphragm) by changes in air pressure that create sound waves. This is the "Sound Pressure" model.

Figure A3.2 Standing waves – sound pressure versus particle velocity

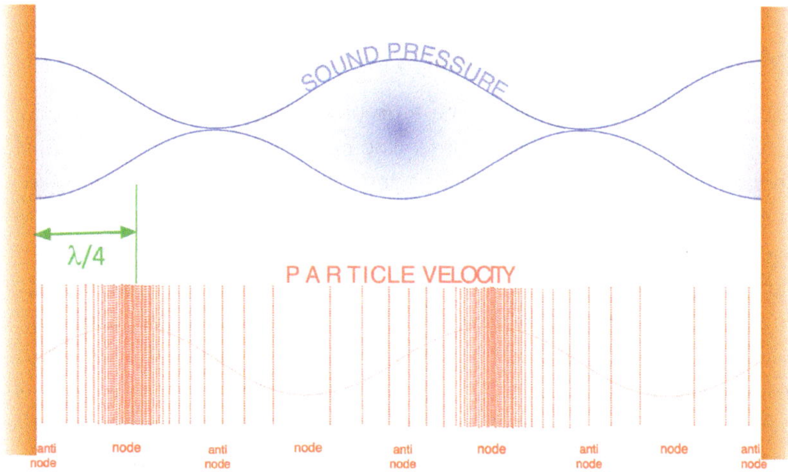

These models do not always tally; however, in "standing waves" at the room boundary the *sound pressure must be at its highest* because it is trapped and can go no further. *Because the wave is now static, its velocity must be zero*, and so changes in air particle density must be happening elsewhere in the room (Figure A3.2). These points of squash or expansion of the velocity of air particles happen from 1/4λ (one quarter wavelength) away from the wall and are called nodes. It is at these points where the sound wave pivots on its axis. Points where the air particles have no velocity are called antinodes, including the surfaces of the walls. Particle velocity is constantly changing; it is a mistake to consider the wave at the wall is pinned like the end of a guitar string; the reality is that the wave is free moving, pivoting on the nearest node.

Let's talk about the two most popular types of absorbers used by acousticians in room treatments. 1) The *porous* absorber works through the friction of movement (converting sound to heat) and is efficient with *high particle movement* or velocity, and it is at its best at medium/high frequencies. For low frequency absorption we need a *resonant* device like Helmholtz or membrane absorbers; these need to be sited where the *sound pressure is highest* to operate well. So, different absorbers need different siting, placing a sound pressure (resonant) absorber where there is a node for the frequency is a waste of time, and equally placing porous absorbers where there is no change in velocity at that frequency, i.e. hard against end walls and corners, is inefficient to say the least! In Figure A3.3, a reminder of the size of a wavelength at different frequencies. In Chapter 8, "Introduction to Acoustics", we used metres on our scale; for imperially minded readers here is a version of the same table in feet.

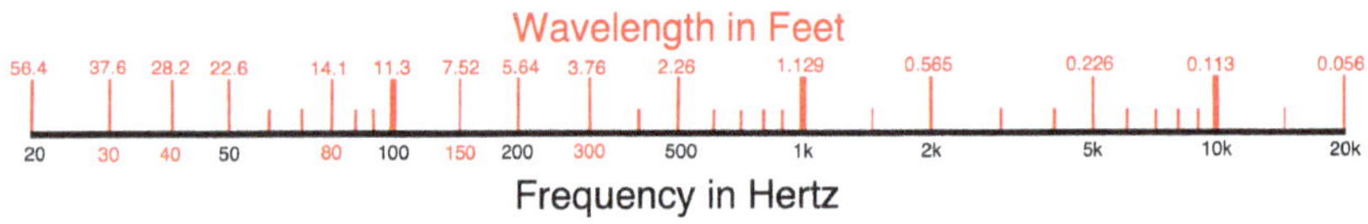

Figure A3.3 Wavelength in feet against frequency

TYPES OF ABSORBERS – VELOCITY BASED

This is the "porous absorber", the type of absorber that occurs naturally in nature. It might be grass, trees or fencing outside; inside it is a carpet on the floor, a sofa, curtains, people in the room or even (to a small extent) the air in the room or the plastered brick wall itself! In a porous absorber the sound energy enters a maze of loose material reflecting off multiple obstacles, growing weaker in the collisions and so giving up energy to heat through friction. The greater the depth of the porous material, the wider and lower the range of frequencies that are absorbed. So felt-backed carpet is more effective than carpet alone, and pleated curtains distanced from a window are better at absorbing the middle frequencies than those close to it. The materials most often encountered in practical porous absorbers are either open cell acoustic foam or low-density mineral wool slabs (as used in home insulation). Increasingly naturally grown fibres such as grasses are used, known as PLA, although dust free containment, whilst maintaining an open porous entry, is problematic. These absorbers work well with sound arriving either directly front on or at an angle; indeed, simple geometry suggests a slanting arrival of sound wave encounters a greater depth and so is more effective at lower frequencies. However, remember porous absorbers are at their least efficient where there is low to no velocity of sound at lower frequencies, principally at the surface of the wall!

Seeking massive amounts of porous absorption is not clever; too much of the same absorber affecting a certain band of frequencies will ruin your goal. The rule is to *balance the absorption equally across the frequency spectrum* to create a space with a naturally pleasing *even* reverberation (RT_{60}) throughout the frequency range. In short, more is often not better, unless of course you are in the business of selling the panels! High frequency absorption is typically the predominant sound loss in a normally furnished room (as most natural absorbers like carpets soak up these frequencies), which is why so many treated studios sound dull and unexciting.

TYPES OF ABSORBERS – RESONANCE BASED

We have learnt when it comes to bass frequencies, sticking foam profiles or other commercial porous panels on your wall will do nothing for a low frequency eigentone and the early overtones up to 300 Hz where it matters. For a porous absorber to be effective it must have the material depth of at least a quarter of the wavelength it is expected to tackle. This is popularly called the

"Quarter Wavelength Rule", and if you think about it, that's just plain sense! But what does that mean practically? Take a typical nasty standing wave of 35 Hz (the "fundamental" or 1st harmonic and 1/2λ) in one long axial plane of a well sized bare parallel walled room. There will be further frequency response peaks seen at the multiples of 35 Hz; namely 70 (2nd hamonic and the wave frequency), 105 (3rd), 140 (4th), 175 (5th) and 210 Hz (6th), etc. With respective wavelengths of approximately 10 m (fundamental or 1st harmonic), 5 m (2nd harmonic), 3.3 m (3rd harmonic) and 2.5 m (4th harmonic) and so on.

Using the quarter wavelength rule, a generous 8 inches (200 mm) of porous acoustic slab will fully absorb from 430 Hz, but barely scratch the surface of the example given above. However, it will certainly dull the reverberation of the higher frequencies where you possibly already have more than enough sound soak. To be fair, I do appreciate there may be "hidden" value in the grazing (or glancing) effect of low frequencies passing over acres of porous absorbers in a room, but I have seen no credible science to measure and catalogue such an effect and so ensure it is a predictable outcome in a known space. So, please beware of buying the dubious product of purely porous material "bass traps" for your corners! Point made, read on about how bass reflections such as standing waves can be effectively tamed.

HELMHOLTZ OR SLOT ABSORBERS

We have learnt inconsistencies are far greater at low frequencies in small/average rooms than they are at higher frequencies. As stated, porous absorbers can easily be built to manage medium and high frequencies but are inefficient in normal sized rooms (at tenable sizes) as bass absorbers. It needs a different approach. There are two main types of purely bass absorber, the Helmholtz resonator and the membrane absorber. The Helmholtz or slit resonator was first developed in the 1850s and has had a commercial revival in recent years, but frankly I'm not sure why. To understand the science, a simple home test is to blow across the open neck of a bottle. The mouth (width and depth) of the bottle resonates at a particular frequency; the smaller the neck, the higher the frequency. Try it! The frequency you hear is the fundamental one where the bottle will absorb sound.

Tuned "bass traps" are designed to the millimetre to isolate and control a single frequency (and to an extent its harmonics) by this resonant effect and can be manipulated to have a given Q (sharpness of trough) which may or may not match your eigentone peak. On the upside, they are efficient on space; however, they are an incredibly nuanced tool which needs to be placed in expert acoustic hands where, with much experimentation, they can be sincerely incredible! The British would say it's all a bit of a "faff", and even the BBC have traditionally only resorted to them when other solutions fail. That leaves the practical user with "membrane"-based bass absorbers.

PANEL OR MEMBRANE ABSORBERS

The first great point to make about well-designed damped membrane absorbers is that, with good design, they leave the middle and top frequencies virtually unchanged, so they can be used independently of porous absorbers in an acoustic design. Rather like loudspeakers have a woofer for the low notes and a tweeter for the high, the amount of absorption can be adjusted to balance between the two types for a naturally wide-frequency acoustically controlled area. In a membrane absorber a heavy "skin" is sealed across an airtight box, so a low frequency sound pressure wave drives the skin to move in sympathy with it, and the trapped air acts as a spring. The energy of the sound is dissipated in creating the physical movement of the skin. An internal porous material is often placed just behind the membrane to widen the range of absorbed frequencies (although it is technically tuned to one), and usefully the waves with most energy (and therefore needing most control) will win in the race to shift the membrane the most.

At its simplest the membrane is a mass loaded vinyl (MLV) sheet sealed across the face of a wooden frame with an airtight back (old school was roofing felt sheets). This solution is fine for the limited bass down to 100 Hz produced in a Talks Studio. If lower frequencies are required to be tamed (say in the control room of a music studio), additional mass in the membrane and/or deeper voids in the airtight frame will be needed. The membrane mass can be doubled with a second sheet of MLV or roofing felt well bonded to the first, or further by a plywood or hardboard sheet thoroughly bonded to the flexible sheet (or the board can be used on its own). A BBC favourite option is to use slightly perforated hardboard, which has a wider absorption characteristic yet will still flex with the bass. In extremis whole walls can be finished with flexible three- or five-ply sheets of plywood, and the depth between the battened sheets and the wall varied to meet the resonance of the room. Frankly that's what many home plasterboard partition walls achieve anyway; bet you didn't think of that! The flexible wall solution is tenable in large structures, but not for our needs here, and our solutions will stick with what's best in small and medium room designs. The membrane absorbers can still be faced with an "acoustically transparent" covering of a favourite colour, so the looks remain in keeping with a quality installation. The efficiency of an absorber, whether it is porous or membrane, is shown as a number between zero and one. Zero is a fully reflective surface whilst one is 100% absorbent (although more than 100% is possible!). Whilst porous absorbers come close to "1" in their frequency range, membrane absorbers are less efficient. In Figure A3.4, we see an absorption coefficient of 0.75 at the resonant peak of 100 Hz, which is excellent for this design.

Figure A3.4 Resonance in a membrane absorber

A popular way to show the action of an absorber is to show its electrical equivalent. For instance, the band stop filter mirrors the action of a membrane absorber, and using this and other equivalents can be used to create a large model of air impedance inside a space.

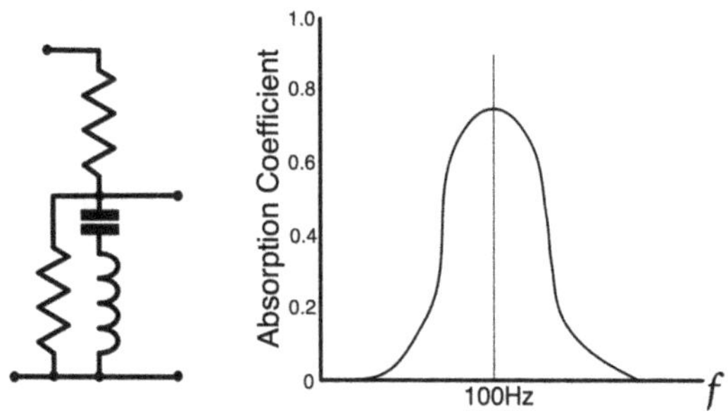

A WORD ON DECORATIVE FACINGS

We'll most likely want our product to look smart after spending good money on the project. Ideally the covering should be acoustically transparent; that is, 100% of the sound should pass through to the absorbent material behind. Sadly, even the most open weave cloth does have some resistance in air; however, we can take precautions, and they are surprisingly simple. Take a piece of the cloth, hold it up to your mouth and breathe through it. If you need to expend effort, the cloth will not work; only where you can breathe effortlessly will the material satisfy the requirement. Next, even though we have found the perfect cloth, it will predictably change the characteristic of the absorber in two ways. If the cloth is stretched across a frame without touching the material behind it, it will have its own "membrane" resonant peak, which will depend on the size of the absorber (but often lies between 100 Hz and 150 Hz). Further, the cloth is a natural porous absorber and will start to become effective from about 4 kHz, improving on its capability as the frequency rises. I note these two problems are identical in loudspeaker cabinet grills. Again, I say, beware of *designing in* too much high frequency attenuation, as in a typical furnished area, the other objects in the room will be achieving it for you.

PRACTICAL DESIGN OF ABSORBERS

I realise I'm setting myself up to fail here, as industry data is rare, differs, is often approximate, and other properties within a room can modify the end measurements of a built structure. The intent behind the graphs and absorbers is an honest one, to help readers understand the physical limitations of certain uses of treatment set against the size/depth required by the frequency that we wish to control. For instance, as you learnt above, if we wanted to gaurantee to absorb a standing wave of 50 Hz and above with a porous absorber it would need to be two metres thick! A proper Schroeder Diffuser needs a depth of 8 inches (200 mm) to tackle frequencies down to 1 kHz (as the BBC design does, yet even that can only diffuse to 3.5 kHz at the high end!). That's very big and heavy on the wall or ceiling (I note many commercial ones are a pretty looking

2 inches [50 mm] deep, missing out on mid band). In this conundrum we have the comfort to choose 23-degree angled wood panels on a rear wall to assist the breakup of a 50 Hz standing wave by dispersion at only 12 inches (300 mm) deep! This is a design that is affordable and practical!

As I write, my kitchen table is piled high with books, tables and research papers from up to half a century ago. The approach has barely changed, Live End, Dead End™ (LEDE) design concepts, microperforation and "Diffusors" apart, and the physics certainly hasn't! The decision is clear: manage the high frequencies with a suitable simple high-frequency module, matched to a module (or modules) appropriate to manage low frequency absorption. The hesitation is module depth; am I designing the minimum to suit in a small home sitting room or going deeper for a serious music studio? The problems are twofold: how deep to absorb, bearing in mind what the room can physically produce, and how to seamlessly (as possible) match between the absorption types at a critical 200 Hz or so. The decision is in: neat shallow devices to improve sound field imaging in a lounge, a slimmer one to assist in light commercial premise problems and a separate comprehensive, adventurous design for the studio.

We have identified two practical, reliably predictable modes of absorber: the membrane absorber for low frequencies and porous for middle and high frequencies. The porous blanket is the easier to describe; we have two simple options: the density of the slab and the depth. Several charts exist in specialist books which place the amount of absorbency against the frequency of commercial mineral wool slabs, some based on theoretical formulae and others on practical testing. There are quite large variations in places, but in general more agreement than disagreement. A professional studio may hand over cubic metres of footprint for deep porous absorption, but this is not practical in the home or smaller studio space. A cost-effective porous design must not overpower a small room with deep "boxes" on the wall, yet we must reach down to where it is practical for a low frequency absorber to take over the work. Over time I have tried the effectiveness of differing slab densities of 45 kg/m^3 to 100 kg/m^3, which by their nature have differing "flow densities" (i.e. the ease of absorption). At one end I found 45 kg/m^3 is difficult to work with as it is too flimsy to sit in a frame, yet it absorbs well; at 100 kg/m^3 low frequency absorption is slightly better; however, at high frequencies particles start to glance off the denser material, so it is less effective.

Long ago I chose the compromise of 60 kg/m^3 as it is efficient, is easy to work (does not sag) and can provide a decent match to a design solution for low frequency panels. To recap, porous slabs can be stacked thicker, reduced in fibre density or mounted over an air gap, all improving the absorption at the lower end of the medium frequencies. I should add porous absorbers are often faced with perforated board (typically 25% open, 75% board for mid-band and under 1% surface perforation for low frequency absorption) which changes the absorption characteristic by allowing an element of reflection/portage from the panel. This is effective for custom diffusers and clearly has its place in the commercial acoustics industry, in buildings where we are using a sound level meter with "A Weighting" (see Chapter 1, "Introduction to Sound"), as the goal is the subjugation of annoying mid-band chatter.

Our goal is very different; complex panels are an unnecessary complication and, in fact, a distraction if applied to our smaller spaces. *We want equal absorption across the usable frequency spectrum.* I have chosen a 50 mm slab thickness in the design as it is flexible in use (I use the Rockwool RW3 slab; other 60 kg/m^3 slabs perform similarly). At just 50 mm, it can create a slim ceiling panel, with an air gap a useful lounge panel, or, when doubled up to 100 mm, it will approach a low-end absorption frequency reliably deep to match to suitable membrane units in a studio setting.

THE MODULAR LOW-END ABSORBER

Figure A3.5 Example material weights for a membrane absorber Photo by the author

Modular membrane absorbers have been around in sound studios for 80 years or so and have gone through many iterations. Early anomalies in performance are now explained, and theories reconsidered, and there is a well-proven mathematical expression to tie the required frequency spot to the surface mass of the chosen membrane against the depth of the box (does not consider porous infill). The physics of the membrane absorber is a return to the *spring and mass* of Chapter 8, "Introduction to Acoustics". The membrane provides the mass (membrane samples in Figure A3.5), and the pocket of air trapped behind it in the sealed box is the spring. Complex spring infills other than air have been successful; however, this is still an emergent highly specialist field. The efficiency of the air-filled membrane boxes can be as poor as a third of the porous variety, particularly if the choice is made to make the material thicker (and so often stiffer) for a lower resonant peak; however, efficiency can be doubled by changing the air impedance simply by placing a porous mat close behind the membrane in the airtight cavity. In addition, the peak frequency of absorption in a porous material filled box is also usefully lowered by up to 30% compared to the air-filled formula. A no-brainer! The membrane can be doped to improve the width of the low frequency response peak, and this further improves as the RT_{60} is curtailed (inverse function). I note they are not currently so popular as there is concern on getting the right siting, but common sense prevails that they are placed where the sound pressure is highest.

Predictable room solutions can be designed using the performance of a particular panel, set against the combined surface area and plugging those numbers in with what is known about the other materials and contents of the room. Poorly designed membrane absorbers can have a narrow spike of resonant absorption, which is unhelpful at best. My solution of using a combination of two damped modules increases the effective frequency spread and boosts the absorption to "1" at critical frequencies to match a third porous unit. There is a frequency offset of the two membrane modules, so anomalies are largely tamed. Painstakingly careful choice of the two resonant frequencies, set against calculated panel efficiency allows an even spread that otherwise is unobtainable. Figure A3.6 shows the predicted absorption coefficient of the three working together. Of course it needs a catchy name: Three Element Wide Band Absorber (TEWBA). In the real world the response will be more of a roller coaster; they always are! And despite the theory, there is likely to be a slight loss of efficiency at the top end of the porous absorber, perhaps by 10% at 8 kHz. It's not a problem in a practical room. At the bass end a theoretical absorption of 0.4 at 25 Hz, rising to 0.9 at 40 Hz in a 120 mm depth system is outstanding.

The decision to keep the box depths the same for ease of building, and at just 120 mm deep to preserve space, could have proved problematic. There was a concern that I would need a range of membrane materials creating a wide and expensive shopping list; but perhaps just by dumb luck the maths all fell into place. I hope you enjoy making them!

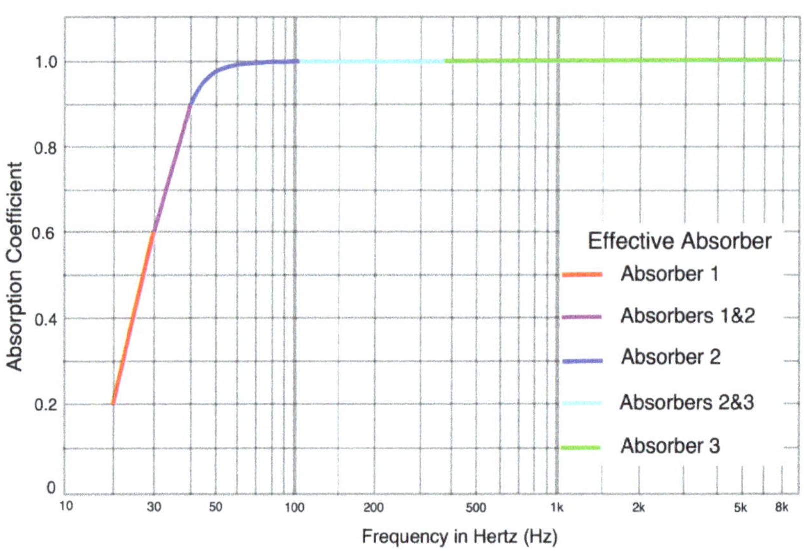

Figure A3.6 Overall absorption coefficient of TEWBA studio panels

PHYSICAL CONSTRUCTION OF THE MODULES

Economic Acoustic Absorber Shopping List

12 mm thick medium-density fibreboard (MDF) sheet for frame sides cut into 75 mm plank width (home systems), 50 mm plank width (commercial) and for the studio boxes, 120 mm width planks.
9 mm MDF sheet backs for the modules that need it (600 mm × 600 mm).
Mineral Wool 60 kg/m^3 density in 50 mm slabs (Rockwool RW3 or similar).
Glue or sealant to make airtight enclosures, screws or pins to hold.
Strong Fine Nylon mesh to seal in fibres (similar to bridal veil material).
Weldmesh 13 mm × 13 mm to secure mineral wool (secure to frame by U staples).
Heavy duty corrugated cardboard boxing for spacer for air gaps.
MLV sheet (5 kg/m^2).
Plain hardboard sheet 3.3 mm (⅛ inch).
Wall and ceiling fixings to suit.

Clarification: the box size of 600 mm × 600 mm has no science behind it. You can make these absorbers round, oval, hexagonal or mermaid shape if you like. The depth matters, the mass matters, airtightness matters (on membranes); however, the frontal face size is irrelevant for each unit as long as the sum of total area covered in an installation meets the required RT_{60}. The recommended size suits whole slabs, with just one cut each for the studio project, and no cuts on the others. If you want to mount them on a stud partition plasterboard wall, the studs are normally at 600 mm spacings, so you may wish to go with a slightly smaller frame to make sure you hit the timber verticals successfully!

Let's tackle the home first. Whether the home system is stereo or surround sound, *in tight spaces*, we have learnt the first reflection of the front pair, off the near side wall(s) must be dampened between 1 kHz and 8 kHz, before it reaches the listener to stop comb filtering in the image. At the higher end this happens naturally as the speakers become more directional. (You already have the front pair turned in to face you, don't you!) My design for this uses 50 mm 60 kg/m^3 Rockwool, placed on one inch of airspace. The surrounding frame or "cassette" depth is 75 mm. I have shown a back in the picture, but this "velocity" module does not benefit from it, so the frame can be affixed to the wall simply with mirror brackets if you want, or any other method of your choosing. Wherever the mineral wool is exposed it should be covered by a strong light nylon mesh to stop fibres peeling off into the air. The air space is created by inch strips of corrugated box cardboard on edge, either in a loose spiral or (my favourite) strips cut half through with slots and created as a square matrix, like stumpy wine glass separators from a new box (on the left of Figure A3.7).

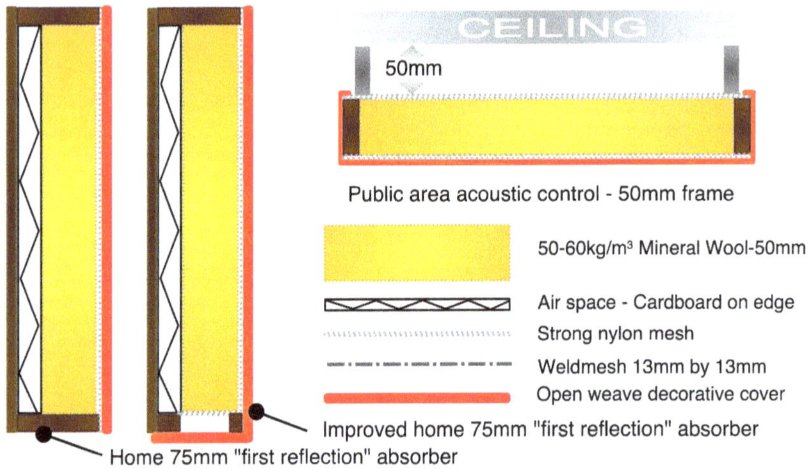

Figure A3.7 Home and public area absorbers

This cassette is slab size, so 1200 mm × 600 mm, and is positioned as explained in the test case, vertically, halfway up the side walls, between the front face of the loudspeakers and the listener's ear. A refinement for lower-ceilinged rooms is an additional pair of these absorbers mounted longitudinally directly on the ceiling (no gap required), again placed to catch the first reflection between the speakers and listening position. The rear wall should be diffuse rather than absorbent, as indicated elsewhere in the book. A small front frame can have open-weave canvas or hessian wrapped around it like a removable loudspeaker grille and in a colour of your choice, finally pinned through to attach to the main frame. A neat improvement, if you have the skills, is to expose the side facing the speaker to the mineral wool edge using large holes or a wide slot in the frame. I find this improves the performance in this critical application.

Some years ago, I was asked to consult for an architectural firm based in Victoria, London. One of their clients was a nearby restaurant, loved by sitting Members of Parliament. The MPs had worked out that if they could hear conversations at the other end of the restaurant, it was likely they were being listened to as well! It was a familiar problem, hard sound reflective flooring, walls and ceilings. Clients might as well have been eating in a bathroom! Further, as mentioned in Chapter 1, "Introduction to Sound", the "Cocktail Party Effect" is at play here. The multiple reflections of sound muddy the table conversation an older diner is trying to hear, often beyond comprehension! If they can't join in with their friends it will reduce their enjoyment of the venue, and next time they will go elsewhere and spend their money with a competitor. There is a saving grace in that in most cases in finding a solution in public spaces, we only need to consider the middle speech frequencies in our reverberation control.

The ceiling unit (on the right of Figure A3.7) suits mixed public spaces such as music rooms, community rooms and restaurants and is designed to create a pleasing and supportive acoustic for various activities. It is a simple 50 mm deep frame around a whole RW3 mineral wool slab; the Rockwool is loosely wrapped in fine strong nylon mesh to stop shedding, and the front and sides can be faced with open-weave canvas in a colour of your choice (washable where use requires it!). On the top side (to the ceiling) the wool is left exposed, just with a nylon mesh protection, and the unit is suspended on brackets 50 mm from the ceiling. Around a quarter to a third of the ceiling area should be covered for most purposes, with the units spaced well apart from each other. In tall echoey rooms increase the suspension distance to 250 mm. Remember additional safety chains or suspension wires where required in public access spaces.

A WIDE-BAND STUDIO ABSORBER SOLUTION

Legal Notice – the Three Element Wide Band Absorber (TEWBA) is an economic solution for use in small-scale studio design, whether it be for a studio performance area or a control and monitoring space. Publication in this book establishes copyright of the concepts and construction of the "TEWBA" in the UK and internationally through the Berne Convention.

I freely grant the right to individuals to embrace DIY construction and experimentation of the TEWBA design, including reaching out to colleagues and local professionals to construct and complete personal projects. The design is offered with experience, diligence and good faith; but I can't guarantee the performance of an individual unit or combined use of units, due to the diverse uses it will be asked to fulfil, and my inability to provide local quality control of the room, the build and the positioning.

A reminder, sound control is not soundproofing; this must have been achieved by eradication of the smallest of air gaps and, where necessary, additional layers of structural mass separated from each other and the structure by pliable materials (isolation, spring and mass systems – see Chapter 8, "Introduction to Acoustics").

In our high-grade studio design, the uniform broadband reduction of reverberation is achieved by multiple installations of groups of *three identical box frames*. I have designed the units to use the same "box" or cassette, so these can be built in bulk without giving thought to the end purpose and easily repurposed to rebalance the acoustics where required up the way. The frame (width and height) is externally 600 mm square to snugly fit half blocks of RW3. The frame depth is 120 mm (without back) and it is made from strips of 12 mm thick MDF panel. The back is made from 9 mm MDF panel, or in rare cases where it is to be battened away from the wall surface increase this to 12 mm MDF. This casing must be glued or silicone sealed to be airtight, and then either securely pinned or screwed.

Here is the detail of the TEWBA absorbers, as shown in Figure A3.8. The three identical box frames each take two layers of RW3, but installation does differ within the box. They are in order of complexity, so first is last. Box 3 is the high frequency absorber.

1) Box 3. Above 100 Hz, a simple pure porous or velocity absorber of 120 mm depth builds its sound reducing ability to an absorption coefficient of 1 by 400 Hz. The RW3 is spaced 20 mm from the rear panel (using a corrugated cardboard matrix as above) and a depth of 100 mm brings the mineral wool to near flush where 1/2 inch Weldmesh (13 mm × 13 mm) holds it in place and fine strong nylon mesh is used to retain loose fibres. The box maintains a uniform absorption of "1" to at least 4 kHz and may have a small but insignificant reduction above it.
2) Box 2. The first of our two membrane absorbers manages the tricky upper bass region where there are many opportunities for resonance issues. Two layers of RW3 fill an airtight box. Again, weldmesh secures the wool, so it can't flop forward onto the membrane, leaving a 20 mm airgap at the front of the unit. The unit has a simple single membrane of ⅛ inch (3.3 mm) unperforated hardboard. The membrane is carefully pinned and glued to seal it to the frame.
3) Box 1. This membrane absorber manages the deep lower bass region and so requires a higher mass solution (as the box depth remains 120 mm). Two layers of RW3 fill an airtight box as before and 1/2 inch Weldmesh secures the wool again, leaving a 20 mm airgap at the front of the unit. This is the complex bit. There are two membranes working as one; the first is a 5 kg/m^2 MLV sheet, and the second membrane is a repeat of our 1/8-inch hardboard. If we just pinned them on the frame together it would be hard for the air pressure to move, not (just) because of the mass, but because of the stiffness of two thick layers at the surround, giving poor compliance where the membrane meets the frame. To get the absorption coefficient we need, we must get the mass moving; the solution is an old one of my own.

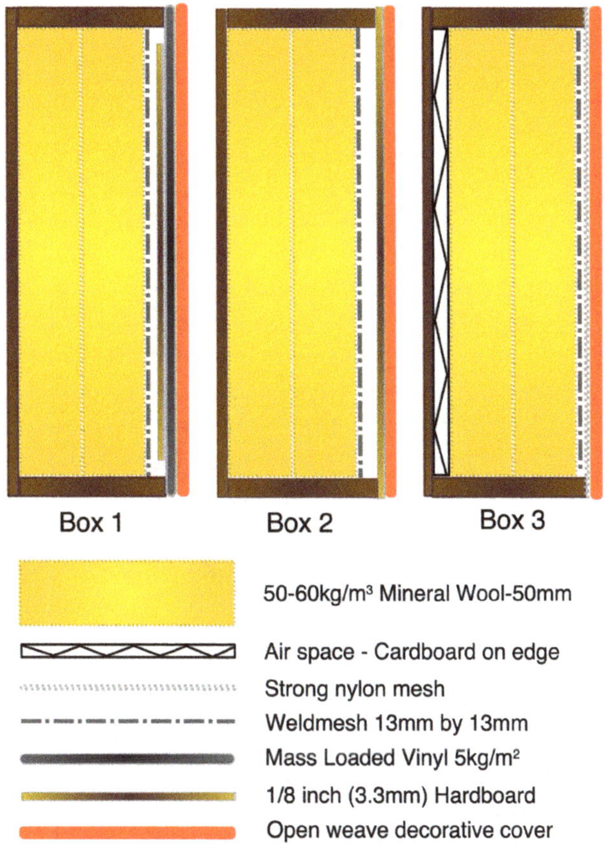

Figure A3.8 The TEWBA for studio installation

Each of the boxes can be finished with colour cover of either open weave material such as cloth or canvas, or the bargain basement but beautiful natural hessian. This can be tucked over on a thin frame and pinned or use a commercial product such as the brilliant FabriTrak. Be careful to avoid materials classed as "soundproof" like felt, as they are the opposite of what is needed here.

The hardboard is trimmed to be 1/2 inch (13 mm) smaller *than the inside of the frame* on all sides, and the corners are rounded off (use the stem of a small wine glass or tumbler as a pattern). Another way to look at it, is the hardboard is 1 inch (25 mm) smaller, on all sides, than the MLV. The hardboard is thoroughly glued all over (quality aerosol spray fixative is best) and then suck on the centre of the rough side of the MLV square. When the combination is thoroughly set, the MLV is glued and pinned to be airtight on the box frame (hardboard inside-smooth MLV side to the outside), so it is just the flexible MLV that attaches onto the frame, the hardboard does not touch. The change of membrane mass with the reduction of hardboard size is insignificant on a 600 mm × 600 mm box at around 1/80. It now moves at its resonant frequency with less hindrance, and a healthy absorption coefficient is achieved.

POSITIONING OF THE FINISHED MODULES

In a Studio Setting

Before considering box positioning let's recap on other aspects of the room. A wise start is to use carpet (with underlay) on the lower traffic areas of the floor and wood strip or carpet tiles glued to the door at high traffic places (castor chairs are terrible for wearing flooring). Don't work with a bare table; cover it with a sheet of thick (3 mm) felt, and the same on a script stand. Windows (if there are any) could have old-fashioned heavy pleated curtains or quality material blinds which can be drawn to make a useful change on the acoustics to suit various uses. Many larger instruments, and certainly the male voice, extend down to 100 Hz, so our area has to have control down past that goal. Too many talk studios, old BBC included, showed a steep rise in bass around that frequency that encouraged a nasty "boomy" colouration, and the smaller the space, the bigger the problem! I have learnt over time that the Eyring modified Sabine formula becomes unreliable under 200 m^3, a size which is often likely with sound "booths". That said, experience has shown it is easy to apply too much treatment, making a dull and uninspiring workspace, which must be avoided at all costs!

On paper, Boxes 1 and 2 together are designed to match the absorbency of Box 3, however in my experience of *real* rooms, membrane absorbers can seem to surprisingly overachieve. I believe this may well be the plasterboard partitions and perhaps the ceiling joining in the membrane game (often around 70 Hz), so, particularly with partition walls, it will need a measure of trial and error in your space. As box sizes are identical, some can be repurposed at small cost and inconvenience; the ratio can be adjusted if problems occur in odd-shaped spaces or with awkward dimensions or again a lot of glass and/or doors compared to wall space.

All three absorbers will be needed opposite glass in a performance area. Box 3 works well at ear height but don't use more than six panels (three large) without interspacing with plain wall or Box 1,2 combinations. Don't use Boxes 1 and 2 at ear height, where you can avoid it. Don't use Box 3 directly opposite other Box 3 panels on the flank walls but do use the same number of Box 1 and 2 panels on facing walls.

To recap:

- Deploy absorbers opposite the hard surfaces of the walls, but don't place absorbers opposite other absorbers in the first experimental installation.
- Use equal numbers of Boxes 1, 2 and 3 in your first installation. Adjust later.

- Do not cover more than 40% of the studio wall area with absorbers without pausing to carefully test the result. It should be enough in many applications.
- Mount Box 3 at ear height, do not mount Boxes 1 and 2 at ear height.
- In general, do not mount the membrane boxes directly opposite others but do use an equal number on facing walls.
- Don't clump many Box 3 together; a maximum of six, as the overall performance will be compromised.

TEST CASE – A MONITORING ROOM

A hot topic, "listening rooms" and attendant sound cubicles are a subject of much speculation on the internet, and I'm pleased to see that many people are taking the subject of modifying a room to optimise the sound of the installed equipment seriously! The point that is often missed by sound enthusiasts is that the sound *should* be modified by the room in which it is received; that's the point of "phonic" delivery over "aural" delivery, as described in Chapter 1, "Introduction to Sound". It is also the basis of auditorium designs going back centuries. It is the *management* of the changes that is the expertise; whatever the area is used for, and monitoring areas are no exception. I don't think anyone would argue with positive terms like "airy" or "balanced" and that is what should be expected of a good design. A constricted sound comes from overuse of porous absorption, by which its very nature is likely to hammer the RT_{60} at the higher frequencies. A honking sound (like a squash court) will come from sound created between two hard parallel surfaces, a brittle one from sound in hard corners and a muddled one from reflection between multiple hard surfaces. And don't get me started on eigentones!

As in all interfaces of art and science, there are many shades of grey, and as you might eagerly consult the internet on skin rashes, investment opportunities and the best recipe for chicken soup, there is sadly poor advice and useless products mixed in with the good. I feel a little defensive here, as the aim of the book and particularly "Appendix 3 – A Guide to Home-Built Absorbers" is to help people to gain an affordable universal foothold in a new subject, not redesign the domes in the Albert Hall!

I find there is a current fascination with corners, as if trolls might lurk there, and whilst there is a proven build-up of sound pressure at the boundaries, it's not necessary to attempt to smite all life with hulk-sized behemoths. And that's before the $1/4\lambda$ rule is followed as explained above. After all, the corners are not where you sit to listen, are they? I note that we were fine without them for over 100 years of sound history; you didn't hear George Martin wailing it was impossible to get the iconic Beatles sound unless pillar boxes of wadding were stashed in every studio orifice. Fashion is what fashion does... In this test case I pay homage to Don Davis, Chip Davis and LEDE™; it uses some of the key findings in their published paperwork, however it is not intended to copy the rigorous and refined research that is used in each LEDE™ installation, or to imbue in any way that the results are comparable. These guys are legends! We'll use the TEWBA system as a basis for this test case, as an aid to your further experimentation. Nothing would please me more than for you to take this base offer and build on it in your own way, surpassing what has been achieved to date.

I said earlier that you should put your money into the best monitors you can afford, and obviously these should be sited carefully. I don't understand why so many squeeze the sound desk virtually up to the studio glass, either mounting speakers in the walls or on tottering stands pressed against the wall. I heard some great advice from a sage studio owner years ago. Others were present, but I've not heard of it again and it may well have been overlooked as his method wasn't shiny enough for some. He started with the empty room; he then sited the monitors facing him at the corners and placed his "listening position" chair in between them. He then moved the chair back down the centre of the room listening to the sound, all the way. He hit a sweet spot just about (unsurprisingly) the middle of the room. He then did the same again with the monitors still facing him, now a half metre out from the corners and repeated the test of moving the chair. He then moved the monitors out further and so on for an entire afternoon. Time well spent, he was keen to say, and I agree with him. He eventually found a sweet spot where he heard beautifully neutral sound and had formed a stereo friendly equilateral triangle with his speakers; he marked it out with tape. Only then did he start to think about absorption!

Speakers radiate uniformally in all directions below around 200 Hz. As the frequency rises, they become increasingly directional, until you will miss a lot of top information if you fail to turn the monitors in to face your ears. This phenomenon is a major reason why there is additional bass (compared to other frequencies) behind a typical speaker and why it's traditional, and it's right to use bass absorption with membrane absorbers, Helmholtz bass traps or literally cubic metres of wadding in studios behind them to tame it. With TEWBA, use Box 3 at ear height and alternate Box 1 and 2 below and above Box 3, in vertical strips of the three behind the speakers (either side of the studio window, if it exists). Above the window add Box 3s and below it, alternate with Box 1s and 2s. A useful experiment in addition is to hang a ceiling unit (the earlier public area absorption slab is fine) just above at the window height and slanting up towards the ceiling as it nears the listening position at the mixer. This is in addition to the Box 3s.

Now the side walls. With your favourite monitor position and seated position marked, take a seat and get a friend to walk the currently naked side wall with a decent mirror flat against it. Mark the wall where the mirror shows the reflection of the front of this side's speaker, and then do the other flank. Now we break the rule of not treating opposite walls with identical absorbers, using two vertically stacked Box 3s either side at the centre of the wall markings. This is to dull first reflections of the direct sound reaching by a second path likely (in most rooms) to be delayed by 25 mS or more (1.5 m or greater path distance). Add a further TEWBA Box 1 and 2, above and below, ensuring 1) same units are not opposite each other and 2) there is an equal number of each box on both flanking walls. Do this up the side walls to the back to the mixer's chair *and allow the walls and ceiling behind and beyond to remain untreated with acoustic absorption.*

Remember, don't lose the goal here. We are taking down the reverberation time uniformly across the range of required frequencies, to get an RT_{60} for the total area. The RT_{60} is assessed to meet the predominant use of the area. The control room area must be slightly less reverberant than the studio to hear the presenters/musicians in the correct acoustic.

Please use diffusion on the rear wall. You may well wish to build Schroeder's brilliant invention of a 2D or 3D diffuser. I'm fine with it if you follow the correct design criteria and philosophy, use premium materials to ensure it does not resonate and be aware of the lowest frequency that will be dispersed. Is it low enough to achieve your objective? I would suggest a different solution in a cost-conscious application. F. Alton Everest describes an effective sound dispersion design in his excellent book Acoustic Techniques for Home and Studio. He uses triangles in a pleasing geometric design which he placed in the centre of the rear studio wall. It's roughly 3 m^2 in size, although the overall shape is hexagonal. In the heading "Dispersion" near the start of this appendix, I laid out some ground rules for breaking up large sound waves with relatively small panels, and at the end you will find a table which (in purple) gives the minimum panel dimension to break up frequencies across the band. Why not have a go at your own design?

COMMUNITY HALLS

This may seem a strange end to this appendix, and I guess it's here because it's a loose end that readers will need an answer to. We have witnessed evolutionary changes in the uses of our community spaces, often former church halls. Activities are now predominantly out on the floor, rather than the original design criteria of sound projected from a raised stage. The removal of old curtains and carpets has created problems which were sorted in the original design. These halls are now often far too reverberant! My answer to these modern problems lies in large porous 100 mm panels (60 kg – open on both faces) fixed to the ceiling on ties with an air gap of between 800 mm and 1600 mm (ceiling height allowing) to increase the effective surface area and low frequency effectiveness. Don't overdo it! Evenly spaced across a quarter of the ceiling area should do it for most, rising to a third in hard-faced buildings.

If the stage is used regularly, the upper surface of the rear wall (furthest from the stage) should be treated to stop standing waves. Simple large membrane absorbers of enclosed plywood panels help in tricky situations, and angled diffusers are also an excellent tool, taking little depth away from the room, yet breaking up eigentones. These should be used in combination with some absorption on the side walls (often between tall or high windows). These porous panels high on the side walls are useful, both in acoustic noise control of daily activities and as a support to quelling standing waves through the grazing effect (replacing the absorption formally provided by the curtains).

It's all cheap materials, the main problems being in finding an acceptable finish, adequate framing of the fibre matting and of course safety whilst working at height. Don't forget the fire regulations! The ideal is to bring the RT_{60} down to around 1.5 seconds, too much treatment will mean the sound will no longer make it down the length of the hall from the stage to the back seats! A perfect simple job for the Eyring modified Sabine formula.

This chapter has been an introduction for the reader to have confidence to take further study if you found it interesting (see References). Finally, I have created Figure A3.9 as another single page "catchall" diagram which provides the theoretical sizes from which different acoustic treatments should work effectively. Note, absorbers will absorb, and diffusers diffuse at lower frequencies, but the effectiveness is unpredictable. I hope this diagram proves a useful tool.

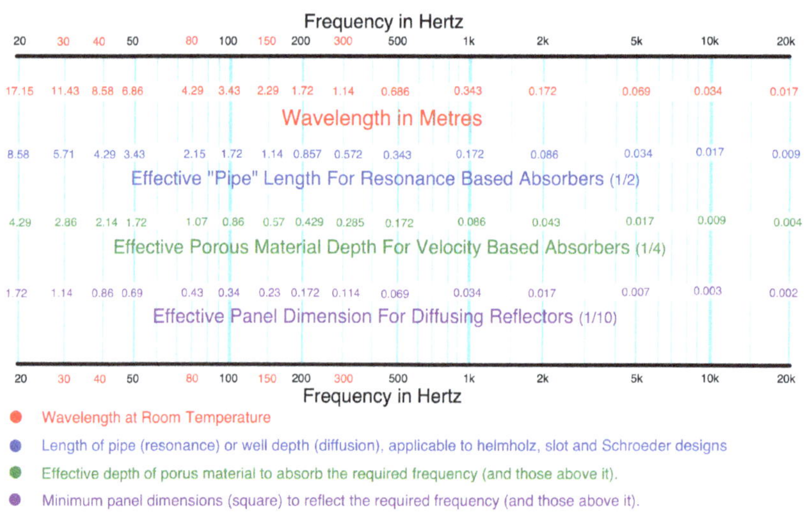

Figure A3.9 Effective sizing of materials for acoustic control

BIBLIOGRAPHY

Alton, F. (1984) *Acoustic Techniques for Home & Studio*, 2nd edition, 4th printing. United States of America: Tab Books Inc.

Architectural and Civil Engineering Department BBC Engineering (1980) *Guide to Acoustic Practice*. BBC Engineering Training Department.

Bronkhorst, Adelbert W. (2000) The Cocktail Party Phenomenon: A Review on Speech Intelligibility in Multiple-Talker Conditions. *Acta Acustica United with Acustica*, 86, 117–128.

Burd, A.P., Gilford, C. and Spring, N. (1966) *Data for the Acoustic Design of Studios*. BBC Engineering Monograph No. 64. London: BBC Engineering Division.

Cox, T. and D'Antonio, P. (2017) *Acoustics and Diffusers: Theory, Design and Application*, 3rd edition. Boca Raton: CRC Press (Taylor & Francis Group).

Davis, D. and Davis, C. (1980) The LEDE™ Concept for the Control of Acoustic and Psychoacoustic Parameters in Recording Control Rooms. *Journal of the Audio Engineering Society*, 28(9).

Fletcher, J. (1991) *The Interaction of Acoustic Treatment in an Experimental Sound Control*. BBC Research Department, Engineering Division.

Fletcher, J. (1992) *The Design of a Modular Sound Absorber for Very Low Frequencies*. BBC Research Department, Engineering Division.

Mackenzie, G.W. (1964) *Acoustics*. Bungay, Suffolk: Focal Press Limited, Richard Clay and Company Ltd.

Moore, J.E. (1978) *Design for Good Acoustics and Noise Control*. Macmillan Press Ltd.

Parkin, P.H., Humphreys, H.R. and Cowell, J.R. (1979) *Acoustics, Noise and Buildings*, 4th edition. London and Boston: Faber and Faber.

Sengpiel, A. and Sengpiel, E. (2025) *Standing Waves*. https://sengpielaudio.com/StandingWaves.htm.

Taylor, E. (1987) *Theoretical and Practical Aspects of the "Functional Absorber" Method of Arranging Sound Absorbing Panels*. BBC Research Department, Engineering Division.

Walker, R. (1990) *The Design and Application of Modular, Acoustic Diffusing Elements*. BBC Research Department, Engineering Division.

Unlock additional features at the book's companion website:

https://www.soundknowledge.org/

Visit the Audio Vault to find unique free downloadable colour pictures and charts to support your journey into professional sound. Many are drawn from the book; others are added as they occur or readers request them. Write to: info@soundknowledge.org.

Index

Note: Page numbers in *italics* indicate figures; page numbers in **bold** indicate tables

For Product Safety Concerns and Information please contact our EU
representative GPSR@taylorandfrancis.com
Taylor & Francis Verlag GmbH, Kaufingerstraße 24, 80331 München, Germany

www.ingramcontent.com/pod-product-compliance
Lightning Source LLC
LaVergne TN
LVHW081311190926
848399LV00036B/367
* 9 7 8 1 0 4 1 2 0 5 7 9 1 *